Horizons in World Physics, Volume 262

ASTROPHYSICS AND CONDENSED MATTER

Horizons in World Physics Series

Volume 261 **Superconductivity Research Advances,** Edited by James E. Nolan

Volume 260 **Instabilities of Relativistic Electron Beam in Plasma,** Edited by Valery B. Krasovitskii

Volume 259 **Self-Focusing of Relativistic Electron Bunches in Plasma,** Edited by Valery B. Krasovitskii

Volume 258 **New Topics in Theoretical Physics,** Edited by Henk F. Arnoldus and Thomas George

Volume 257 **Boson's Ferromagnetism and Crystal Growth Research,** Edited by Emerson D. Seifer

Volume 256 **Horizons in World Physics,** Edited by Victor H. Marselle

Volume 255 **Frontiers in General Relativity and Quantum Cosmology Research,** Edited by Victor H. Marselle

Volume 254 **Optics and Electro-Optics Research,** Edited by Albert V. Berzilla

Volume 253 **Trends in General Relativity and Quantum Cosmology,** Edited by Albert Reimer

Volume 252 **Chemical Physics Research Trends,** Edited by Benjamin V. Arnold

Volume 251 **Quantum Dots: Research Developments,** Edited by Peter A. Ling

Volume 250 **Quantum Gravity Research Trends,** Edited by Albert Reimer

Volume 249 **General Relativity Research Trends,** Edited by Albert Reimer

Volume 248 **Spacetime Physics Research Trends,** Edited by Albert Reimer

Volume 247 **New Developments in Quantum Cosmology Research,** Edited by Albert Reimer

Volume 246 **Quantum Cosmology Research Trends,** Edited by Albert Reimer

Volume 245 **Horizons in World Physics,** Edited by Tori V. Lynch

Volume 244 **Horizons in World Physics,** Edited by Albert Reimer

Volume 243 **Theoretical Physics 2002, Part 2,** Edited by Henk F. Arnoldus and Thomas F. George

Volume 242 **Models and Methods of High-T_c Superconductivity: Some Frontal Aspects, Volume 2,** Edited by J. K. Srivastava and S. M. Rao

Volume 241 **Models and Methods of High-T_c Superconductivity: Some Frontal Aspects, Volume 1,** Edited by J. K. Srivastava and S. M. Rao

Volume 240 **Horizons in World Physics,** Edited by Albert Reimer

Volume 239 **Theoretical Physics 2002, Part 1,** Edited by Henk F. Arnoldus and Thomas F. George

Volume 238 **Theoretical Physics 2001,** Edited by Henk F. Arnoldus and Thomas F. George

Volume 237 **Generalized functions in mathematical physics: Main ideas and concepts,** Edited by A.S. Demidov

Volume 236 **Edge Excitations of Low-Dimensional Charged Systems,** Edited by Oleg Kirichek

Volume 235 **On the Structure of the Physical Vacuum and a New Interaction in Nature (Theory, Experiment, Applications),** Edited by Yu. A. Baurov

Volume 234 **Electron Scattering on Complex Atoms (Ions),** Edited by V. Lengyel, O. Zatsarinny and E. Remeta

Volume 233 **An Introduction to Hot Laser Plasma Physics,** Edited by A.A. Andreev, A.A. Mak and N.A. Solovyev

Volume 232 **Electron-Atom Scattering: An Introduction,** Maurizio Dapor

Volume 231 **Theory of Structure Transformations in Non-equilibrium Condensed Matter**, Edited by Alexander L. Olemskoi

Volume 230 **Topics in Cosmic-Ray Astrophysics**, Edited by M. A. DuVernois

Volume 229 **Interactions of Ions in Condensed Matter**, Edited by A. Galdikas and L. Pranevièius

Volume 228 **Optical Vortices**, Edited by M. Vasnetsov and K. Staliunas

Volume 227 **Frontiers in Field Theory**, Edited by Quantum Gravity and Strings, R.K. Kaul, J. Maharana, S. Mukhi and S. K. Rama

Volume 226 **Quantum Groups, Noncommutative Geometry and Fundamental Physical Interactions**, Edited by D. Kastler, M. Russo and T. Schücker

Volume 225 **Aspects of Gravitational Interactions**, Edited by S.K. Srivastava and K.P. Sinha

Volume 224 **Dynamics of Transition Metals and Alloys**, Edited by S. Prakash

Volume 223 **Fields and Transients in Superhigh Pulse Current Devices** Edited by G.A. Shneerson

Volume 222 **Gaussian Beams and Optical Resonators,** Edited by A. N. Oraevsky

Volume 221 **Laser Cathode-Ray Tubes,** Edited by Yu. M. Popov

Volume 220 **Quantum Electrodynamics with Unstable Vacuum,** Edited by V.L. Ginzburg

Volume 219 **The Muon Method in Science,** Edited by V.P. Smilga and Yu.M. Belousov

Volume 218 **Radiative and Collisional Characteristics of Ions in Hot Plasmas,** Edited by I.I. Sobel'man

Volume 217 **Nonlinear and Quantum Optical Phenomena in Nonequilibrium Media** , Edited by V.A. Shcheglov

Volume 216 **Semiconductors and Insulators: Optical and Spectroscopic …,** Edited by Yu.I. Koptev

Volume 215 **Relativistic Theory of Gravity,** Edited by A.A. Logunov

Volume 214 **Research in Quantum Field Theory**, Edited by V.I. Man'ko

Volume 213 **Nonlinear Theory of Strong Electromagnetic Wave-plasma{xe "plasma"}…,** Edited by O.N. Krokhin

Volume 212 **Dynamics of Elementary Atomic-Molecular…**, Edited by O. S. Scheglov

Volume 211 **Optics and Lasers,** Edited by G.G. Petrash

Volume 210 **Theory and Technology of High Temperature Superconductors,** Edited by Yu.I. Koptev

Volume 209 **Theory of Interaction of Multilevel Systems…**, Edited by V.I. Man'ko

Volume 208 **Physical Properties of High Temperature Superconductors,** Edited by Yu.I. Koptev

Volume 207 **Generalized Mayer Series in Classical Statistical Mechanics**, Edited by Igor J. Ivanchik

Volume 206 **Mathematical Physics, Applied Mathematics and Informatics,** Edited by Yu.I. Koptev

Volume 205 **Squeezed and Correlated States of Quantum Systems**, Edited by M.A. Markov

Volume 204 **Sakharov Memorial Lectures in Physics**, Edited by L.V. Keldysh and V.Ya. Fainberg

Volume 203 **Fiber Optics: Research and Development,** Edited by Ye.M. Dianov

Volume 202 **Gas Dynamics**, Edited by Yu.I. Koptev

Volume 201 **Generation of Nonlinear Waves and Quasistationary …**, Edited by L.M. Kovrizhnykh

Volume 200 **Laser Thermonuclear Targets and Superdurable Microballoons**, Edited by A.I. Isakov

Volume 199 **Spatiotemporal Characteristics of Laser Emission,** Edited by M.V. Pyatakhin and A.F. Suchkov

Volume 198 **Theory of Nonstationary Quantum Oscillators**, Edited by M.A. Markov

Volume 197 **Reliability Problems of Semiconductor Lasers**, P.G. Eliseev

Volume 196	**Pulsars,** Edited by A.D. Kuzmin
Volume 195	**Physical Characteristics and Critical Temperature of High Temperature Superconductors** , Edited by M.M. Sushchinskiy
Volume 194	**Biophysical Approach to Complex Biological Phenomena,** Edited by E. Volkov
Volume 193	**Research on Chemical Lasers** Edited by A.N. Orayevskiy
Volume 192	**Atomic and Ionic Spectra and Elementary Processes in Plasma** , Edited by I.I. Sobelman
Volume 191	**Polarization of the Vacuum and a Quantum Relativistic Gas ...**, A.Ye. Shabad
Volume 190 , (Supp. Vol.)	**Superconductors with A15 Lattice and Bridge Contacts Based ...**, Edited by M. M. Sushchinskiy
Volume 190	**Metal-Optics and Superconductivity**, Edited by A.I. Golovashkin
Volume 189	**Clusters of Galaxies and Extragalactic Radio Sources**, Edited by A.D. Kuz'min
Volume 188	**Integral Systems, Solid State Physics and Theory of Phase Transitions. Pt. 2. Symm. and Alg. Structures in Physics**, Edited by V.V. Dodonov and V.I. Man'ko
Volume 187	**Quantum Field Theory, Quantum Mechanics and Quantum Optics: Pt. 1. Symm. and Alg. Structures in Physics,** Edited by V.V. Dodonov and V.I. Man'ko
Volume 186	**Photoproduction of Pions on Nucleons and Nuclei**, Edited by A.A. Komar
Volume 185	**Injection Lasers in Optical Communication and Information Processing Systems**, Edited by Yu. M. Popov
Volume 184	**Electron Processes in MIS-Structure Memories**, Edited by A.F. Plotnikov
Volume 183	**Invariants and the Evolution of Nonstationary Quantum Systems**, Edited by M.A. Markov
Volume 182	**Luminescence of Wideband Semiconductors**, Edited by M.D. Galanin
Volume 181	**Pulsed Metal and Metal Halogenide Vapor Lasers**, Edited by G.G. Petrash
Volume 180	**Inelastic Light Scattering in Crystals**, Edited by M.M. Sushchinskiy
Volume 179 , (Supp. Vol. 2)	**Electron-Excited Molecules in Nonequilibrium Plasma**, Edited by N.N. Sobolev
Volume 179 , (Supp. Vol. 1)	**Interaction of Ultrashort Pulses with Matter** , Edited by M.D. Galanin
Volume 179	**X-Ray Plasma Spectroscopy and the Properties of Multiply-Charged Ions** , Edited by I.I. Sobel'man
Volume 178	**The Delfin Laser Thermonuclear Installation: Operational Complex ...**, Edited by G.V. Sklizkov
Volume 177	**Stoichiometry in Crystal Compounds and Its Influence on Their Physical Properties**, Edited by Yu. M. Popov
Volume 176 , (Supp. Vol)	**Quantization, Gravitation and Group Methods in Physics** Edited by A.A. Komar
Volume 176	**Classical and Quantum Effects in Electrodynamics** Edited by A.A. Komar
Volume 175 , (Supp. Vol.)	**Luminescence and Anisotropy of Zinc Sulfide Crystals** Edited by M.D. Galanin
Volume 175	**Luminescence Centers of Rare Earth Ions in Crystal Phosphors** , Edited by M.D. Galanin
Volume 174 , (Supp. Vol. 2)	**Electron Liquid Theory of Normal Metals** Edited by V.P. Silin
Volume 174 , (Supp. Vol. 1)	**Nonequilibrium Superconductivity** , Edited by V.L. Ginzburg
Volume 174	**Thermodynamics and Electrodynamics of Superconductors,** Edited by V.L. Ginzburg
Volume 173	**Quantum Mechanics and Statistical Methods** , Edited by M.M. Sushchinskiy
Volume 172	**Phase Conjugation of Laser Emission** , Edited by N.G. Basov
Volume 171	**Research on Laser Theory** , Edited by A.N. Orayevskiy

Volume 170 **The Theory of Target Compression by Longwave Laser Emission** , Edited by G.V. Sklizkov

Volume 169 **The Physical Effects in the Gravitational Field Black Holes**, Edited by M.A. Markov

Volume 168 **Issues in Intense-Field Quantum Electrodynamics**, Edited by V.L. Ginzburg

Volume 167 **Group Theory, Gravitation and Elementary Particle Physics**, Edited by A.A. Komar

Volume 166 **The Nonlinear Optics of Semiconductor Lasers**, Edited by N.G. Basov

Volume 165 **Solitons and Instantons, Operator Quantization**, Edited by V.L. Ginzburg

Horizons in World Physics, Volume 262

Astrophysics and Condensed Matter

Thomas G. Hardwell
Editor

Nova Science Publishers, Inc.
New York

Library of Congress Cataloging-in-Publication Data

Astrophysics and condensed matter / Thomas G. Hardwell, editor.
p. cm. -- (Horizons in world physics ; v. 262)
ISBN 978-1-60456-332-0 (hardcover)
1. Astrophysics. 2. Condensed matter. I. Hardwell, Thomas G.
QB461.A7754 2008
523.01--dc22
2007049521

Published by Nova Science Publishers, Inc. ✦ New York

CONTENTS

PREFACE

This new book presents selected research in two leading areas of physics.

Expert Commentary - Sulfur-nitrogen heterocycles such as the 1,2,4-triazolo[3,4-*b*]-1,3,4-thiadiazoles have a distinct tendency to establish intermolecular links in a solid state through secondary bonding interactions S...N. The associaton of these molecules was identified by systematic survey of crystallographic data. The interaction between these molecules is dominated by contributions arising from orbitals mixing, which can be interpreted as the donation of a nitrogen lone pair into the sulfur-centered antibonding orbitals. Because of the bulky substituents, 1,2,4-triazolo[3,4-*b*]-1,3,4-thiadiazoles associate in a coplanar fashion forming dimers. In addition to its strength, the coplanar dimer is characterized by being rigid, yet no activation barrier is expected for the association/dissociation process.

Chapter 1 - It is now generally accepted that knowledge of three-dimensional structure at the molecular level is absolutely essential for understanding the function of biological macromolecules and their interactions with other molecules. Indeed, much of modern biological and medical research is aimed at understanding life at the molecular level. Thus, three-dimensional structural information is essential for understanding disease, for developing new drugs, for improving the environment, and for many other applications of importance to society through biotechnological developments by the pharmaceutical and agronomical industries. X-ray crystallography is probably the most effective way of obtaining three-dimensional structural information, although there are a number of limitations including the necessity to produce single crystals that yield good diffraction patterns to a useful resolution. Over the last decade or so, dramatic developments have been made to make macro-molecular crystallography more of a routine scientific tool for structure determination rather than a highly specialised discipline. Automation has been invoked to accelerate crystallisation, data collection, and structure elucidation procedures. One of the major advances has involved the use of synchrotron radiation which provides a very brilliant, highly focused source of X-rays which is also readily tuneable over the wavelength range of most use to structural biologists. This has led, in turn, to the development of methods to overcome the so-called phase problem, a necessary prerequisite for deriving structural information from X-ray diffraction patterns. In parallel, many advances have been made in the problem of obtaining suitable single crystal samples for X-ray studies. Although these have been directed mainly to the cases of soluble proteins, the structures of integral membrane and associated membrane proteins are now being increasingly studied. The information deriving from the large amount of structural data that is now under deposition has also led to advances in the field of bioinformatics. This review will attempt to describe some of the more important technical advances that have been made with respect to crystallization, synchrotron beam lines and data collection, and reflection phase determination utilising the variability of wavelength made possible by the use of synchrotron radiation. It will also indicate some of the more important current limitations and future perspectives.

Chapter 2 - The phenomenon of diamagnetism in which a negative susceptibility emerges as the response of a system under an external magnetic field is intriguing and fascinating from various aspects of physical science. The authors demonstrate explicitly that diamagnetism is a robust property of charged spinless bose systems at finite temperature regardless of their interactions. The authors also point out the connection of this diamagnetism with other dissipative systems and problems in condensed matter physics. Starting from a single particle, the discussion will go on to N(finite) particles system and finally, end on to an interacting charged scalar field theory in an arbitrary spatial dimension d. A novel regularization scheme particularly suitable for magnetic field problem has been utilized to regularize the zero point divergent energy. Finally, as an illustration of this theory, the authors take up two classic examples of charged bosonic type of excitations in condensed matter physics.

Chapter 3 - In this chapter, the authors report electronic and magnetic structure of pure and (As-) doped manganese clusters from density functional theory using generalized gradient approximation for the exchange-correlation energy. Ferromagnetic to ferrimagnetic transition takes place at n=5 for pure manganese clusters, Mn_n, and remarkable lowering of magnetic moment is found for Mn_{13} and Mn_{19} due to their closed icosahedral growth pattern and results show excellent agreement with experiment. On the other hand, for As-doped manganese clusters, Mn_nAs, ferromagnetic coupling is found only in Mn_2As and Mn_4As and inclusion of a single As stabilizes manganese clusters. Exchange coupling in the Mn_nAs clusters are anomalous and behave quite differently from the Ruderman-Kittel-Kasuya-Yosida like predictions. Finally, possible relevance of the observed magnetic behaviour is discussed in the context of Mn-doped GaAs semiconductor ferromagnetism.

Chapter 4 - The authors apply the homotopy group theory in classifying the topological defects in atomic spin-1 and spin-2 Bose-Einstein condensates. The nature of the defects depends crucially on the spin-spin interaction between the atoms. The authors find the topologically stable defects both for spin-1 ferromagnetic and anti-ferromagnetic states, and for spin-2 ferromagnetic and cyclic states. With this rigorous approach the authors clarify the previously controversial identification of symmetry groups and order parameter spaces for the spin-1 anti-ferromagnetic state, and show that the spin-2 cyclic case provides a rare example of a physical system with non-Abelian line defects, like those observed in biaxial nematics. The authors also show the possibility to produce vortices with fractional winding numbers of 1/2, 1/3 and their multiples in spinor condensates.

Chapter 5 - The technology necessary for constructing muon colliders is currently under investigation around the world. These machines would employ muon beams rather than electron beams, and would require new cooling and acceleration methods to succeed. If the technical obstacles to construction of these novel colliders can be overcome, one can envision the possibility that muon colliders will vastly extend for decades to come the energy frontier that can be probed experimentally in particle physics.

Muon colliders can also play a unique role in determining the properties of Higgs bosons and fully understanding the Higgs sector. The much larger mass of the muon compared to the mass of the electron means the s-channel production rates are much larger at a muon collider.

A muon collider with energy tuned to the mass of a Higgs boson would offer possibilities in measuring parameters of the Higgs boson that is highly complementary to Higgs studies at other planned colliders such as the Large Hadron Collider (LHC) or a future Linear Collider (LC). In this article, I detail the precision measurements of the Higgs boson mass and total width, and probe the ability of the muon collider to distinguish between the Standard Model Higgs boson and the lightest Higgs boson of supersymmetric extensions. The ability to determine the properties of a general extended Higgs sector, possibly with CP-violating couplings, is explored. The detection and the mass separation of heavy supersymmetric Higgs bosons is described. These supersymmetric Higgs bosons satisfy mass sum rules and I analyze the attractive possibility that very precise tests of radiative corrections in the Higgs sector can be performed. Finally the method of performing threshold scans of *ZH* production at a muon collider to determine the mass and spin of a Higgs boson is described.

Chapter 6 - This article will provide a review of the current methodologies in use for the identification and classification of protein crystallization experiments, covering computer-aided identification, the use of cross-polarization, UV-stimulated fluorescence and the addition of fluorescent markers. The current state of the field will be reviewed and potential new directions will be explored.

Chapter 7 - It is well-known that one of the most challenging problems of general relativity, which has not yet yield a satisfactory solving is the energy-momentum localization. For giving an answer to the problem of energy and momentum localization, many attempts have been made in the past but this remains an important issue to be settled. The difficulty relies in the lack of a generally accepted expression for the energy density. Since Einstein has given his important expression for the energy-momentum complex, the localization of energy is also connected to the use of other energy-momentum complexes, including the prescriptions of Landau-Lifshitz, Papapetrou, Bergmann-Thomson, Weinberg and Møller. The Einstein, Landau-Lifshitz, Papapetrou, Bergmann-Thomson and Weinberg energy-momentum complexes give meaningful results only if the calculations are restricted to quasi-Cartesian coordinates. The energy-momentum complex of Møller gives the possibility to determine the expressions for energy in any coordinate system. In his recent paper, Lessner formulated his viewpoint that the Møller definition is a powerful concept of energy and momentum in general relativity. The concept of energy-momentum complex was reconsider by Virbhadra and his collaborators, and since then many interesting studies were done in this area. Chang, Nester and Chen showed that the energy-momentum complexes are actually quasi-local and legitimate expressions for the energy-momentum. They pointed out the existence of a direct relationship between energy-momentum complexes and quasi-local expressions because every energy-momentum complex is connected to a legitimate Hamiltonian boundary term.

The authors evaluate the energy distribution of a charged black hole solution in heterotic string theory in the Møller prescription. This solution was given by A. Sen and is an exact classical solution in the low-energy effective field theory, which describes a black hole carrying a finite amount of charge and angular momentum. The authors consider that the angular momentum vanishes and the authors have two metrics ds^2 and ds^2_E. The first metric describes the transformed solution and the second is the Einstein metric, which is given by $ds^2_E = e^{-\Phi'} ds^2$.

For both metrics the authors obtained meaningful results for the energy distribution in the Møller prescription. The energy distribution depends on the mass *m* and the arbitrary number

α. If we consider only the first term in the expressions for the energy distribution, we obtain for the energy the expression $E=m$ in the both cases. For the Einstein metric the authors perform some calculations and the case $\alpha=0$ corresponds to the value $E=M$, the energy is equal to the mass M of the black hole.

The authors result sustain the Lessner opinion that the Møller energy-momentum complex is a powerful tool for evaluating the energy and momentum in general relativity, and the very important Cooperstock hypothesis which states that "the energy and momentum are confined to the regions of non-vanishing energy-momentum tensor of the matter and all non-gravitational fields". This result, and many others obtained by several authors support the idea that the Møller energy-momentum complex yield acceptable results for different space-times.

Chapter 8 – Starting from the crystallographic viewpoint, the chemical bonding theory of single crystal growth was proposed and introduced into the morphology prediction and control of inorganic crystals and powders, which can also help us to further understand the process of single crystal growth. In this chapter, the inorganic crystals within seven crystal systems were respectively selected to quantitatively predict their ideal morphology. The predicted morphologies by calculating the vertical growth rate of selected planes, can agree well with experimental observations. The authors took $Mg(OH)_2$ as an example to comprehensively analyze the effect of both crystal structure and reaction conditions on the powder morphology. The authors' results indicate that the current theory can lead to better understanding of the single crystal growth and the morphology control of various powders, which can guide the control of crystal and powder morphology from a kinetic viewpoint.

Chapter 9 - It has recently been pointed out that the 511 key emission line detected by INTEGRAL/SPI from the bulge of our galaxy could be explained by annihilations of light dark matter particles into $e^+ e^-$. The authors present the possibility that dark matter could be made of scalar candidates, namely, of the Higgs bosons in the models based on $SU(3)_C \otimes SU(3)_L \otimes U(1)_N$ (3-3-1) gauge group.

These particles are singlet of the $SU(2)_L \otimes U(1)_Y$ group, so they do not interact with the ordinary particles, except the Higgs boson of the standard model. The Spergel-Steinhardt condition for self-interacting dark matter gives a bound on the mass of the candidates to be a few MeVs. Beside the scalar candidates, which exist in both non-SUSY and SUSY 3-3-1 models with right-handed neutrinos, the spin ½ candidate exists in a variant 3-3-1 version with exotic neutral lepton.

In contrast to the singlet models, where an extra symmetry must be imposed to account the stability of the dark matter, in this case the decay of the candidates is automatically forbidden in all orders of perturbative expansion. This is because of that, these scalars are singlets, i.e., in bottom of the Higgs triplet. Therefore, the standard model fermions and the standard gauge bosons cannot couple with them. The 3-3-1 models are favor of the light dark matter.

In: Astrophysics and Condensed Matter
Editor: Thomas G. Hardwell, pp. 1-9
ISBN: 978-1-60456-332-0

Expert Commentary

THE NATURE OF THE DIMERS ASSOCIATION OF 1,2,4-TRIAZOLO[3,4-*B*]-1,3,4-THIADIAZOLES

Kang Zhang[1], Li-xue Zhang[*,1], Jian-yu Jin[2], An-jiang Zhang[1], Yu-zhen Liu[1] and Hai-le Zhang[3]

[1]College of Chemistry and Materials Engineering, Wenzhou University, Wenzhou 325027, P.R. China,
[2] College of Education, Wenzhou University, Wenzhou 325027, P.R. China
[3]Dept. of Molecular Biology & Chemistry, Scripps Research Institute, San Diego, CA 92037,USA

Abstract

Sulfur-nitrogen heterocycles such as the 1,2,4-triazolo[3,4-*b*]-1,3,4-thiadiazoles have a distinct tendency to establish intermolecular links in a solid state through secondary bonding interactions S…N. The associaton of these molecules was identified by systematic survey of crystallographic data. The interaction between these molecules is dominated by contributions arising from orbitals mixing, which can be interpreted as the donation of a nitrogen lone pair into the sulfur-centered antibonding orbitals. Because of the bulky substituents, 1,2,4-triazolo[3,4-*b*]-1,3,4-thiadiazoles associate in a coplanar fashion forming dimers. In addition to its strength, the coplanar dimer is characterized by being rigid, yet no activation barrier is expected for the association/dissociation process.

Introduction

We have witnessed remarkable progress in the construction of complex structures by the spontaneous assembly of small building blocks in the last two decades. Most of these achievements in supramolecular chemistry have been accomplished through the use of two types of interactions: hydrogen bonding and coordination of metal ions[1-3]. The use of other supramolecular forces would grant access to unique structures with unusual properties. In this

[*] E-mail address: zhanglixuelz@yahoo.com.cn

context, aurophilic interactions in 1,2,4-triazolo[3,4-*b*]-1,3,4-thiadiazoles are receiving attention[4-7].We synthesed some 1,2,4-triazolo[3,4-*b*]-1,3,4-thiadiazoles[8-10] (Chart 1) and succeded in obtaining their single crystals. Then the SBIs in 1,2,4-triazolo[3,4-*b*]-1,3,4-thiadiazoles are identified by the crystallographic data.

Heavy main-group elements often display short interatomic contacts that can be attributed to secondary bonding interactions (SBIs), attractive interactions between atoms that had previously satisfied their main valences. Some recent publications illustrate the ability of SBIs of heavy p-block elements to tailor the properties of materials and elicit unique chemistry. Dipolar molecules connected by -NO_2...I-Ar SBIs formed a noncentrosymmetric lattice capable of optical second-harmonic generation[11], a chiral fluorocarbon was resolved thanks to the formation of an infinite helix through Br^-...Br-C SBIs[12]▯ chalcogen-chalcogen SBIs in tellurium-and seleniumsubstituted acetylenes assembled nanotubes that host organic solvents,[13-17] and the addition of a selenenyl halide to olefins was stereochemically controlled by the intramolecular SBI of Se with a chiral amine.[18] More examples in organochalcogen compounds are collected in a detailed review.[19]

In search for the SBIs in 1,2,4-triazolo[3,4-*b*]-1,3,4-thiadiazoles, we have examined published structural data, with the intention of identifying the best candidates and applying them in future experimental investigations. Here, we reported a detailed study of the nature of the dimers association of 1,2,4-triazolo[3,4-*b*]-1,3,4-thiadiazoles.

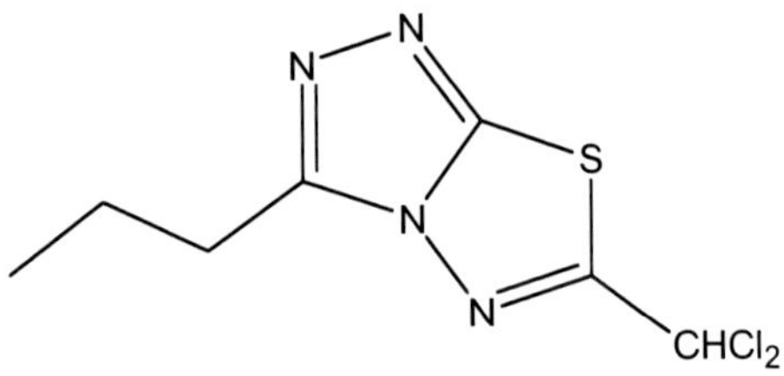

I

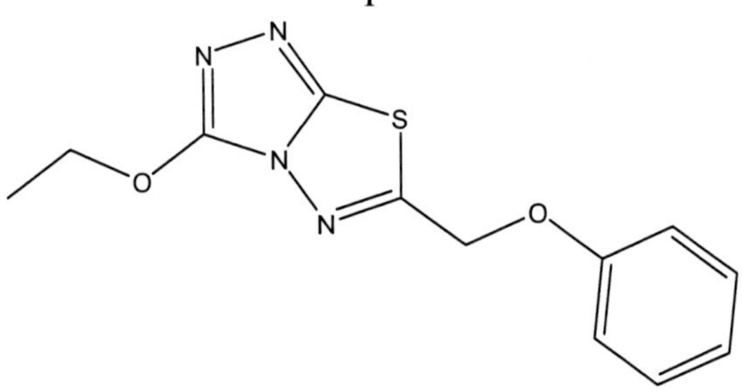

II

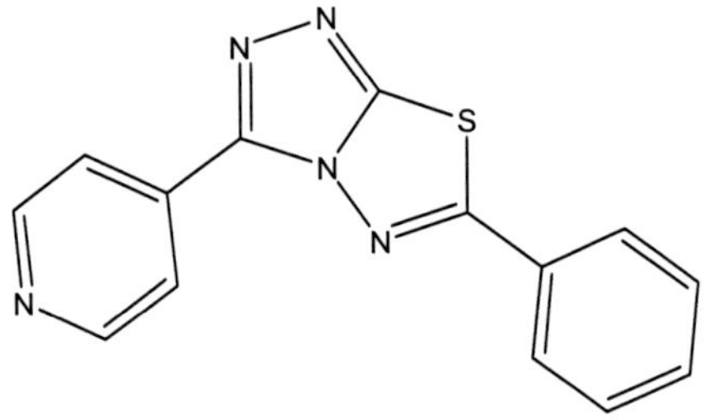

III

Chart 1.

Results and Discussion

The crystal structures of compounds 1,2,4-triazolo[3,4-*b*]-1,3,4-thiadiazoles often display intermolecular contacts that can be ascribed to SBIs. Of particular interest are those cases in which two rings are associated by two antiparallel S...N SBIs, forming a six-membered cyclic supramolecular synthon (Scheme 1).The arrangement of these heterocycles is especially attractive as it leaves room for pendant functional groups. A survey of the Cambridge Structural Database (CSD)[20-21] was performed to identify those crystals where these features are observed.[22-45] This provided the SBIs distances of the 1,2,4-triazolo[3,4-*b*]-1,3,4-thiadiazoles that are compared in Table 1 to the sum of the van der Waals radii.The set of data suggesting that the SBIs distances are much shorter than van der Waals radii. Thus, the SBIs are much stronger than the corresponding van der Waals force. The molecular structure of I, showing ellipsoids at the 30% probability level, is showed in Fig.1.The packing diagram for I assembled though the secondary bonding interactions S...N is showed in Fig.2. The crystal data and summary of data collection and structure refinement of I are given in Table 2. Selected bond lengths and angles are given in Table 3. The geometric calculations were performed using the program SHELXL-97. The Cambridge Structural Database survey also found that thiadiazoles are frequently associated in dimers.

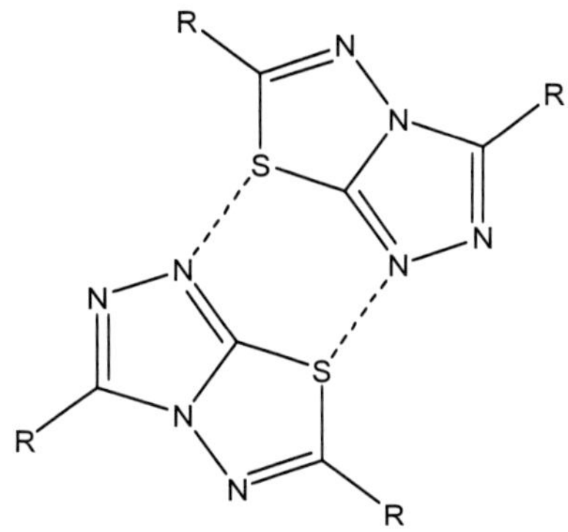

Scheme 1.

Table 1. SBIs Distances (Å) in 1,2,4-Triazolo[3,4-*b*]-1,3,4-thiadiazoles

	I	II	III
$r_{Evdw} + r_{Nvdw}$	3.26	3.26	3.26
S...N	2.85	3.03	2.87

These "monomers" were used to construct models of dimers, which were fully optimized. All three dimers converged to centrosymmetric coplanar structures with the two expected SBIs (Scheme 1), These pairs are in principle stable because their energies are lower than those of the two isolated molecules and all their vibrational frequencies are real. The 1,2,4-triazolo[3,4-*b*]-1,3,4-thiadiazoles in the dimers are compared to those of the isolated molecule in Table 4. The S...N SBI distances are in agreement with the average of the distances experimentally observed.

The total interaction energy between two thiadiazole molecules forming a dimer can be split in three contributions (Eq. 1): reorganization of the structure to adopt the geometry attained in the dimer, binding of the two fragments, and the zero-point energy correction. These contributions are presented in Table 5 for the thiadiazole ring. The total interaction energy of the 1,2,4-triazolo[3,4-*b*]-1,3,4-thiadiazoles is -2.82 kJ/mol. These calculations cannot account for London dispersion forces, but in any case these would only provide a small attractive contribution that would not significantly modify the results. A similar trend has been obtained using HF/3-21+G(dp) calculations for the dimers of chalcogen diamides.[46] Using VT NMR, Tomoda has estimated that Se…N interaction energies lie between -31.8 and -79.5 kJ/mol for intramolecularly coordinated amino selenides.[47]

$$\Delta E_{total} = 2\Delta E_{Reorganization} + \Delta E_{Bonding} + \Delta ZPE \text{ (Eq.1)}$$

The most straightforward interpretation of the binding energy is twice the energy of the individual SBI. A more complete description of the stability of the dimers is obtained by plotting the binding energy as a function of the S–N distance as the rings are displaced along the direction of the S…N axes. This analysis reveals that the thiadiazole pairs sit in well-defined potential minima, and there is no activation barrier expected for the processes of association and dissociation of these molecular aggregates.

Table 2.

Crystal data and summary of data collection and structure refinement	
Compound	$C_7H_8Cl_2N_4S$
Color	Colorless
Formula weight	251.13
Crystal system, space group	Monoclinic, C2/c
Temperature,℃	25(298K)
Cell constants	
a (Å)	21.116(4)
b (Å)	5.1836(9)
c (Å)	19.781(3)
α (°)	90
β (°)	90
γ (°)	90
Volume (Å^3)	2105.8(6)
Formula units	8
Calculated density (Mg/m^3)	1.584
F(000)	1024
Absorption coefficient, mm^{-1}	0.780
Limiting indices	$-20 \leq h \leq 24$; $-6 \leq k \leq 6$; $-23 \leq l \leq 21$
Reflections collected / unique	5254 / 1880 (R(int) = 0.0196)
Absorption correction	Semi-empirical from equivalents
Max. and min. transmission	0.8724 and 0.7720
Refinement method	Full-matrix least-squares on F^2
Data / restraints / parameters	1880 / 0 / 128
Goodness-of-fit on F^2	1.061
Final *R* indices [$I > 2\sigma(I)$]	$R_1 = 0.0394$, w$R_2 = 0.0983$
R indices (all data)	$R_1 = 0.0444$, w$R_2 = 0.1012$
Absolute structure parameter	0.04(10)
Largest diff. peak and hole (e Å^{-3})	0.401 and -0.299

There are two large potential barriers that prevent the lateral movement of the rings with respect to each other because of the repulsion between lone pairs of nitrogen and sulfur. In both cases, the planar conformation was the most stable and steep slopes prevent any distortions. Additional calculations considered the possibility of face-to-face interactions driven by the molecular dipole moments but they failed to produce a minimum of potential energy. All these results provide a semiquantitative view of the stability of the thiadiazole pair which, once formed, is a remarkably rigid structure.

The position of the minima in the potential surfaces is dictated by the balance of three distance-dependent contributions:[48-49] the Pauli repulsion that arises from the interaction of completely occupied orbitals, the electrostatic that results from the local dipole moments created by differences of atomic electronegativities, and the interaction of empty and occupied orbitals. The Pauli repulsion that arises from the interaction of completely occupied orbitals, the electrostatic that results from the local dipole moments created by differences of atomic electronegativities, and the interaction of empty and occupied orbitals. Coulombic interactions in this case take place mainly between the negative nitrogen and the positive sulfur atoms.

Table 3.

Selected bond lengths (Å) and angles (°) for I		
S(1)-C(3) 1.736(3)	N(1)-N(2) 1.368(3)	N(4)-C(4) 1.307(3)
S(1)-C(2) 1.760(2)	N(2)-C(3) 1.365(3)	C(1)-C(2) 1.499(3)
Cl(1)-C(1) 1.775(2)	N(2)-C(4) 1.369(3)	C(4)-C(5) 1.483(3)
Cl(2)-C(1) 1.746(2)	N(3)-C(3) 1.299(3)	C(5)-C(6) 1.516(4)
N(1)-C(2) 1.286(3)	N(3)-N(4) 1.409(3)	C(6)-C(7) 1.513(4)
C(3)-S(1)-C(2) 86.78(11)	C(2)-C(1)-Cl(2) 112.10(17)	N(3)-C(3)-S(1) 140.0(2)
C(2)-N(1)-N(2) 107.21(18)	C(2)-C(1)-Cl(1) 108.72(17)	N(2)-C(3)-S(1) 109.12(17)
C(3)-N(2)-N(1) 118.59(19)	Cl(2)-C(1)-Cl(1) 110.52(14)	N(4)-C(4)-N(2) 108.2(2)
C(3)-N(2)-C(4) 106.2(2)	N(1)-C(2)-C(1) 118.9(2)	N(4)-C(4)-C(5) 128.1(2)
N(1)-N(2)-C(4) 135.2(2)	N(1)-C(2)-S(1) 118.26(18)	N(2)-C(4)-C(5) 123.7(2)
C(3)-N(3)-N(4) 105.6(2)	C(1)-C(2)-S(1) 122.67(17)	C(4)-C(5)-C(6) 113.9(2)
C(4)-N(4)-N(3) 109.1(2)	N(3)-C(3)-N(2) 110.9(2)	C(7)-C(6)-C(5) 112.3(2)

However, the most important contribution to the stabilization of the dimers results is the result of the interaction between empty and occupied molecular orbitals; in other words, the S…N SBI has a strong donor-acceptor character. In the single molecules, the lone pairs of the nitrogen atom are in the vicinity of the HOMO while the LUMO+1 and LUMO+2 consist of combinations of the $\sigma^*_{S\text{-}N}$ orbitals. Therefore, the nitrogen atom can donate a lone pair into low-lying empty orbitals which are polarized toward the sulfur. This interpretation is consistent with the model of three-center four-electron bonding that has gained increasing acceptance in the interpretation of main-group SBIs.[50-52] The previous discussion is primarily focused on σ interactions only and ignores the intense repulsion that exists between the lone pairs of nitrogen and the sulfur atom. Studies of the intramolecular "coordination" of chalcogen atoms in diazenes indicated that π interactions can substantially contribute to alleviate such repulsion and in one case are indispensable to stabilize the secondary S…N interaction.[53] In the I dimers, the σ and π orbitals are easily distinguished by symmetry. The

σ interactions are indeed stabilizing, but their contribution is in all instances I order of magnitude smaller than the σ. All the dimers are stable in calculation where the π virtual orbitals are removed but not when the σ acceptor orbitals are excluded.

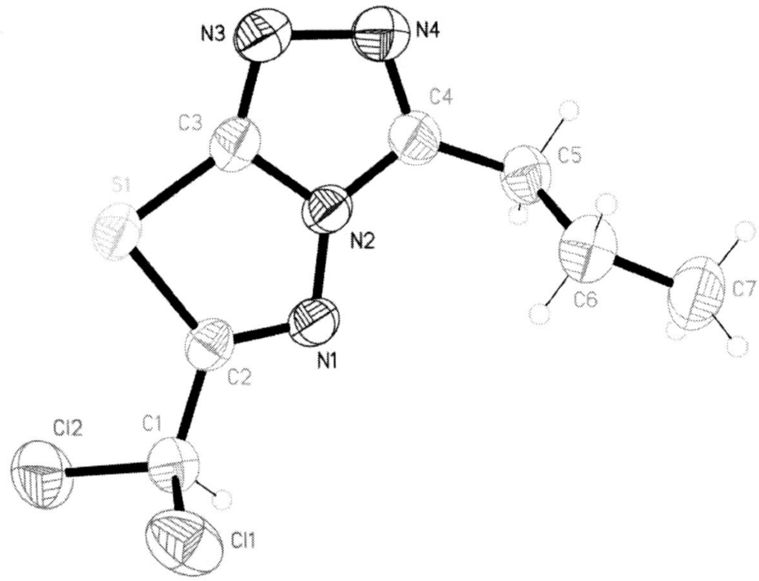

Figure 1. The molecular structure of I , showing ellipsoids at the 30% probability level.

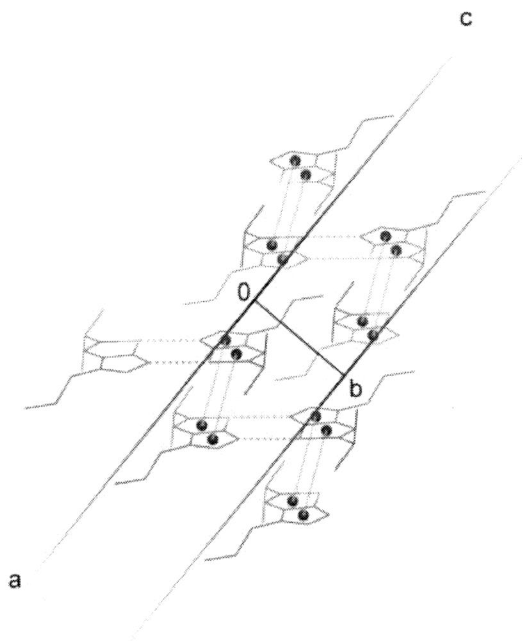

Figure 2. The packing diagram for I assembled though the secondary bonding interactions S…N.

Table 4. Optimized Bond Length (Å) and Angle (°) for the Model Structures of 1,2,4-Triazolo[3,4-*b*]-1,3,4-thiadiazoles and Their Dimers

	I	I_2
C(2)-S(1)	1.769	1.760
C(3)-S(1)	1.730	1.736
C(2)=N(1)	1.302	1.286
C(3)=N(2)	1.357	1.365
N(1)-N(2)	1.369	1.368
S(1)...N(3)		2.85
C(2)-S(1)-C(3)	88.05	86.78
N(1)=C(2)-S(1)	115.72	118.28
C(2)=N(1)-N(2)	108.66	107.19
N(1)-N(2)=C(3)	118.39	118.58
N(2)=C(3)-S(1)	109.16	109.14

Table 5. Contributions to the Energy of Interaction between Two Thiadiazole Molecules

	contribution	
reorganization	binding	ZPE
-0.08	-5.05	2.39

Conclusions and Outlook

The nature of the dimers association of 1,2,4-triazolo[3,4-*b*]-1,3,4-thiadiazoles is attributed to SBIs, which are a result of both electrostatic and orbital contributions; most prominent is the donation of a long pair of the nitrogen into the antibonding orbitals of the sulfur. The S...N distance is much shorter than the van der Waals radii, so the SBI in 1,2,4-triazolo[3,4-*b*]-1,3,4-thiadiazoles is much more than a van der Waals contact.

The thiadiazole dimmer constitutes a promising supramolecular synthon that, in addition to being stable, is fairly rigid yet should be labile enough to form and dissociate reversibly. An additional feature of sulfur-nitrogen SBIs is that they enable electron communication between molecules, which would be advantageous in conducting materials. The results of these investigation call for more synthetic and structural studies of these and related heterocycles and invite the design of supramolecules based on these interactions.

Acknowledgement

We are grateful for financial support from the Natural Science Foundation of Zhejiang Province, China (Project No. M203149).

References

[1] Lehn, J. M. *Supramolecular Chemistry: Concepts and Perspectives*; VCH: New York, 1995.
[2] Leininger, S.; Olenyuk, B.; Stang, P. *J. Chem. Rev.* 2000, 100, 853.
[3] MacDonald, J. C.; Whitesides, G. M. *Chem. Rev.* 1994, 94, 2383.
[4] Pathaneni, S. S.; Desiraju, G. R. *J. Chem. Soc., Dalton Trans*. 1993, 319.
[5] Desiraju, G. R. *Dalton* 2000, 3745.
[6] Mohr, F.; Jennings, M. C.; Puddephatt, R. *J. Angew. Chem*., Int. Ed. 2004, 43, 969.
[7] Vicente, J.; Chicote, M.-T.; Saura-Llamas, I.; Lagunas, M.-C.; De Arellano, M. C. R.; Gonzalez-Herrero, P.; Abrisqueta, M.-D.; Guerrero, R. *Met. Clust. Chem*. 1999, 1, 493.
[8] Yu, M; Zhang, K; Qian, K. D; Zhang, L. X; Liu, Y. Z; *Acta Crystallogr*., Sect. E 2007, 63, o2551.
[9] Lei, X. X; Zhang, A. J; Huang, X. B; Zhang, L. X; *Acta Crystallogr*., Sect. E 2006, 62, o4418.
[10] Huang, X. B; Liu, M. C; Zhang, L. X, Zhang, A. J, Xu, L. Y, Hu, M. L; *Acta Crystallogr*., Sect. E 2005, 61, o2233.
[11] Sarma, J. A. R. P.; Allen, F. H.; Hoy, V. J.; Howard, J. A. K.; Thaimattam, R.; Biradha, K.; Desiraju, G. R. *Chem. Commun*. 1997, 101.
[12] Farina, A.; Meille, S. V.; Messina, M. T.; Metrangolo, P.; Resnati, G.; Vecchio, G. *Angew. Chem*., Int. Ed. 1999, 38, 2433.
[13] Gleiter, R.; Werz, D. B.; Rausch, B. J. *Chem. Eur. J.* 2003, 9, 2676.
[14] Werz, D. B.; Staeb, T. H.; Benisch, C.; Rausch, B. J.; Rominger, F.; Gleiter, R. *Org. Lett*. 2002, 4, 339.
[15] Werz, D. B.; Gleiter, R.; Rominger, F. J. *Org. Chem*. 2002, 67, 4290.
[16] Werz, D. B.; Rausch, B. J.; Gleiter, R. *Tetrahedron Lett*. 2002, 43, 5767.
[17] Werz, D. B.; Gleiter, R.; Rominger, F. *J. Am. Chem. Soc*. 2002, 124, 10638.
[18] Fujita, K.-i.; Murata, K.; Iwaoka, M.; Tomoda, S. *Tetrahedron* 1997, 53, 2029.
[19] Vargas-Baca, I.; Chivers, T. *Phosphorus Sulfur Silicon Relat. Elem*. 2000, 164, 207.
[20] Allen, F. H.; Kennard, O. *Chem. Des. Automat. News* 1993, 8, 31.
[21] Allen, F. H.; Davies, J. E.; Galloy, J. J.; Johnson, O.; Kennard, O.; Macrae, C. F.; Mitchell, E. M.; Mitchell, G. F.; Smith, J. M.; Watson, J. *J. Chem. Inf. Comput. Sci.* 1987, 31, 187.
[22] Suzuki, T.; Fujii, H.; Yamashita, Y.; Kabuto, C.; Tanaka, S.; Harasawa, M.; Mukai, T.; Miyashi, T. J. *Am. Chem. Soc*. 1992, 114, 3034.
[23] Ono, K.; Tanaka, S.; Yamashita, Y. Angew. Chem., Int. Ed. Engl. 1994, 33, 1977.
[24] Esteban-Calderon, C.; Martinez-Ripoll, M.; Garcia-Blanco, S. *Acta Crystallogr*., Sect. C 1984, 40, 80.
[25] Barclay, T. M.; Cordes, A. W.; George, N. A.; Haddon, R. C.; Itkis, M. E.; Mashuta, M. S.; Oakley, R. T.; Patenaude, G. W.; Reed, R. W.;Richardson, J. F.; Zhang, H. *J. Am. Chem. Soc*. 1998, 120, 352.
[26] Yamashita, Y.; Eguchi, J.; Suzuki, T.; Kabuto, C.; Miyashi, T.; Tanaka, S. *Angew. Chem., Int.* Ed. Engl. 1990, 29, 643.
[27] Bagryanskaya, I. Y.; Gatilov, Y. V.; Miller, A. O.; Shakirov, M. M.; Zibarev, A. V. *Heteroat. Chem*. 1994, 5, 561.

[38] Klimasenko, N. L.; Chetkina, L. A.; Ginzburg, S. L.; Neigauz, M. G.; Smelyanskaya, E. M. *Zh. Strukt. Khim*. 1973, 14, 108.
[29] Kane, J.; Schaeffer, R. *Cryst. Struct. Commun*. 1981, 10, 1403.
[30] Gieren, A.; Betz, H.; Hubner, T.; Lamm, V.; Neidlein, R.; Droste, D. *Z. Naturforsch*., *B* 1984, 39, 485.
[31] Yamashita, Y.; Saito, K.; Suzuki, T.; Kabuto, C.; Mukai, T.; Miyashi, T. *Angew. Chem*., Int. Ed. Engl. 1988, 27, 434.
[32] Yamashita, Y.; Ono, K.; Tanaka, S.; Imaeda, K.; Inokuchi, H.; Sano, M. *Chem. Commun*. 1993, 1803.
[33] Yamashita, Y.; Ono, K.; Tomura, M.; Tanaka, S. *Tetrahedron* 1997, 53, 10169.
[34] Hatta, T.; Kawano, M.; Imaizumi, Y.; Tsuge, A.; Moriguchi, T.; Maeda, H.; Tsuge, O. *Heterocycles* 1997, 46, 651.
[35] Fukushima, T.; Okazeri, N.; Miyashi, T.; Suzuki, K.; Yamashita, Y.; Suzuki, T. *Tetrahedron Lett*. 1999, 40, 1175.
[36] Richardson, C.; Steel, P. J. Acta Crystallogr., Sect. C 2001, 57, 197.
[37] Suzuki, T.; Fukushima, T.; Miyashi, T.; Tsuji, T. *Angew. Chem*., Int. Ed. Engl. 1997, 36, 2495.
[38] Mellini, M.; Merlino, S. *Acta Crystallogr*., Sect. B 1976, 32, 1074.
[39] Tian, W.; Grivas, S.; Olsson, K. *J. Chem. Soc*., Perkin Trans. 1 1993, 257.
[40] Klimasenko, N. L.; Chetkina, L. A.; Gol¢der, G. A. *Zh. Strukt. Khim*. 1973, 14, 515.
[41] Bertini, V.; Dapporto, P.; Lucchesini, F.; Sega, A.; Munno, A. D. *Acta Crystallogr*., Sect. C 1984, 40, 653.
[42] Neidlein, R.; Knecht, D.; Gieren, A.; Ruiz-Perez, C. Z. *Naturforsch*., *B* 1987, 42, 84.
[43] Chivers, T.; Gao, X.; Parvez, M. *Inorg. Chem*. 1996, 35, 9.
[44] Hanasaki, Y.; Watanabe, H.; Katsuura, K.; Takayama, H.; Shirakawa, S.; Yamaguchi, K.; Sakai, S.; Ijichi, K.; Fujiwara, M.; Konno, K.; Yokota, T.;Shigeta, S.; Baba, M. *J. Med. Chem*. 1995, 38, 2038.
[45] Mikhno, O. A.; Ezhkova, Z. I.; Zhdanov, G. S. Kristallografiya 1973, 18, 99.
[46] Kontopoulos, C.; Sigalas, M. P. *J. Mol. Struct*. (THEOCHEM) 1999, 490, 125.
[47] Iwaoka, M.; Tomoda, S. J. *Am. Chem. Soc*. 1996, 118, 8077.
[48] Ziegler, T.; Rauk, A. *Inorg. Chem*. 1979, 18, 1755.
[49] Ziegler, T.; Rauk, A. *Inorg. Chem*. 1979, 18, 1558.
[50] Landrum, G. A.; Hoffmann, R. *Angew. Chem*., Int. Ed. 1998, 37, 1887.
[51] Tani, K.; Kato, S.; Kanda, T.; Inagaki, S. *Org. Lett*. 2001, 3, 655.
[52] Iwaoka, M.; Komatsu, H.; Katsuda, T.; Tomoda, S. J. *Am. Chem. Soc*. 2004, 126, 5309.
[53] Chivers, T.; Krouse, I.; Parvez, M.; Vargas-Baca, I.; Ziegler, T.; Zoricak, P. *Inorg. Chem*. 1996, 35, 5836.

In: Astrophysics and Condensed Matter
Editor: Thomas G. Hardwell, pp. 11-69

ISBN: 978-1-60456-332-0

Chapter 1

CRYSTALLOGRAPHIC RESEARCH DEVELOPMENTS

Gordon Leonard[a], Didier Nurizzo[a], Christoph Mueller-Dieckmann[a], Edward Mitchell[a], Joanne McCarthy[a], David Flot[b], Isabel Bento[c] and Peter Lindley[c]

[a] European Synchrotron Radiation Facility, Grenoble Cedex, France.
[b] European Molecular Biology Laboratory, Grenoble Cedex 9, France.
[c] Instituto de Tecnologia Química e Biológica, Universidade Nova de Lisboa, Oeiras, Portugal.

Abstract

It is now generally accepted that knowledge of three-dimensional structure at the molecular level is absolutely essential for understanding the function of biological macromolecules and their interactions with other molecules. Indeed, much of modern biological and medical research is aimed at understanding life at the molecular level. Thus, three-dimensional structural information is essential for understanding disease, for developing new drugs, for improving the environment, and for many other applications of importance to society through biotechnological developments by the pharmaceutical and agronomical industries. X-ray crystallography is probably the most effective way of obtaining three-dimensional structural information, although there are a number of limitations including the necessity to produce single crystals that yield good diffraction patterns to a useful resolution. Over the last decade or so, dramatic developments have been made to make macromolecular crystallography more of a routine scientific tool for structure determination rather than a highly specialised discipline. Automation has been invoked to accelerate crystallisation, data collection, and structure elucidation procedures. One of the major advances has involved the use of synchrotron radiation which provides a very brilliant, highly focused source of X-rays which is also readily tuneable over the wavelength range of most use to structural biologists. This has led, in turn, to the development of methods to overcome the so-called phase problem, a necessary prerequisite for deriving structural information from X-ray diffraction patterns. In parallel, many advances have been made in the problem of obtaining suitable single crystal samples for X-ray studies. Although these have been directed mainly to the cases of soluble proteins, the structures of integral membrane and associated membrane proteins are now being increasingly studied. The information deriving from the large amount of structural data that is now under deposition has also led to advances in the field of bioinformatics. This review will attempt to describe some of the more important technical advances that have been made with respect to crystallization, synchrotron beam lines and data collection, and reflection phase determination utilising the variability of wavelength made possible by the use of synchrotron radiation. It will also indicate some of the more important current limitations and future perspectives.

1. Introduction

During the last 10-15 years macromolecular crystallography (MX) has become the predominant tool for the investigation of structure-function relationships in biology. Nowadays, it is almost considered a right by those working in the life sciences to be able to deduce the detailed structures of interest by using this method. This is a significant contrast to the preceding decade or two, when determination of these macromolecular crystal structures was a very difficult, lengthy procedure requiring expert and specialist knowledge. This leap forward is due in part to the advent of new biological tools for providing purified proteins and for producing them in a crystal form suitable for X-ray diffraction studies, but it is primarily due to the tremendous advances in technology which have not only improved the quality, sensitivity and speed of the measurements made, but have also allowed these powerful experiments to be automated. In this regard the use of synchrotron radiation has played a crucial central role. The result is a truly golden age for MX where the combination of new specimen preparation and handling methods with the full automation of the MX experiment at synchrotron sources is providing an unprecedented throughput of new structures.

In this chapter a brief account will be given of crystallization methods, but the bulk of the material will be oriented towards the major advances brought about by the availability of synchrotron radiation. These include MX beam line design and automation, data collection, and structure elucidation utilizing the tunability of the synchrotron source.

2. Protein Crystallization

The developments that have been achieved in molecular biology in the genomics era have been extended today to the three dimensional structural level, as genes encode for proteins that show specific functions which are determined by their three-dimensional structure. A major effort has been driven towards the elucidation and understanding of protein structure and the most widely used technique is single crystal X-ray crystallography. This technique is completely dependent on the production of highly ordered crystals with a suitable size to allow collection of precise X-ray diffraction data. With all the developments that have been implemented concerning X-ray sources and software algorithms, obtaining such ordered crystals constitutes the bottleneck in the determination of a protein three-dimensional structure. This is particularly true for integral membrane or membrane associated proteins, although technology is developing rapidly as more investigators enter the field, but insoluble proteins still present formidable problems and have to be transformed to soluble versions or complexed with soluble components.

In crystallisation a protein solution is brought into supersaturation by varying several parameters such as pH, temperature, ionic strength and specific concentration of salts, organic additives and detergents [Ataka, 1993; McPherson, 1999; Ducruix & Giegé, 1999]. It appears to be of major importance to work with a pure protein sample in contrast to small organic molecules, for example, where crystallization has long been a means of providing purity. Impurities in protein solutions are known to affect the crystallisation process preventing it from happening or causing contamination of the crystal lattice and consequently poor quality crystals.

Usually, protein crystallisation comprises two steps; screening and optimization. The objective of the first step is to screen a wide range of parameters that affect crystallization. Once some promising conditions are found, these are then optimised. Whereas the screening step is more a trial and error process, the optimisation of crystals is more a rational process based on a phase diagram. The phase diagram is divided in four zones; a zone of high degree of supersaturation where proteins will precipitate; a zone of a moderate degree of supersaturation where nucleation occurs; a metastable zone with a low degree of supersaturation where crystals grow in a well ordered manner and no further nucleation occurs; a zone of undersaturation where the protein is in solution and crystallisation does not occur [Chayen, 2004]. In a crystallisation experiment it is necessary to drive the system containing the protein solution into a moderate degree of supersaturation for nucleation to take place and subsequently go back to metastable conditions for nuclei to grow, avoiding in this way the formation of more nuclei.

There are four standards methods used to a obtain protein crystals; batch, free interface diffusion, dialysis and vapour diffusion. In the batch method the protein solution is mixed with the crystallisation agents at conditions where supersaturation is achieved immediately. In the other three methods at the beginning of the experiment the protein solution is mixed with the crystallisation agents at undersaturated conditions and, as the system reaches the equilibrium with the reservoir solution, saturation is gradually achieved. The vapour diffusion method uses evaporation and diffusion of water between solutions of different concentration to drive the protein solution into the supersaturation state. Using the sitting drop or the hanging drop technique a drop containing protein and reservoir solution, diluted from their initial concentration, is suspended and sealed over a reservoir solution containing the precipitant agents at the desired final concentration. The difference in precipitant concentration between the reservoir solution and the drop is the driving force that makes water evaporate from the drop until the precipitant concentration reaches that of the reservoir. In the free interface diffusion method layering two solutions of different density, the concentrated protein solution onto the concentrated precipitant solution, makes them slowly mix over time leading to protein nucleation and crystal growth at the interface between the two layers. This method has the disadvantage of consuming large amounts of protein. In the dialysis method diffusion through a semi-permeable membrane is used to slowly equilibrate the protein solution against the precipitant solution to achieve supersaturated conditions and eventually crystallization.

The first step of crystallization is the screening of a large number of different conditions with the aim of obtaining an initial hit. This is often performed neglecting any knowledge of the physical characteristics of the protein, although use of this knowledge can save time and effort. One reason for this is that considerable effort has been made in the past decade towards the automation of this initial step and today crystallisation plates support screening of thousands of conditions and automated evaluation enables the results to become readily available [Berry *et al.,* 2006; Korczynska *et al.,* 2007 and references therein]. Crystallization robots, which dispense nanolitre volume samples have been designed to perform crystallization experiments using microbatch and vapour diffusion methods involving hanging and sitting drop techniques. Frequently, crystals are mounted and used for X-ray experiments without any further optimisation. Indeed, the direct exposure of plates containing crystals to the X-rays [Jacquamet *et al.,* 2004; Yadav *et al.,* 2005; Eisenstein, 2007] has been reported. Nevertheless, obtaining crystals of suitable quality and size for X-ray experiment is still problematic for a large number of proteins. Therefore it is of major importance to be able to manipulate the phase diagram in a way that crystal growth is promoted [Chayen, 2004] whereas the formation of new nuclei is avoided.

A common method used in crystal optimisation, that separates nucleation from crystal growth, is seeding. Two different types of seeding are normally used, macro- and micro-seeding. The main objective is to use a few nuclei as seeds and transfer them to new drops previously equilibrated at a low degree of supersaturation. The results obtained can be further improved by simultaneously fine-tuning some of the crystallization parameters, such as protein and precipitant concentration, buffer pH and the use of additives [Bergfors, 2003; Stura, 1999].

Crystallization in gels is a method also used in crystal optimisation. Crystals can grow either in a small volume of gel inside capillaries using counter diffusion techniques [Garcia-Ruiz *et al.,* 2002], or using the microbatch method with gelled drops in high-throughput mode [Chayen & Saridakis, 2002]. Oils can be also used to improve crystal quality. In the vapour diffusion method covering the reservoir of the trial sitting or hanging drops with a paraffin/silicon oil mixture will slow down the rate of equilibration avoiding in this way large amounts of nucleation [Chayen, 1997; Barends & Dijkstra, 2003; Loll *et al.,* 2003]. This method has also been adapted and automated for high-throughput mode [Chayen & Saridakis, 2002] and proved to be successful in crystallising membrane proteins [Chayen, 2003].

In the vapour diffusion method a simpler way of achieving the separation between nucleation and crystal growth is to transfer the coverslip with the crystallization drop to a reservoir containing a solution with lower precipitant concentration. The same effect can also be attained in the microbatch method by adding buffer to the crystallisation mixture [Chayen, 2004 and references therein]. There are other methods that have proved to be effective in improving crystal quality; evaporative microdialysis which is a simple variant of conventional microdialysis in which the reservoir solution is allowed to evaporate slowly, thus increasing the precipitant concentration and allowing crystals to grow without increasing nucleation, [Bunick *et al.,* 2000]; the application of magnetic fields to aid molecules that are incorporated onto the growing crystal faces to adopt the same orientation [Ataka & Wakayama, 2002]; the temperature effects on crystal quality, quantity and size has also been studied, in particular, for those proteins that exhibit temperature-sensitive solubility [Blow *et al.,*1994; Zhu *et al.,* 2006 and references therein].

Dynamic light scattering has been employed as a tool for studying pre- and post-nucleation events and for predicting favourable crystal conditions [Wilson, 2003]. This technique allows the selection of pH, buffer, temperature and precipitant type by measuring the polydispersity of the sample and the presence of protein aggregates that might interfere with the crystallisation process, when varying these parameters [D'Arcy, 1994; Bergfors, 1999]. Other approaches have also been proposed with the common objective of making protein crystallization a more successful process. McPherson and Cudney (2006) have searched for "silver bullets" by accessing the impact of 200 different chemicals on the propensity of 81 different proteins to crystallise. This study was based on the hypothesis that certain small molecules might establish stabilising, intermolecular, non-covalent interactions with protein crystals promoting in this way lattice formation [Mcpherson & Cudney, 2006]. Holyoak *et al.*, (2003) demonstrated the advantages of using malonate as a cryoprotectant and as a versatile stabilising solution for crystal manipulation and soaking for salt-grown protein crystals.

Improvement in crystal quality can also be achieved by post-crystallization treatments such as soaking, cross-linking, crystal annealing and controlled dehydration [Heras & Martin, 2005 and references therein]. In crystal annealing a flash-cooled crystal is warmed to room temperature and flash-cooled again [Harp *et al.*, 1998]. Three different protocols were reported; the macromolecular crystal annealing method [Harp *et al.*, 1998], the flash annealing method [Yeh & Hol, 1998] and the annealing on the loop method, which is a variation of the flash annealing method. However a study using all three protocols in crystals of six different proteins proved that the more effective and reproducible one was the macromolecular crystal annealing [Harp *et al.*, 1999]. A fourth annealing protocol has been reported more recently, although this approach requires a system that allows the regulation of the temperature in the cryostream because it restricts the warming temperature to be under 250K [Kriminski *et al.*, 2002].

Crystal dehydration is another post–crystallisation treatment that has been proven to produce a significant improvement in the resolution of protein crystals [Heras *et al.*, 2003; Abergel, 2004]. Several protocols of crystal dehydration have been reported including those for use with membrane proteins and dehydration can be combined with other post-crystallisation treatments giving spectacular improvements in the diffraction quality and resolution of protein crystals [Heras & Martin, 2005 and references therein]. Post-crystallisation soaking involves soaking protein crystals in a solution containing a higher concentration of precipitants or cryoprotectants. The procedure is similar to that used for crystal dehydration, but in this case dehydration does not occur and changes in the unit cell dimensions or solvent content are not observed. Nevertheless, a significant improvement of the diffraction quality of the crystal has been also reported [Heras & Martin, 2005 and references therein]. Another strategy to improve crystal diffraction properties is to cross-link protein crystals. Usually this procedure is used when crystals do not tolerate being transferred to cryoprotective solutions. In this case, using glutaraldehyde or another cross-linking agent the crystals are chemically cross-linked before transferring them to the cryoprotectant solution. Considering all four post-crystallisation treatments and the quality of the protein crystals at room and cryo temperatures Heras & Martin have proposed a protocol for the selection of the most appropriate post-crystallisation treatment to use in order to improve the diffraction quality and resolution of the protein crystals [Heras & Martin, 2005 and references therein].

Crystallization protocols are continually under refinement and development, but crystallization still has an element of "green fingers" rather than strict logic.

3. The Value of Synchrotron Radiation; 40 years of Advancements in Synchrotron Beam Lines

Ever since the first applications of synchrotron radiation to biological problems in 1971 [Rosenbaum *et al.*, 1971] and the subsequent use specifically for macromolecular crystallography for which the first experiments were made in the mid-1970s [Harmsen *et al.*, 1976; Phillips *et al.*, 1976], structures determined at synchrotrons have formed a growing fraction of all those produced annually, Figure 3.1.

The initial catalysis for growth was the construction of sources dedicated to scientists wishing to use the "annoying" radiation given off by the first generation cyclotrons dedicated to particle physics studies. This second generation of machines, the first of

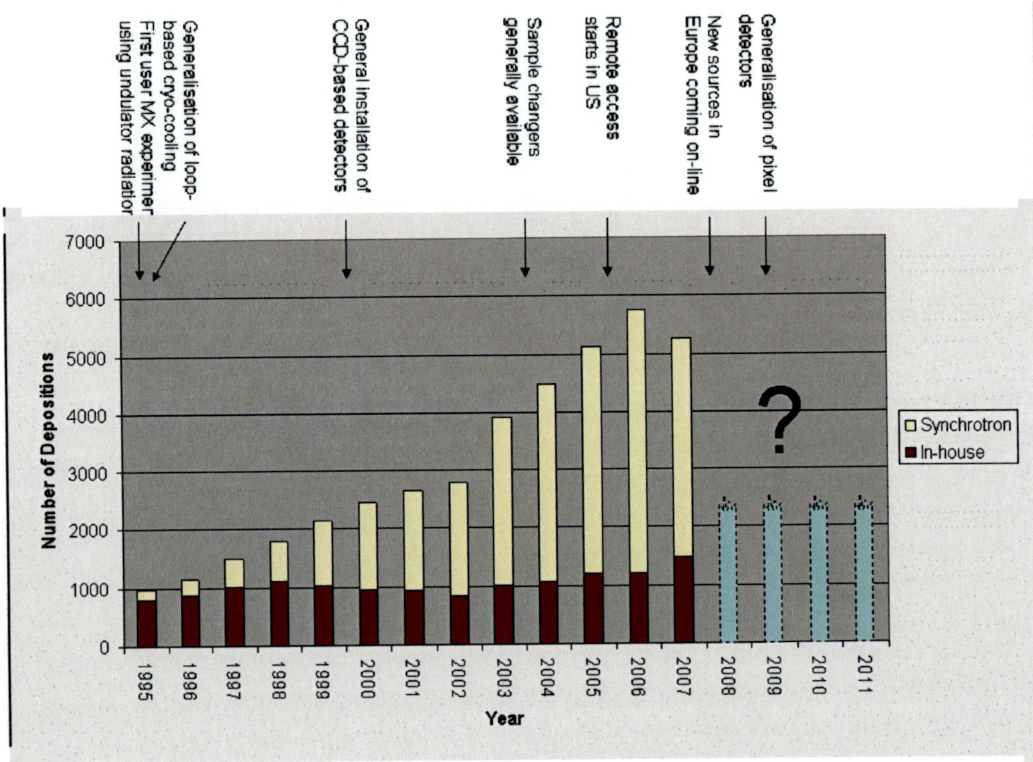

Figure 3.1. A graph of PDB depositions by time showing the ratio determined using synchrotron and in-house X-ray sources [data taken from http://biosync.rcsb.org/BiosyncStat.html; Jiang & Sweet, (2004)]. The numbers for 2007 are only for part of the year. In addition, the key technology developments are shown over the time period.

which was the Synchrotron Radiation Source (SRS) constructed at Daresbury (UK) completed in 1980 [Munro, 1997], then gave way to the third generation of synchrotrons that started with the European Synchrotron Radiation Facility (ESRF) based in Grenoble, France and was closely followed by the Advanced Photon Source (APS) in Chicago, USA and SPring8 in Japan. The third-generation facilities make use of undulators inserted in straight sections of the particle storage ring to produce highly intense beams of collimated X-rays. The growth has also been due to the improved technologies developed over the last four decades since the first synchrotron-based biological diffraction experiments took place. These decades have seen numerous developments but the main advances opening the field of synchrotron macromolecular crystallography are the following;

- Availability of dedicated synchrotron radiation sources with tunable, stable and reliable beam lines [Lindley, 1999; Helliwell, 2001].
- Development of high quality detector systems [Eikenberry *et al.*, 1992; Tate *et al.*, 1995; Broennimann *et al.*, 2006)]
- Advent of cryo-cooling to protect samples from radiation damage [Teng, 1990; Garman, 2003]
- The "industrialisation" of the synchrotron MX experiment with automation [Deacon *et al.*, 2004; Arzt *et al.*, 2005, Ueno *et al.*, 2005; Beteva *et al.*, 2006; Girard *et al.*, 2006].

Synchrotron MX beam lines have moved beyond the mere collection of simple diffraction data. Other on-line techniques are becoming available as standard, such as micro-spectrophotometry to look, for example, at metal oxidation states during data collection. The modern MX beam line is a comprehensive structure determination machine, with quality control points, high levels of automation, and ease of use – even to the point of remote data collection, permitting users to remain at home to run experiments [see McPhillips *et al.*, 2002; Beteva *et al.*, 2006; Skinner *et al.*, 2006].

The application of experiments complementing MX (such as time-resolved diffraction, powder diffraction, EXAFS, SAXS and X-ray imaging) has broadened and continues to enlarge the armoury of Life Scientists wishing to look at a system over many scales of length (angstroms to microns and up to whole organs) and time (hundreds of picoseconds upwards) to harvest different, but complementary information. Together, these techniques make the synchrotron an especially potent centre for Structural Biology, especially when combined with on-site ancillary techniques and expertise (such as electron microscopy and specialized sample preparation laboratories).

The demand for synchrotron beam time still outstrips supply and, despite the availability soon of around 30 end-stations dedicated to MX in Europe alone, the market for MX beam time will remain strong. Increased use will be made to routinely screen hundreds or even thousands of samples on a routine basis to identify the few crystals of sufficient quality to help in determining the macromolecular structure concerned. New opportunities will arise with a range of end-stations able to answer demands for specialized experiments. Above all, scientists now require regular, rapid access to synchrotron beam lines to gather high quality diffraction data; the use of synchrotron beam lines is very much now the norm rather than the exception.

3.1. Developments in the Synchrotron X-ray Source

Third-generation light sources make use of undulators (often several placed in series on a straight section) to produce highly intense and brilliant beams of X-rays. Radiation is emitted by bending ("undulating") the relativistic (around 99.99999% of the speed of light) particle beam (usually electrons) in a strong magnetic field made by an array of permanent magnets. The radiation is emitted at each peak of the sinusoidal particle path and is reinforced with each undulation following a power law of N^2 (where N is the number of periods). In this way incredibly intense beams of light can be produced. Undulators do not produce a broad continuous spectrum of light, but narrow bandwidth X-rays at well defined energies, *i.e.* a spectrum consisting of a series of harmonics. The energy of the harmonics can be tuned by varying the magnetic field (by opening the jaws of the undulator around the electron beam). In this way the output photons can be changed in energy – one of the most valuable properties of synchrotron light and one critical to experiments using an optimised anomalous signal (for example multi-wavelength anomalous diffraction, MAD, measurements).

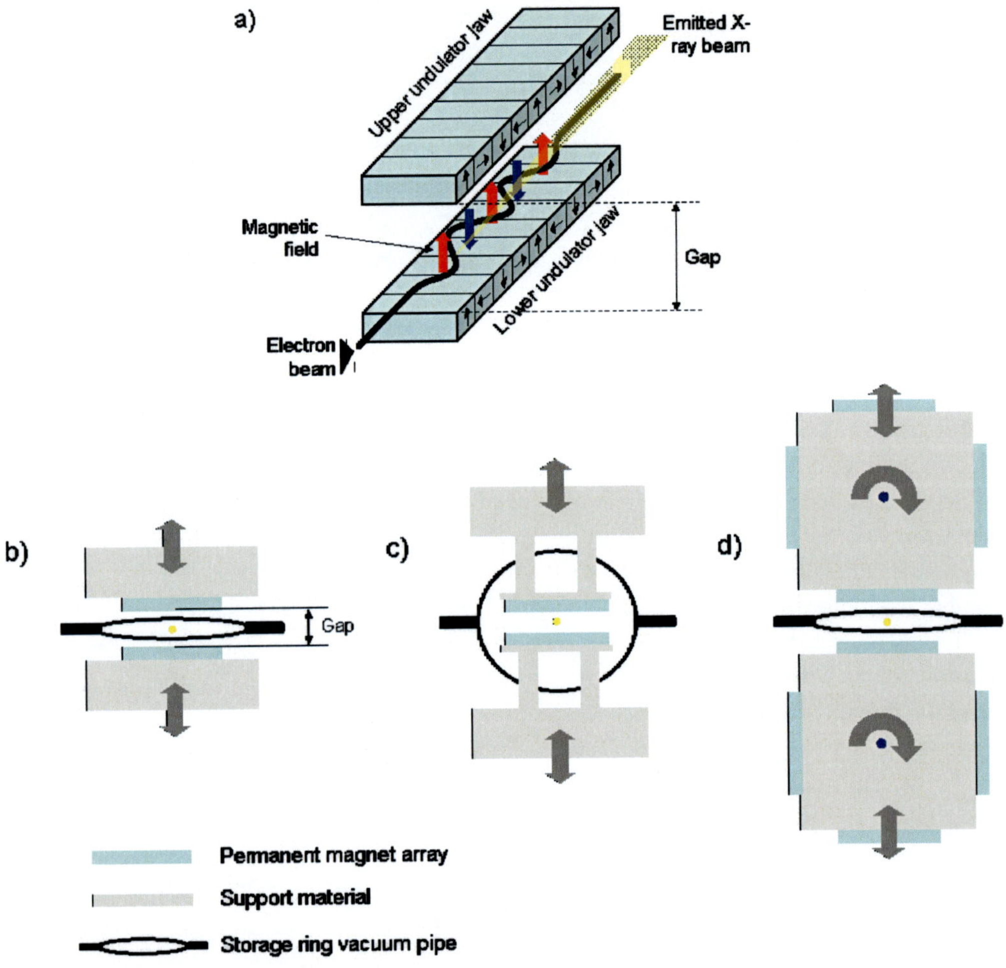

Figure 3.2. a) Schematic view of an undulator (in reality it would consist of many more magnetic periods with a total length of 1-2 m, although much longer undulators have been constructed. Schematics of three basic undulator types shown in cross-section; b) an ex-vacuum undulator c) an in-vacuum undulator with smaller minimal gap d) a revolver-type undulator (the magnetic construction can only be rotated with the jaws in the open position).

Undulator technology is evolving and the minimal jaw gap has decreased markedly in the last few years with the advent of “in-vacuum” undulators installed on third-generation sources [Hara *et al.*, 1998; Schmidt *et al.*, 2001] where the magnetic array is contained inside the vacuum vessel holding the circulating electron beam rather than above and below it outside, Figure 3.2. At the ESRF, for example, this permits the smallest gap to be 5 mm, giving a much higher magnetic field than achievable previously and in turn generating unprecedented beam intensity. Other innovations include the use of a "revolver carriage" [Isoyama *et al.*, 1989] to allow the undulator(s) to be changed on a particular beam line (Figure 3.2d), thereby allowing a broader range of energies to be applied. One application for this is to install a revolver with one undulator optimised for low energy X-rays (around 6 keV for sulphur-SAD phasing) and a second undulator optimised for a spectrum from 8 to 20 keV for more standard experiments. The use of "canted" undulators with two undulators arranged at a small angle (several mrad) to each other on a single straight section is increasing and allows two X-ray beams to be generated and used on two independent beam lines, thereby making the use of straight sections highly efficient. Examples of beam lines using canted undulators include beam line ID23 at the ESRF [Nurizzo *et al.*, 2006] and beam line GM/CA-CAT at the APS [Fischetti *et al.*, 2007].

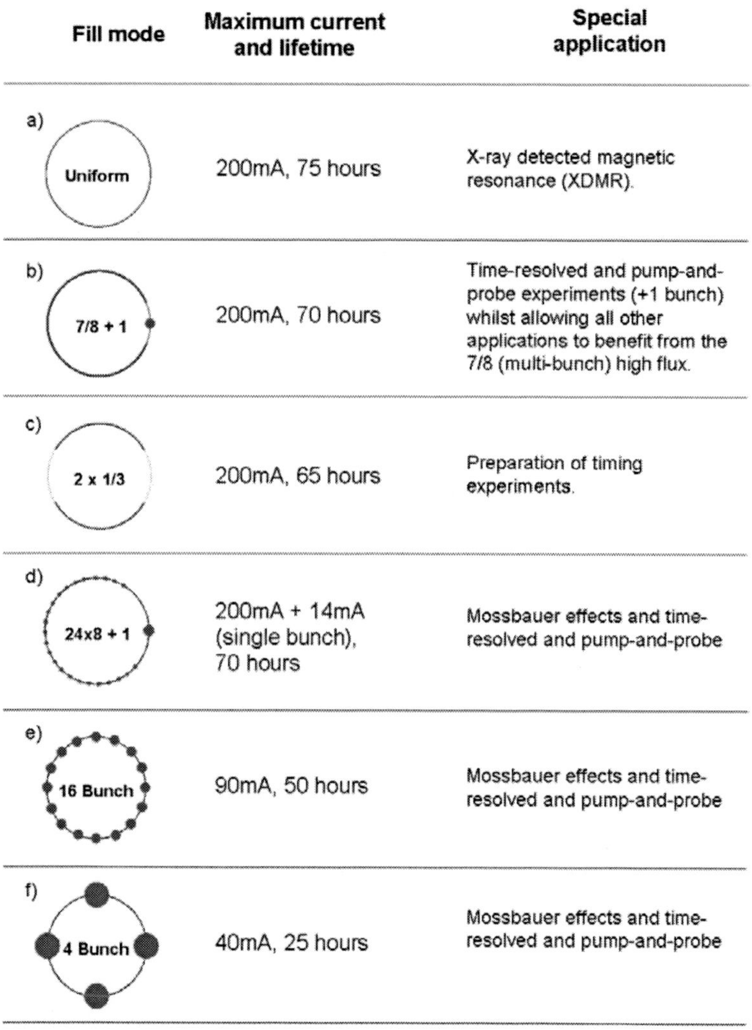

Figure 3.3. Storage ring filling modes and bunch structure used at the ESRF. The 7/8 +1 mode is the newest fill implemented at the ESRF and is used to generate strong X-ray beams with the 7/8 fill whilst still allowing time-resolved experiments to be performed with the "+1" single bunch. Pictures by courtesy of J-L Revol (ESRF).

A further key feature of synchrotron radiation is the use of the time structure that can be delivered in the X-ray beams. Though not used for standard macromolecular crystallography experiments, this time structure is used to probe reactions where structural changes take place. Several examples have been published, one of which concerns the changes that myoglobin undergoes as it releases carbon monoxide [Schotte *et al.*, 2003] and another looking at the reaction pathways in photoactive yellow protein [Ihee *et al.*, 2005]. Although these reactions take place on the picosecond scale, the stored particle beam can be given a very precise time structure with "bunches" of electrons which are used to generate X-ray pulses, Figure 3.3.

By precise synchronization with an initiating ("pump") laser pulse to the crystalline sample, these X-ray pulses can be used to perform "pump-and-probe" experiments on suitable systems. In the future, this technology may also be extended to spectroscopic techniques, allowing pump-and-probe X-ray absorption and emission spectroscopy to examine rapid changes during reactions, for example of photosystem II [Barry *et al.*, 2006; Yano *et al.*, 2005].

It should be noted that in-house X-ray sources have also evolved. Nowadays "microsources" available to be purchased off-the-shelf are able to supply up to 1.6 x 10^{11} photons/mm^2/second into a focal spot of less than 100 microns in size. The newest machines also have dual wavelength capability. Larger "compact light sources" which use a stored electron beam interacting with a laser beam, via inverse Compton scattering, to produce hard X-rays [Huang & Ruth, 1998] are coming on-stream. These sources have become feasible and cost effective with the availability of high power lasers and optimized very low emittance electron beams.

3.2. Optics Instrumentation

A synchrotron beam line can be divided into three basic sections; the X-ray source, the optics hutch and the experimental hutch, Figure 3.4. The general design is partially driven by the properties of the X-ray source. The optics hutch components are used to tailor the beam to the requirements of the experiment, whilst the experimental hutch components deal with the sample positioning in the beam and data collection. The source and the optics hutch are separated by a "front end" whilst the optics and the experimental hutches are separated by a substantial safety shutter allowing safe interventions in the experimental hutch whilst the X-ray beam is contained in the optics hutch. Front end components consist of water cooled X-ray absorbers, carbon filters and beryllium or diamond windows. Today, the optics hutch components are generally (from source to sample) a monochromator, a focusing element and various sets of beam defining slits (water cooled if in the white beam) and diagnostic components.

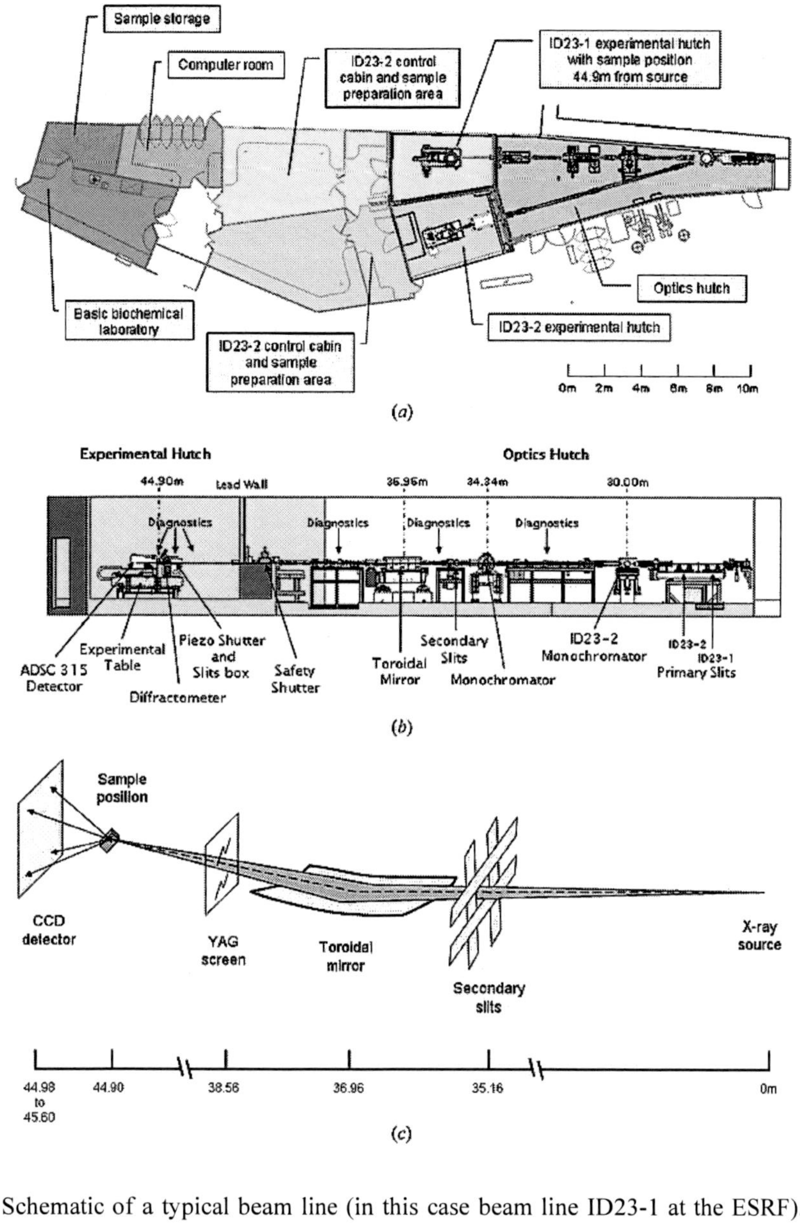

Figure 3.4. Schematic of a typical beam line (in this case beam line ID23-1 at the ESRF). This beam line has two independent end-stations through the use of canted undulators. Only the optical components of beam line ID23-1 (tunable) are shown. Reproduced with permission of the International Union of Crystallography from Nurizzo *et al.*, 2006.

3.2.1. Wavelength selection - monochromator. The X-ray source supplies a certain X-ray energy spectrum. The role of the monochromator is to provide a beam of narrow energy range selected from this spectrum. The monochromator is very often therefore a high quality single crystal with the monochromatic beam produced by diffraction. The variation of the incidence angle between the white beam and the monochromator crystal by rotation allows the beam energy to be selected, following Bragg's law. The part of the incident beam not diffracted passes through the crystal, being partially absorbed and then finally stopped by a beamstop (a tungsten block). The most common material for the monochromator crystal is silicon using the [111] reflection but Si[311], Ge[220] and diamond are also commonly used. The higher order reflections of silicon can be useful for improved energy resolution (the energy band-pass of the Si[311] reflection is smaller than Si[111] with a $\Delta E/E$ of around $5x10^{-5}$ against $1x10^{-4}$), which can be more effective for anomalous measurements despite the consequently lower intensity beam.

A number of designs exist for the monochromator: "single bounce" (one diffracting crystal), "double crystal" (two crystals) or "channel cut" (one large block where two parallel pieces are cut). In the case of a single-bounce monochromator, the rest of the experimental setup must track the reflected beam as the wavelength is changed (by rotation of the monochromator crystal). However, large energy changes require a considerable beam line realignment (up to 30 minutes). In a similar setup, a translation between two sets of two vertically stacked (but different) crystals has been used to obtain discrete energies but in a fixed geometry as implemented on SRS Daresbury beamlines 14.1 and 14.2 [Duke *et al.*, 1998]. The choice of a single crystal monochromator diffracting horizontally is an option to allow a good separation between experimental stations as with beam lines with a wiggler X-ray source or with canted insertion devices (for example beam lines ID23-1 and ID23-2 at the ESRF).

With a double-crystal monochromator, the simultaneous rotation of the two crystals allows the diffracted beam to be maintained parallel to the incident beam. To keep the beam at a fixed exit (when the rotation axis is horizontal) over all the energy range covered by the monochromator, it is necessary to translate the second crystal. A channel-cut monochromator consists of a single crystal specifically cut to have two diffracting faces parallel to each other thereby keeping the beam parallel at the cost of a few millimetres displacement over the whole energy change but with increased simplicity in alignment. As the beam is shifted vertically when the monochromator is rotated, the focusing mirror and the experimental table have to be moved vertically to recatch the beam. To limit the contamination by higher energies, the parallelism of the two diffracting crystals can be slightly detuned at the cost of beam flux.

With the monochromators described above, only one small energy range is used and the rest of the beam is lost. The Troika concept [Grübel *et al.*, 1994] is based upon the use of several transparent diamond crystals positioned in series, allowing a number of beam lines to function from one straight section [*e.g.* beamline ID14 with four stations at the ESRF; Wakatsuki *et al.*, 1998]. Three *double* crystal diamond monochromators in series (forming a "trichromator") have been used at the RIKEN beam line 1 (BL45XU) at SPring-8 for MAD data collection [Kumasaka *et al.*, 2002]. The trichromator consists of three pairs of transparent diamond monochromators simultaneously producing a single trichromatised X-ray beam of three different wavelengths. The switching and combining of any three wavelengths can be achieved in few seconds by the rotation of the beam choppers located between two diamond crystals of an individual monochromator. Three-energy MAD datasets may be therefore collected successively and rapidly without beam line realignment or indeed could be recorded on a single oscillation image.

The main part of the white X-ray beam is not diffracted by the monochromator; the non-diffracted energy is partially absorbed by the crystal and transformed into heat. Typical undulators installed on a third-generation ring straight section at minimum gap (*i.e.* highest field strength) produce powerful synchrotron radiation with a power density of about 300 kW $mrad^{-2}$ and a total power of 11 kW [Biasci *et al.*, 2002]. A careful design is therefore required to cool the monochromator crystal so that there is no vibration or drift. The cooling must cope with changes in the beam intensity linked with stored current changes and through the various fill modes of the machine. Ring operation in a top-up mode (*i.e.* frequent but small injections of electrons to top-up the stored current) is becoming more common as it maintains a constant heat load on the monochromator giving improved stability. The heat load has increased massively with the third-generation of machines. For the first-generation sources with a bending magnet, monochromator cooling with water was generally sufficient. With a wiggler X-ray source, the crystal is often set in a bath of liquid alloy such as Ga-In-Sn which is also water cooled. The advent of the third-generation storage rings with undulator light sources with higher and more concentrated heat loads (the beam is about 10 times less in size) required a new approach for the cooling of the monochromator. Whereas cooling can be made efficient to prevent melting, the main difficulty arises from the thermal distortions generated by the temperature gradients in the first monochromator crystal (*i.e.* creation of a "thermal bump" on the crystal surface). The effect of a thermal bump is to increase the divergence of the X-rays, blowing up the beam size and creating instabilities. Thus the challenge in the mid- to late-1980s was to engineer a solution capable of preserving the brilliance of the third-generation ring X-ray beams. One solution was to take advantage of the particular properties of materials such as silicon (and other diamond-like group III-V and II-VI mixed semiconductors) where the coefficient of thermal expansion passes through zero at ~125 K and the thermal conductivity is greatly enhanced at cryogenic temperatures. This development led to cryogenically cooled (using liquid nitrogen) monochromators [Bilderback *et al.*, 2000]. Cooling is either indirect *via* a copper mounting block or directly through channels in the crystal (though this technology still requires perfecting). Current increases in the existing storage rings and the development of fourth-generation (XFEL) sources will create new challenges in terms of heat load dissipation whilst maintaining the beam properties.

3.2.2. Focussing - mirror elements. The X-ray source emits the beam with a certain size and divergence (emittance). As the sample is rather far from the source itself (usually tens of metres), it is necessary to compensate the natural beam expansion by focussing it with the monochromator or by another optical element such as a mirror, generally located after the monochromator. In an ideal world, the beam size should match the sample size in order to allow the incoming X-rays to illuminate the whole sample and minimising illumination of the cryoprotectant and the sample holder whose scattering will merely be a source of noise and increased background. The beam size corresponds to the source size (defined by the storage ring properties) divided by the demagnification ratio of the focusing components (defined by the beam line); the demagnification ratio is the ratio between the optical element to source distance and optical element to sample distance, *i.e.* the relative position of the focussing element will define the beam size. A beam larger than the crystal can be reduced to match the sample size. However, in addition to loss of intensity, slits can generate adverse effects that reduce resolution and prevent measurements close to the direct beam [Le Bolloc'h *et al.*, 2002]. To achieve a high quality beam size smaller than 20 μm in diameter it is therefore necessary to focus the beam and not to simply resize it using slits. A small beam size can be achieved if the focusing element is near to the sample but the beam divergence will be increased. The beam is usually focused at the sample position but it may be focussed on the detector (*e.g.* beam lines XS06 at Swiss Light Source (SLS) and GM/CA-CAT at Advanced Photon Source).

Various systems exist to focus the X-ray beam including Fresnel zone plates [Hart *et al.*, 1966], beryllium lenses [Snigirev *et al.*, 1996, Baron *et al.*, 1999], glass capillaries [Bilderback *et al.*, 1994], but the most popular on MX beam lines is probably a simple mirror. Typically a piece of special glass or single crystal silicon coated with at least one thin metal layer (*e.g.* rhodium or platinum), a mirror can be initially flat and then mechanically bent or shaped in the horizontal or vertical or both directions. The best shape optically-speaking is ellipsoidal as there are no aberrations created but such mirrors are very difficult to make; toroidal mirrors, approaching an ellipsoidal figure, are more common as they are less difficult to make. The focus is achieved by a proper positioning of the mirror in the beam and a fine tuning by bending the mirror longitudinally. Due to the large meridionial radius of curvature (in the kilometre range) describing the mirror shape, the bender has to be very precise and stable. A strong advantage of the simple mirror is that the focusing is not function of the incoming X-ray energy, it can be large enough to accept the whole beam and reject high energy harmonics. The quality of mirrors has improved considerably over the last few years. The best commercially available mirrors now have a 0.3 μrad slope and 3 nm shape error over typically 500 mm length and a micro-roughness of 0.1 to 0.2 nanometres rms, [Freund, 2003]. With toroidal mirrors a spot size less than 100 μm in diameter can be quite easily obtained [for example 50 by 60 μm^2 full-width half-maximum for a 780 mm long mirror at ESRF beamline ID23-1; Nurizzo *et al.*, 2006].

3.2.3 Microfocus technology. The average size of crystalline samples has decreased markedly over the last few years; 50 micron samples are not uncommon and the trend is to smaller micro-sized crystals. To collect good quality useful data from such samples, beam lines with microfocus X-ray beams are being designed. Poor quality, larger crystals, often being grown from large multimeric protein complexes, may also benefit from small sections of the sample being exposed to an equally small beam in the search for the best quality data. In the case of X-ray mirrors, surface roughness plays a dominant role in their reflectivity and errors on large mirrors prevent a very small beam size being attained. One solution is to use two smaller (easier to produce) crossed mirrors; one to focus the beam horizontally and one to focus the beam vertically in the so-called Kirkpatrick-Baez geometry [Kirkpatrick and Baez, 1948]. Beam line ID13 at the ESRF is one of the pioneers in the field, producing a 30 micron beam in 1994 [Engström *et al.*, 1995] and demonstrating the benefits of collecting data from micro-crystals with a matched and focused X-ray beam. Initially devoted to small molecule crystallography, pioneering experiments were made on ID13 to see if it was possible to collect data from very small protein crystals. This work allowed one of the first structures of membrane proteins, that of bacteriorhodopsin, to be determined [Pebay-Peroula *et al.*, 1997; Luecke *et al.*, 1999] and the advantages of such a method to be evaluated [Cramer & Müller, 1997; Cusack *et al.*, 1998].

Although it was possible to use the ID13 environment successfully for macromolecular crystallography, the setup has many inherent limitations such as difficult crystal mounting, tricky crystal centering and easy misalignment of the collimator-sample-beam stop chain of elements. An improved monolithic diffractometer, with a more friendly user environment and allowing far easier crystal alignment, on-axis viewing, etc, has been in operation since the end of 1998 at ID13 [Perrakis *et al.*, 1999]. However the demand for microfocus stations dedicated to macromolecular crystallography has led to the design of a number of new beam lines. ID23-2, a fixed energy beam line at the ESRF, has been in operation since 2005 with a beam of around 5 microns in diameter. Microfocus stations with MAD capabilities are in construction in the new third-generation sources (*e.g.* beamline I24 at Diamond, Proxima II at Soleil, MX2 at Petra-III) and plans also exist in existing sources (Structural Biology Center at APS, BL41XU at SPring 8). Submicron beams of 100 nm or less can be generated by X-ray optical elements such as crossed mirrors in Kirkpatrick-Baez geometry, crossed Fresnel systems or crossed refractive lenses; the first experiments on macromolecular crystal systems making use of such submicron beams ("nanofocus") have already taken place [Riekel *et al.*, 2005].

3.3. Technology for Automation

Fortunately, synchrotron radiation macromolecular crystallography (SRMX) facilities are particularly amenable to automation as they specialise in one technique (data collection from a single crystal of a macromolecule) that is repeated many times a day and consists of a simple sequence of operations, Figure 3.5. Several years ago the idea of streamlining the process of determining macromolecular crystal structures into a "pipeline" was conceived [Stevens & Wilson, 2001]. Of course to streamline a series of events into a pipeline allowing automatic structure determination requires first that each of the individual events themselves be automated. Much of the pioneering work in this area began at the Stanford Synchrotron Radiation Laboratory (SSRL) in the United States at the beginning of the current decade [Abola *et al.*, 2000, Kuhn & Soltis 2001]. Automation of the data collection part of the full structure determination pipeline has been adopted by almost all the other synchrotron facilities worldwide since then, culminating at the present time in rather remarkable progress towards a fully automatic data collection pipeline (DCP).

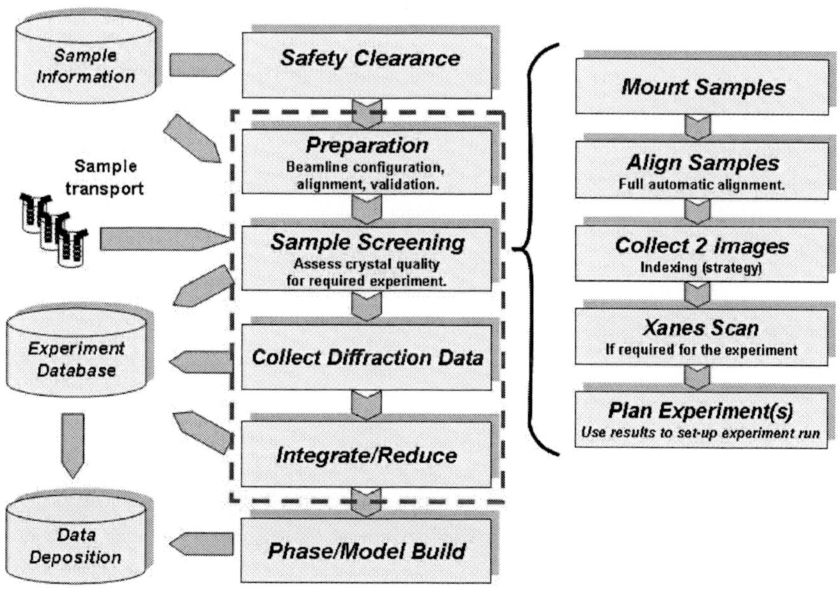

Figure 3.5 Schematic view of a typical MX experiment which follows a well-established layout permitting a systems approach to integrating and automating the steps.

Included in any automation procedure towards a DCP at synchrotron facilities should be: administrative and safety clearance for the sample to be studied, the provision of a suitable X-ray beam onto the sample, the mounting, dismounting and storage of the sample, the reduction (integration and scaling) of any data collected, initial structure solution and refinement processes and the linking of SRMX experiments with home laboratories through the exchange of information between remote beam line sources and home data tracking systems. The implementation of automatic routines for each of these steps relies heavily on the advanced technology available for SRMX instrumentation and instrument control software. To fully comprehend the impact that progress in technology has had on the SRMX experiment, one only needs to look at how these experiments were carried out in the era of the first protein structure determination beam lines.

Although the first X-ray diffraction experiment on a biological sample (muscle fibres) was conducted at DESY in Hamburg [Rosenbaum *et al.,* 1971], the first purpose-built beam lines for protein crystallography were not built until more than a decade later. They included stations 7.2 [Helliwell *et al.*, 1982], 9.6 [Helliwell *et al.*, 1986] and 9.5 [Thompson *et al.*, 1992] at the SRS, Daresbury Laboratory, UK. These first beam lines, built over 20 years ago, used an experimental set-up based on the use of an Arndt-Wonacott rotation camera [Arndt *et al.,* 1973] along with photographic film for data acquisition. Black paper inserted between several layers of photographic film allowed, for a single oscillation, the measurement of intensity over the large dynamic range required, resulting in something like six films for a single oscillation step. These had to be chemically developed and then scanned by densitometry to obtain the diffracted spot intensities which were subsequently stored on large quantities of tape. However, the quest to automate areas of these difficult and time-consuming experiments was evident even then since the camera included a turn-table allowing the use of up to eight photographic films to be exposed without intervention. In addition, Helliwell and Thompson, (1983), introduced the first remotely controlled movement of the beam stop into and out of the beam, allowing automatic measurement of the direct beam position without having to enter the hutch.

Due to the relatively low photon flux available at these first SRMX facilities, data collections lasted typically hours, if not days. Most of the earliest beam lines were built for applications similar to those conducted on home sources and therefore featured bent single-crystal monochromators (*e.g.* SSRL VII-1, CHESS A1, DESY X11, PF 6A2, SRS 9.6) which produced focussed beams but were not readily tunable. Also the use of photographic film was not well-suited for high precision. A few beam lines were developed with readily tunable double crystal monochromators and accurate electronic area detectors called multiwire proportional counters, notably at LURE D23 [Kahn *et al.,* 1982] and at SSRL I-5 [Phizackerley *et al.,* 1986] but due to detector count-rate limitations these beam lines were used without focussing. These improvements in monochromator and detector technology allowed the first completed multi-wavelength anomalous dispersion (MAD) experiments to be undertaken using synchrotron radiation in 1985. One of these was a study on a parvalbumin from the fish *Opsanus tau*, where the calcium had been replaced with terbium, at LURE D23 [Kahn *et al.,* 1985]. Freezing crystals and maintaining them at cryo-temperatures during the MX experiment only became an integral part of the technique in the early 1990's - before that, crystals were held in capillaries and measured at room temperature. This of course resulted in significant radiation damage and meant that multiple crystals were needed to collect full datasets. For these same reasons, crystal screening was almost non-existent and data were collected from the first crystal which gave decent diffraction. The MX technique was therefore expensive in materials (the crystal samples themselves, photographic film, tape for data storage, etc) and in manpower and was extremely tricky and time-consuming to perform. In the early 1990's at around the same time that cryo-preservation techniques were being refined [reviewed by Rodgers, 1994; Garman & Schneider, 1997], several other important technological developments occurred, namely the implementation of imaging plate area detectors [Amemiya *et al.,* 1988], selenomethionyl proteins and new beam lines dedicated to MAD applications, the first of which was NSLS beam line X4A, followed by NSLS X12C, CHESS F2 and SRS 9.5. The new image plate detector systems and sample cryo-cooling, combined with beam lines purpose-built to deliver focussed radiation that could be readily tuned at high resolution signalled the beginning of a new era for MAD experiments and, from around 1994, the production of MAD structures began to grow substantially [Hendrickson, 1999]. However the readout times of several minutes for the early image plate systems combined with exposure times per image of around fifteen minutes from the most recent sources and insertion devices of that time meant that it could still take around one week for a full three-wavelength MAD dataset to be completed. In the second half of the 1990's, CCD detectors replaced image plates on most synchrotron beam lines and revolutionized diffraction data quality and data acquisition speeds. The timing was perfect for it coincided with the construction of the first protein crystallography beam lines at the ESRF, the first third-generation high brilliance synchrotron radiation X-ray source in the world. From now on, it could be hoped that neither source intensity nor detector speed or sensitivity would be the limiting factors in the MX experiment.

The sample environment at a modern SRMX facility now consists typically of a single- or multi-axis diffractometer, a fast X-ray shutter, a CCD detector, a cryo-cooling device and, for tunable energy beam lines, a fluorescence detector for performing X-ray absorption near edge structure (XANES) scans. Prior to the push for fully automating all aspects of the MX experiment which began some ten years ago or so, the diffractometers were typically rather cumbersome mechanical rotation stages with a sphere of confusion of 20 μm or more. Installed in the very crowded area around the sample itself were the fluorescence detector, various fibre optic tubes and lamps for front and backlighting of the sample, the backstop and its associated motors, the cryo-cooling system, and various cameras with their associated monitors for viewing both the sample itself and the general sample area. The result was a very congested sample environment with many individual elements needing individual care and attention. Mounting and aligning the sample in the X-ray beam was a slow, laborious process involving determining firstly the centre of the axis of rotation of the diffractometer using a sharp pin or wire and then determining the beam position using X-ray sensitive paper on a special holder mounted on the diffractometer and the sample viewing camera mounted perpendicular to the beam axis and therefore prone to introducing a parallax error. The fast X-ray shutters were typically either stepper-motor driven solenoids or mechanical shutters controlled *via* a TTL electronic signal, the opening and closing times of which were typically of the order of 3 to 20 ms or more, depending on the age of the shutter. The evolution of the experimental end-station of the SRMX facility over the last twenty five years or so can be clearly seen in Figure 3.6.

Figure 3.6. The evolution of the protein crystallography beam line experimental end-station. (a) the Arndt-Wonacott and photographic film cassette of beam line 7.2 at Daresbury, UK in 1982 (Reprinted from Helliwell *et al.*, 1982, J. Phys. E **15**, 1363-1372. With permission from IOP Publishing Ltd.), (b) the congested sample environment of the modern but pre-automation end-stations of the ESRF in 2001, (c) the current automation-oriented experimental end-stations of the ESRF, 2007.

The astonishing leap forward in detector technology for X-ray diffraction experiments over the past two decades has been one of the key elements in pushing the advances made in other areas concerning the MX experiment. Detector development has seen greatly increased sensitivity and dynamic range, and cycle or readout times which have dropped from the days required for photographic film to the few minutes of image plate detectors and then to the tens of seconds of the first CCD detectors. The most recent CCD detectors available on the market allow around forty images to be collected per minute using a one second exposure time per image. Advances in detector technology have allowed the MX experiment to capitalise on the high photon flux available at modern synchrotrons and reduce the exposure time per image to the sub-second range. As a result, fast X-ray shutter and diffractometer technologies have had to remain abreast of this progress. Modern fast shutters use piezoelectric actuators to obtain opening and closing times of the order of 2 ms or below whilst the most modern diffractometers use an air-bearing based spindle and are both fast (maximum rotation speed of 180 degrees per second) and highly precise (sphere of confusion radius of 2 μm). These modern diffractometers also comprise much of the equipment required around the sample in a very compact and elegant way. They have integrated backstops, sample illumination and scintillating screens, which convert the X-ray beam to visible light allowing the beam to be "seen" in real time, along with a high-resolution video microscope co-axial with the X-ray beam which allows sample visualisation along the beam axis and thus without parallax error. The sample environment is now much more spacious, allowing room for the fluorescence detectors and cryocooling systems to be inserted as and when needed using pneumatic actuators. Faster diffractometers and shutters have also meant that the synchronisation between spindle movement and shutter opening and closing is extremely critical and adapted electronics cards have been developed to allow this to be done in a precise and reliable way. These state-of-the-art diffractometers, shutters and associated electronics comprise encoders feedback loops which allow the position of the phi spindle and the state of the shutter to be monitored and plotted, meaning that synchronisation may be constantly monitored via software written specifically for these diagnostics. However, further developments will be made and the latest detectors use gating technology to completely remove the need for a X-ray fast shutter. These modern detectors, such as the Pilatus 6M pixel detector [Hülsen *et al.*, 2006] developed at Paul Scherrer Institut (PSI) for the SLS, use gated electronics to turn the detector acquisition on and off in synchronisation with the movement of the phi spindle. For such a detector, when the data are written directly to memory the data collection time for a full 180 degree dataset is of the order of fifteen seconds! The current computer and network technology increases the data collection time back up to 40 images per minute when disk writing is included.

Data transfer and storage is another area where advancing technology has played an integral role in defining the modern day SRMX experiment but where perhaps the beginning of a bottle-neck is apparent. As mentioned previously, data from the earliest SRMX experiments were stored on copious numbers of reels of tape. Around a decade ago the media for data backup had evolved to the use of much more compactly-sized tapes such as DAT (~2 Gb capacity) or DLT (~80 Gb capacity) tapes. However with high-speed data collection and increasing detector sizes, huge amounts of raw image data can be generated. The most technologically advanced beam lines of the present day are capable of producing up to 1Tb of raw data over a 24 hour period! The current procedure for data backup is for users to copy raw image data to their own data storage devices (*e.g.* USB or FireWire disks, portable laptops, magnetic or optical media) or to transfer *via* FTP the images directly to their home laboratories. Even when data can be reduced reliably and sufficiently rapidly at the synchrotron site, the majority of users still do not choose to leave the synchrotron without their raw data. Nonetheless, disk space is becoming cheaper and currently moving ahead of the ability to produce raw images at synchrotrons. This allows user data to be stored on-line at synchrotron facilities, *e.g.* during several weeks at the ESRF, and therefore gives users a possibility to recuperate images if there have been data processing problems. Secure national archiving allowing data to be stored essentially indefinitely with near-line access is becoming available [Rajasekar *et al.,* 2003; Hey & Trefethen, 2003]. Data storage and curation projects exist in Europe, as part of the eScience project (http://www.rcuk.ac.uk/escience/) in the UK, and in the United States, at the San Diego Supercomputing Center (http://www.sdsc.edu/grid_cluster/index.html) which, amongst other clients, manages the diffraction data generated by the Joint Center for Structural Genomics at Stanford. However, as can be seen from the use of the Pilatus 6M gated pixel detector, the potential few seconds total data collection time of such a detector is hampered by current computer and network technology, bringing data collection times back to those currently achieved with the most modern CCD detectors. However the superior sensitivity, signal-to-noise ratio and dynamic range of such a detector, along with its ability to suppress fluorescence via an energy threshold set individually for each pixel, means that pixel detectors will undoubtedly be the technology of the future.

As data collection times decreased, sample handling began to become one of the main stumbling blocks in the MX experiment. Using modern techniques for SRMX data collection, once crystals become available the aim is to screen and characterise them to find the one most suitable for data collection. This used to require hours of sample manipulation in nitrogen, manual mounting, centring and experimental hutch searches and the process was expensive both in manpower and beam time and was also error prone, particularly at SRMX facilities where experiments continue through the night. It became quickly clear that a robotic sample changer coupled with automatic sample characterisation and screening software could greatly enhance the efficiency. Over the last five years or so a number of sample changers of differing styles have been developed around the world (Figure 3.7). The sample changers at the Stanford Synchrotron Radiation Laboratory (SSRL) (Cohen *et al.,* 2002), the APS, the EMBL Hamburg Outstation at DESY [Pohl *et al.,* 2004] and the French beam line for

Investigation of Proteins (FIP) at the ESRF [Roth *et al.,* 2002; Ohana *et al.,* 2004] are based on standard industrial robots offering a high degree of flexibility at moderate costs. Here, the robot can select samples from different positions in a storage dewar. In contrast, the systems developed at the ESRF and the European Molecular Biology Laboratory, Grenoble, [Cipriani *et al.,* 2006], at DORIS [Karain *et al.*, 2002], the Advanced Light Source (ALS) [Earnest *et al.,* 2002; Rupp *et al.,* 2002] and at SPring-8 [Ueno *et al.,* 2004] are based on fixed custom-made transport systems where the cryo-frozen samples are moved in the dewar or the entire dewar is moved to the loading/unloading position. In the meantime, commercial sample changers produced by three different companies have become available: the ACTOR by Rigaku/MSC, developed in collaboration with Abbott Laboratories (http://www.msc.com), and BruNo by BrukerNonius (http://www.bruker-axs.com) are based on industrial six-axis robots whereas the cryogenic sample changer by MarResearch consists of a very compact transport mechanism specially designed for the Mar desktop beam line (http://www.marresearch.com).

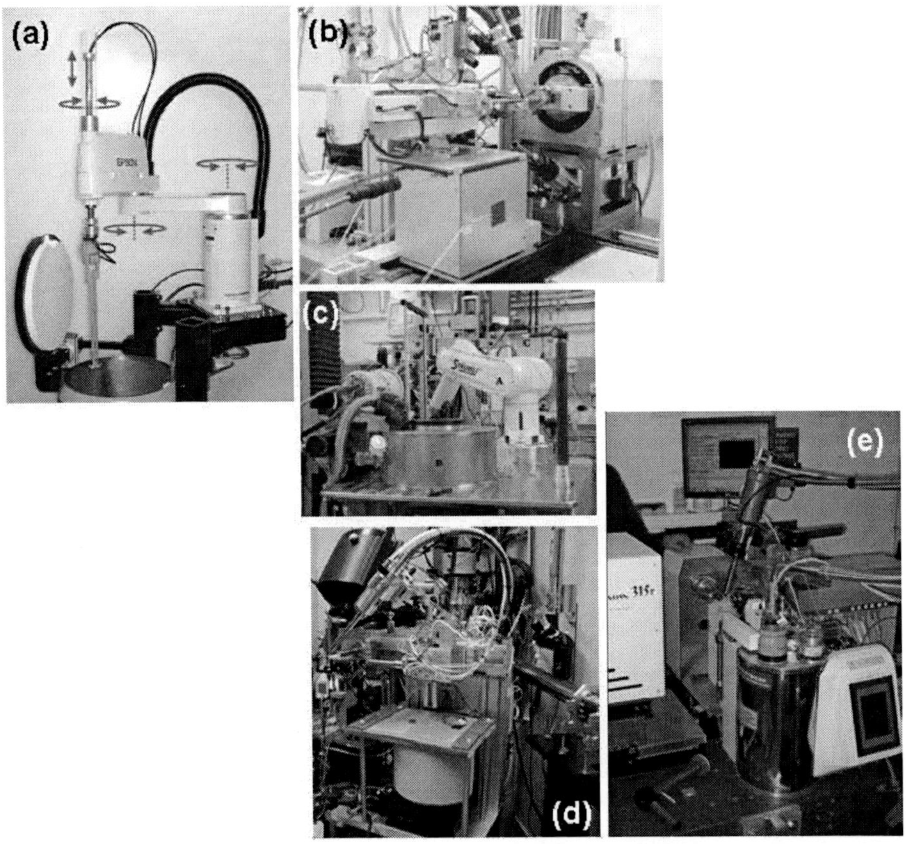

Figure 3.7. A selection of some of the sample changer robots in operation at various synchrotrons around the world. (a) SSRL, USA (Reproduced with permission of the International Union of Crystallography from Cohen *et al.*, 2002); (b) SPring-8, Japan (Reproduced with permission of the International Union of Crystallography from Ueno *et al.*, 2004); (c) FIP beam line, ESRF (Reproduced with permission of the International Union of Crystallography from Ohana *et al.*, 2004); (d) ALS, USA (Reproduced with permission of the International Union of Crystallography from Rupp *et al.*, 2002); (e) ESRF.

In addition to hardware advances, huge progress has also been made with control software, which pervades all steps and features of the SRMX beam line and is central to reliable operation and high levels of automation. Given the intrinsic optical simplicity of MX beam lines and the increasingly high throughput required of them, the most natural future goal from a software point of view is the automation of the entire structure solution process with feedback possible from data processing to data collection [Minor & Otwinowski, 1996; Lamzin & Perrakis, 2000]. This means linking beam line control software, packages written for automatic sample characterisation and screening and databases. Beam line control software is typically organised in three layers. The first, lower level is comprised of a collection of instrumentation programs including drivers and device servers to communicate with individual distributed equipment. The second is constituted by services including sequencers, an expert system holding the intelligence to operate the beam line automatically and database or web servers. Sequencers used at the majority of synchrotrons include SPEC (Certified Scientific Software, PO Box 390640, Cambridge, MA 02139, USA) and EPICS (Experimental Physics and Industrial Control System, http://lansce.lanl.gov/lansce8/Epics/epics.htm). The third software level is a Graphical User Interface (GUI) for users. The goal of a GUI is to hide the details of the low-level instrumentation control during both beam line configuration and data collection such that an inexperienced user is able to confidently and correctly carry out experiments. In the USA, the Blu-Ice software [McPhillips *et al.*, 2002] written at the SSRL and MADNES/OptiX [Skinner and Sweet, 1998] used by the National Synchrotron Light Source (NSLS) at Brookhaven are good examples of improved user and beam line staff environments that have been operational for several years and have set the tone for the integrated systems now being assembled. Both permit a certain level of remote interaction, such as data collection monitoring and chat facilities, thereby allowing synchrotron staff or supervisors to survey from home or office the experiment status or to offer assistance during data collection.

Initiatives to write automatic sample characterisation and screening software include the DNA project [Leslie *et al.*, 2002], born from a European collaboration initially involving the ESRF, CCLRC Daresbury Laboratory and MRC-Cambridge and since expanded to include the Diamond Light Source, EMBL, Global Phasing Ltd, MRC-France (BM14) and Soleil. Another example is the DISTL package [Zhang *et al.*, 2006] used at the SSRL at Stanford, USA. The DNA system includes the collection of two reference images collected 90 degrees apart, detection of blank images or ice rings and determination of unit cell dimensions, space group, diffraction limits and mosaicity. Once these are available the software then proposes a pertinent strategy for the collection of a complete dataset including an optimum exposure time calculated using the program BEST [Popov & Bourenkov, 2003]. The information gathered by DNA then allows for an automatic ranking of samples in terms of their quality defined according to various criteria which may be decided by the user. This is passed to the beam line control software for data collection. If desired, DNA can take into account the wishes of users via communication with an experimental database. Results which include the (optional) integration and scaling of diffraction data collected are also stored in the experimental database. Such software packages now allow the optimisation of data collection strategies and are particularly valuable for projects requiring high levels of crystal screening. They also allow data from at least the more straightforward projects to be collected in an automatic fashion.

The recent advances in high-throughput technology and automation imply the dynamic handling of large amounts of information for sample tracking before, during and after beam line visits. Inevitably this involves the use of databases. The aim of these information management systems is to provide an efficient and coherent infrastructure for data storage and retrieval, thereby making it easier to track physical samples and their associated characteristics and experimental results. Many Laboratory Information Management Systems (LIMS) have been developed that allow sample tracking and reporting of sample information and these have already had a positive impact on structural biology research. For example, a LIMS for protein production (PIMS) is currently in development [Pajon *et al.*, 2005] and a LIMS has been developed for protein crystallisation [Mayo *et al.,* 2005]. Completing the pipeline series is a LIMS for diffraction data collection which has been developed at the ESRF in collaboration with the SPINE (Structural Proteomics in Europe) and the BBSRC funded e-science project e-HTPX [http://www.e-htpx.ac.uk; Allan *et al.*, 2003] and given the acronym ISPyB (Information System for Protein Crystallography Beamlines). ISPyB, as currently deployed at the ESRF, allows the logging of information concerning sample data, sample shipping and data collection and allows the harvesting of data-reduction statistics from DNA. At a modern synchrotron such as the ESRF the beam line database is linked to the User Office database and hence to safety information concerning the samples.

Looking upstream of the experimental end-station, a prerequisite for any automated beam line, whatever the measurements to be made, is a dependable and stable X-ray beam with the correct characteristics to allow the data collection experiment to come to a successful conclusion. Advances in storage ring technology have played a major part in making the present generation of synchrotron sources highly reliable with reproducible beam characteristics, but the source stability and reliability are immediately lost if the optical elements of a beam line are not of equivalent quality and if the procedures to optimally align them are badly implemented. Alignment is where automation can play a key role. Fine-tuning the relative positions of the optical elements of a beam line can be time-consuming and, when performed manually, should not be done by non-experts. For a modern protein crystallography experiment, where changing the wavelength of the incident beam is essential for experiments based on the analysis of the anomalous dispersion signal from a heavy atom present in the protein and can occur many times a day, protocols and software for the automatic alignment of the optical elements of the beam line are therefore primordial. Beam line alignment involves the following principle steps; selecting and maintaining the energy of the X-ray photons, focusing the X-ray beam and fixing the X-rays in an optimised and stable configuration. The optical components required to achieve this alignment can be split into the following discrete units; primary slits, monochromator and secondary slits, focussing element (e.g. toroidal mirror, multilayer, lenses, *etc*). In the automation procedures under development at the ESRF MX beamlines [Nurizzo *et al.,* 2006; Arzt *et al.,* 2005], which will be used as the example here, each of these units is treated separately and functions autonomously, with appropriate diagnostic tools between each element for monitoring and feedback.

The calibration of the position and size of the primary slits is accomplished by measuring the scattering from a pyrocarbon foil inserted into the white beam. A software algorithm deduces the zero position of each blade making up the slits and then moves the slit gaps to a pre-determined size. This foil remains in the white beam and the resultant scattering can be used as a continuous diagnostic tool during experiments. All three tunable ESRF MX beam-lines (ID14-4, ID23-1 and ID29) use channel cut silicon [111] monochromators to define the energy of the X-ray beam impinging on a sample. Such monochromators have been shown to be robust and highly stable with the only real disadvantage being an intrinsic vertical movement of the beam when changing energy. However, this can be readily compensated since on the ESRF standard channel-cut monochromators the stepper motor which drives the rotation of thc monochromator is mounted with an absolute encoder. The encoder has been initialised with a "home position" corresponding to the beam line reference energy (11.563 keV; Pt L_{III} absorption edge). When necessary, the monochromator Bragg angle can automatically be driven close to this reference position and, once there, is properly calibrated using the XANES spectrum of the Pt L_{III} absorption edge of a Pt foil inserted into the monochromatic X-ray beam. The spectrum is automatically measured and analysed. Cross-checking of the calibration can also be carried out using other metal foils held on the same actuator as the Pt foil. The same diode that is used for monochromator calibration also allows the calibration of the four-bladed secondary slits carried out in the same fashion as already described for the primary slits. In this case a kapton foil is inserted into the X-ray beam. This foil can also remain in the X-ray beam and monitor the beam intensity after the secondary slits during user experiments.

The majority of the ESRF MX end-stations are equipped with a toroidal mirror for focusing the monochromatic X-ray beam. The first stage of the alignment of a toroidal mirror uses the same philosophy as that implemented for the monochromator. Five absolute encoders are able to bring the mirror to a flat initial reference position. From this initial position the mirror is tilted to its operational grazing angle (typically 3 mrad). An optical camera visualises the fluorescence generated by the X-ray beam passing through a YAG screen positioned after the mirror and the signal intensity and the centre of gravity of the beam are calculated from this image. A scan of the intensity versus mirror height or horizontal translation then allows the geometrical centre of the mirror to be aligned to the incoming X-ray beam. In the future, the second stage of the mirror alignment will include a wave-front analysis method to automatically align X-ray mirrors. The rotations (yaw and incident angle) and bender tension will be optimised using a camera mounted on an actuator in front of the X-ray diffraction detector in the experimental hutch thus allowing the focus to be either at the sample position or on the detector.

Many protein structure determination experiments require the exploitation of the anomalous scattering properties of a few heavy atoms present in the crystal and so the measurement and analysis of a X-ray absorption near edge structure (XANES) scan around the absorption edge of the anomalous scatterer is a crucial step. From this, the X-ray energies of the absorption peak (f''_{max}) and its inflection point (f'_{min}) may be determined precisely. It involves measuring the fluorescence signal from the heavy atom in question whilst scanning over the energy of the beam. The complexity of changing the energy of the beam, realigning the beam line at this energy and setting up parameters on the fluorescence detectors meant that such a procedure was the cause of much apprehension amongst the beam line users! Performing an absorption edge scan is yet another procedure comprised of a sequence of well-defined events (Figure 3.8) and was therefore clearly a candidate for automation. The principle steps include changing the undulator gap, changing the monochromator angle, adjusting the focussing element and experimental table position, selecting the appropriate attenuation for the sample, recording the absorption edge and deducing the anomalous scattering factors f' and f'' from analysis of the edge scan. Current automation routines use lookup tables for determining the undulator gap size, monochromator angle and focussing element and table positions, and then a final scan is used to fine-tune those positions. The fluorescence detector is inserted near the sample (and removed once the scan has been completed) and its parameters set electronically, and then the monochromator angle is scanned over a pre-defined range using state-of-the-art mechanics and electronics to perform this scan in only a handful of seconds. Once this has been completed, the anomalous scattering factors *f'* and *f"* are automatically obtained using CHOOCH [Evans & Pettifer, 2000] and passed into the data collection software and experiment database. The automation of the XANES scan has been one of the most visible and appreciated automation developments over the last five years. The first MAD experiments performed at the ESRF, on the bending magnet beam line BM14 which opened in 1995, lasted around 24 hours. Since then, automation of the experiment and state-of-the-art detectors, coupled with the latest insertion devices and optical components which allow the full source brilliance to be exploited, means that on beam lines such as ID23-1 and ID29 of the ESRF, a three-wavelength MAD data collection can be done in less than one hour!

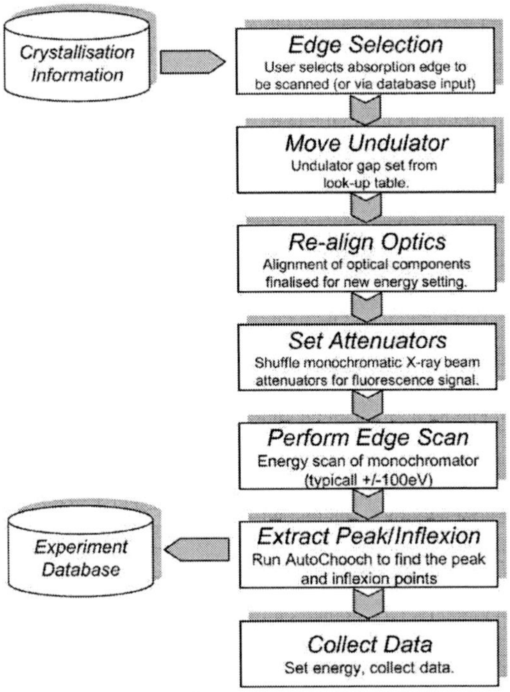

Figure 3.8. Flowchart of automatic absorption edge scans as implemented at the ESRF. Reprinted from Progress in Biophysics and Molecular Biology, 89, Arzt et al., Automation of macromolecular crystallography beamlines, 124-152, Copyright (2005), with permission from Elsevier.

The remarkable advances in hardware and software technology made over the last nearly three decades have allowed the individual steps of the typical SRMX experiment to be automated, Figure 3.5. More importantly though, what it has also done is allow each of these individual automated steps to be linked smoothly together in a sequence to form the data collection pipeline (DCP). In its fully refined form, the DCP will automatically carry out all the steps usually performed manually by experimenters; crystals may be loaded onto (and unloaded from) the host goniometer and centred in the X-ray beam, the software package for automatic data collection will characterize samples, provide detailed information on individual crystals (unit cell parameters, resolution limits, mosaicity, spot shape etc), choose an optimum data collection strategy and exposure time based on experimental requirements contained in the LIMS, use a ranking mechanism if required to choose the best of several crystals and automatically collect, integrate and scale diffraction data. Fully automated DCPs are already in use at some synchrotron sites such as the ESRF [Beteva *et al.*, 2006] and at the SSRL [Van den Bedem *et al.*, 2003].

Taken together, the components of the DCP provide a platform for external user groups to access SRMX facilities remotely since, apart from the actual loading of the samples into the sample changer robot, the whole process may be run *via* the software interfaces. In principle there will no longer be any need for external user groups to be physically on the beam line while data from their crystals are being collected. Indeed remote access to beam lines is already possible at some synchrotron sites; for example, SSRL or the SGX CAT at APS, and preliminary experiments have also been made at the ESRF.

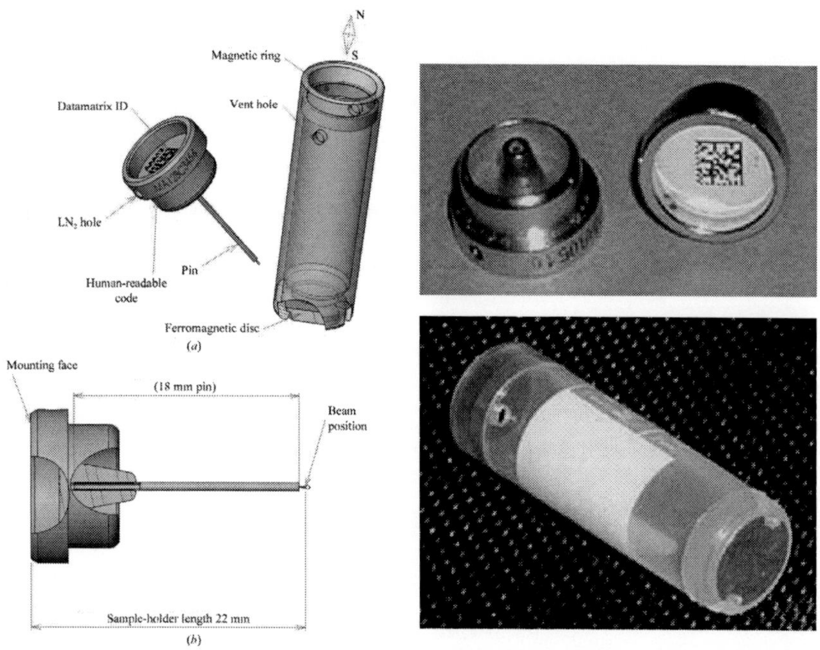

Figure 3.9 The European standard sample holder with data matrix and vial resulting from the EU-funded SPINE project. Reproduced with permission of the International Union of Crystallography from Cipriani *et al.*, 2006.

One of the key issues in facilitating high throughput and automation has been the implementation of standard equipment and procedures. Ensuring the compatibility between user home laboratories and the various synchrotrons worldwide is a vital element in ensuring the smooth transfer of samples and associated data. In the future it is clear that the use of SRMX in structural biology projects will increase from the already high levels of today. An increase in the demand for beam time has been foreseen and considerable resources are being mobilised to construct new storage rings and automate beam lines for MX and other scientific fields. User groups will very likely make use of multiple synchrotron sources and multiple beam lines specializing in complementary techniques, which will require a certain level of cohesion to allow sensible access to and efficient use of beam time. Co-ordination amongst synchrotrons strongly enhances the resource base with beam lines of different specialities complementing each other. The two major EU funded projects, SPINE and BioXHIT, and the UK-funded e-HTPX are targeting partly or wholly these goals in Europe. Elsewhere in the world projects have similar ambitions as part of their structural genomics or proteomic goals; *e.g.* Protein 3000 in Japan [Normile & Jianxiang, 2001] and the Protein Structure Initiative of the United States (http://www.nigms.nih.gov/psi/). In Europe, the SPINE, BioXHIT and e-HTPX projects have so far resulted in, amongst others, the implementation of a standard sample holder with data matrix [Cipriani *et al.,* 2006; Figure 3.9], compatible with the majority of SRMX facilities in Europe, the sample characterisation and data collection software DNA again used by the majority of European SRMX facilities, automatic sample centring software [Lavault *et al.*, 2006], the development of mini-kappa add-ons and the standardisation of the software infrastructure for home and synchrotron LIMS. There is also initiative to extend the IUCR Crystallographic Information File (CIF) concept to cover efficient storage of 2-D area detector data so that images of diffraction data have a standard format making them easier to manipulate and archive (http://www.iucr.ac.uk/iucr-top/cif/cbf/ndb/). In the past detector manufacturers have preferred to develop their own formats but the most modern synchrotrons coming online, such as the Diamond Light Source in the UK, are actively enforcing the use of such a format by including it in their specifications for the purchase of new detectors. Such standards are imperative in allowing future world- or continent-wide MX initiatives to link together in a smooth, and therefore exploitable, fashion.

4. Wavelength Variability and Structure Determination

4.1. The Rise and Fall of MAD

The broad spectral range of SR allows the exploitation of anomalous scattering techniques such as Multiwavelength Anomalous Dispersion (MAD) [Hendrickson, 1991; Smith, 1991; Hendrickson & Ogata 1997; Ogata, 1998] and Single Wavelength Anomalous Dispersion (SAD) [Hendrickson & Teeter 1981; Wang, 1985; Dauter *et al.,* 2002] as methods for solving the phase problem in MX. These techniques coupled with the construction of rapidly tunable synchrotron beam-lines [*c.f.* Biou *et al.*, 1997; Wakatsuki *et al.*, 1998, Walsh *et al.,* 1999b; Pohl *et al.,* 2001, Roth *et al.,* 2002, Nurizzo *et al.,* 2006], the availability of techniques for the routine incorporation of selenomethionine into protein amino acid sequences [Hendrickson *et al.,* 1990; Doublié, 1997], the routine cryo-freezing of crystals of macromolecules [Garman, 2003] and improved software for anomalous scatterer substructure determination [Terwilliger & Berendzen, 1999; Schneider & Sheldrick, 2002], phase probability calculations [Terwilliger & Berendzen, 1999; Terwilliger, 2000; Bricogne *et al.,* 2003], and automatic interpretation of electron density maps [Lamzin *et al.,* 2001; Terwilliger, 2003] have helped fuel the revolution in Structural Biology that has place since the early 1990s.

Both MAD and SAD techniques take advantage of the fact that the X-ray scattering factor ***f*** for an atom has components that are wavelength dependent;

$$f = f_o\,(\sin\vartheta/\lambda) + f'(\lambda) + i\,f''(\lambda) \tag{4.1}$$

A large value of f'' results in significant differences (*anomalous differences*) in the intensities of reflections making up Friedel or Bijvoet pairs (Blow 2003) while the large change in f' around electron shell absorption edges (Figure 4.1) means that the intensity of the same reflection measured at different wavelengths around the absorption edge will be different. Diffraction data collected at different wavelengths in this region thus yield intensity differences known as *dispersive differences.* In the MAD technique diffraction data are collected at two or more wavelengths/X-ray energies around the absorption edge and anomalous and dispersive differences are combined to produce unique values for the phases of reflections in the diffraction data sets.

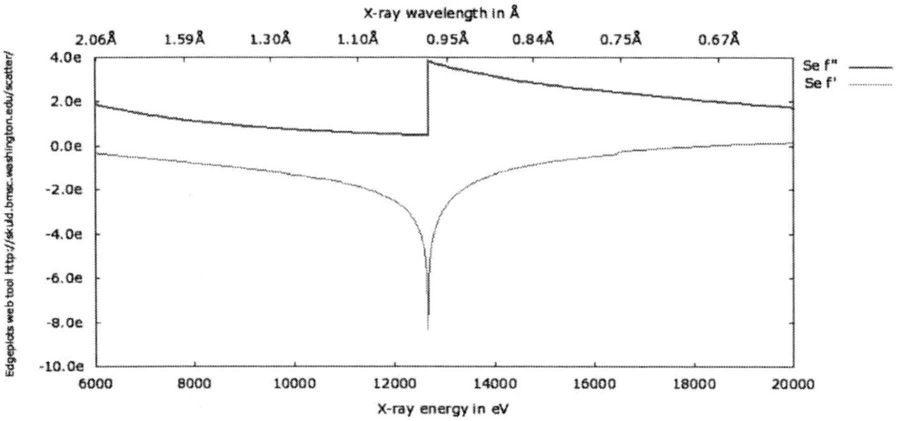

Figure 4.1. The theoretical variation [Brennan & Cowan, 1992] of f' and f'' around the K edge of Se. This figure was generated at http://skuld.bmsc.washington.edu/scatter/

SAD is a simplification of the MAD technique. Here diffraction data are collected at a single wavelength and anomalous differences only are used to derive possible values for the phase. As can be seen from Figure 4.2, SAD does not result in a single value for the phase and following SAD experiments this phase ambiguity needs to be resolved – usually using density modification techniques [Wang, 1985].

In MAD and SAD experiments optimising f'' will give more accurate values for the possible solutions of the phase while (for MAD) maximising the difference in f' will provide better discrimination in the calculation of the correct phase [Woolfson & Fan, 1995]. In order to maximise both f' and f'' many different strategies have been proposed regarding the different wavelengths/energies at which MAD data should be collected [Peterson *et al.,* 1996; González *et al.,* 1999; González, 2003; Burla *et al.,* 2004].

$$<\Delta F/F>_{\text{anomalous}} \approx \frac{1}{\sqrt{2}}\frac{\sqrt{N_A}\,2f''}{<|F_T|>} \tag{4.2}$$

Where for proteins $<|F_T|> \approx \sqrt{3.14 \times molecular \cdot weight}$

$$< \Delta F / F >_{dispersive} \approx \frac{1}{\sqrt{2}} \frac{\sqrt{N_A} \, | f_{\lambda_i} - f_{\lambda_j} |}{<| F_T |>} \tag{4.3}$$

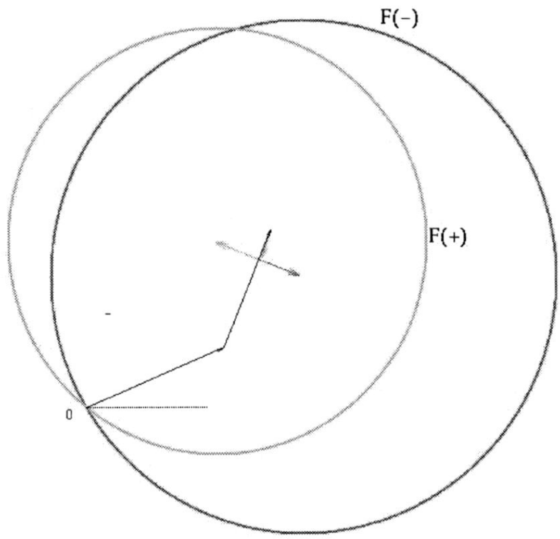

Figure 4.2. A graphical illustration of the phase ambiguity inherent in the SAD technique.

The anomalous (equation (4.2)) and dispersive (equation (4.3)) signals [$< \Delta F / F >$; Smith, 1991] available in a MAD experiments are much smaller than those available *via* isomorphous replacement techniques. Even for data collected at the peak of an absorption edge, the maximum f'' attainable is ~ 6 e^- and ~25 e^- for K- and L-absorption edges respectively. Likewise, the maximum Δf inducible are of the order of 10 e^- (K-edges) and 20 e^- (L-edges). These compare to atomic numbers of ~70 e^- for the 3rd row transition metals that are commonly use to prepare heavy atom derivatives for use in isomorphous replacement (IR) experiments [Blundell & Johnson, 1976]. In principle, therefore, AD experiments should be less useful than those exploiting IR.

However, once cryocooling of crystals of macromolecules became the norm rather than the exception in MX, the data at all the wavelengths required in a MAD experiment could, in most cases, be collected from a *single* crystal. This, by obviating the problems of non-isomorphism that plague isomorphous replacement methods, dramatically increased the chances of success of a MAD experiment to such an extent that the technique moved rapidly from being a 'novelty technique' to being a technique at the forefront of Structural Biology research [Ogata, 1998].

The pre-eminence of MAD as a technique for deriving the phasing information for the *de novo* solution of macromolecular structures was reinforced with the advent of the highly automated, tunable beam-lines constructed at higher intensity, more stable modern synchrotron sources [Walsh *et al.,* 1999b, Wakatsuki *et al.*, 1998; Nurizzo *et al.,* 2006]. The much more intense X-ray beams provided by such sources, coupled with the increased user-friendliness and automation of modern synchrotron beam-lines [Pohl *et al.*, 2004; Gaponov *et al.*, 2004; Arzt *et al.*, 2005; Beteva *et al.*, 2006], allowed data for MAD experiments to be collected extremely rapidly [Walsh *et al.*, 1999a,b; Flot *et al.*, 2006]. However, it soon became clear that the more intense beams provided meant that radiation damage to crystals of macromolecules had reappeared as a serious problem in MX [Burmeister, 2000; Ravelli & McSweeney, 2000; Weik *et al.*, 2000]. This resurgence of radiation damage meant that, in many cases, complete MAD data sets could no longer be collected from a single crystal and that a change of strategy concerning the exploitation of anomalous dispersion in structure determination was needed.

In 2000 a study of MAD datasets collected around the Se K-absorption edge ($\lambda \sim 0.98$ Å) at ALS beam line 5.0.2, NSLS beam line X4A and SSRL beam line 1-5 showed that phases derived using only data from the peak wavelength (*i.e.* the wavelength at which f'' is optimized) in a SAD approach produced, when combined with density modification techniques, electron density maps of comparable quality to those that resulted from MAD phasing, again combined with density modification, using data at all three wavelengths [Rice *et al.*, 2000]. As a result of this study the authors predicted a renaissance of the SAD technique as a means of solving macromolecular crystals structures, particularly in the light of the high throughput structure determination demanded by Structural Genomics initiatives. This has since proved to be the case [Dauter, 2006] particularly for structure elucidation carried out using data collected at undulator based beam-lines at 3rd generation synchrotron sources where radiation damage is severe and particularly targets atoms in the anomalous scattering substructure during MAD and SAD experiments [Rice *et al.*, 2000; Zwart *et al.*, 2004; Weiss *et al.*, 2004; Ravelli *et al.*, 2005, see Figure 4.3].

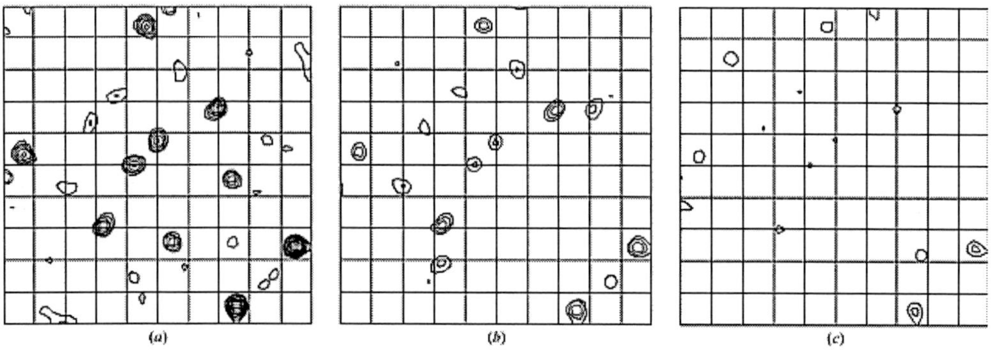

Figure 4.3. The change in occupancy of the Se atoms in selenomethionine as a function of X-ray dose (Reproduced with permission of the International Union of Crystallography from Ravelli *et al.*, 2005).

4.2. Remote SAD

In the study above by Rice *et al.*, the SAD data used were collected at a wavelength corresponding to the peak of the Se K-absorption edge where the anomalous signal available (equation (4.2)) is optimized. However, as it is a single wavelength method, SAD can be used at almost any wavelength where the f'' of an anomalously scattering atom in a crystal is significant (*i.e.* almost any wavelength on the high energy side of an absorption edge). Indeed, it has been demonstrated many times [*c.f.* Hendrickson & Teeter (1981), Wang

(1985), Brodersen *et al.* (2000), Leonard *et al.,* 2005] that structure solution *via* the SAD technique is also possible using the weak anomalous signals available far from absorption edge features (Figure 4.4).

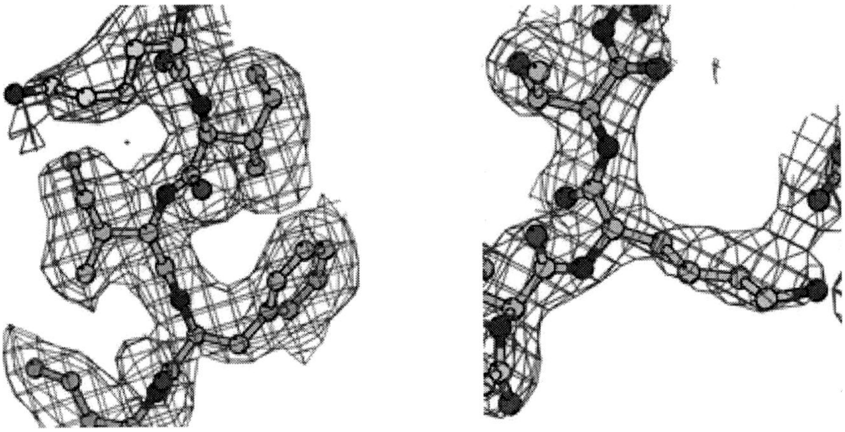

Figure 4.4. Portions of the electron density maps obtained using remote SAD to solve the crystal structures of the two metalloproteins ferrodoxin VI (right) and shrimp alkaline phosphatase (left). The proteins contain 1 Fe^{2+} and 3 Zn^{2+} ions respectively which, at λ= 0.933 Å, yield anomalous signals of 1.4% and 1.9%. Reproduced with permission of the International Union of Crystallography from Leonard *et al.,* 2005.

Such a 'remote SAD' technique has one obvious disadvantage, the signals available in such experiments are smaller than they would be for SAD experiments carried out at the peak of an absorption edge [*c.f.* equation (4.2)]. However, it also has one notable advantage; such experiments can be carried out on fixed wavelength synchrotron beam-lines or even in-house laboratory sources. In fact, fixed wavelength synchrotron beam-lines are ideally suited to the technique as the energies at which these beam-lines operate (> 13 keV) is generally above that of at least one absorption edge of nearly all the elements whose anomalous scattering properties one might wish to exploit in MX.

For example, the energy of the fixed wavelength ID14 beam-lines at the ESRF [Wakatsuki *et al.*, 1998] is ~13.3 keV (λ ~0.933 Å). At this photon energy possible targets (Figure 4.5) for the technique thus include heavy atom derivative crystals produced by soaking native crystals in solutions containing third-row transition metals. Here, using X-rays of λ ~0.933 Å, one would expect (equation (4.2)) anomalous signals of around 4-5% for the series Ta-Ir and around 3.5-4% for the series Pt-Pb assuming one fully occupied heavy-atom site per 300 amino acids. Another series of targets for this technique are crystals of the metalloproteins that make up 30% of all proteins. At 13.3 keV (λ = 0.933 Å), *f"* ranges from ~1.2 e^- for Mn to ~2.3 e^- for Zn. These values are significantly smaller than those available at the peak of the absorption edges of these atoms (~4 e^-) but would be, in many cases, sufficient for successful structure solution using the remote SAD technique [Leonard *et al.,* 2005; Figure 4.4]. Crystals of selenomethionyl derivatives of proteins are very attractive targets for this technique, since at λ = 0.933 Å, f''_{Se} = 3.2 e^- resulting in an anomalous signal of 4.0% for the average protein, which contains around one methionine every 42 amino-acid residues.

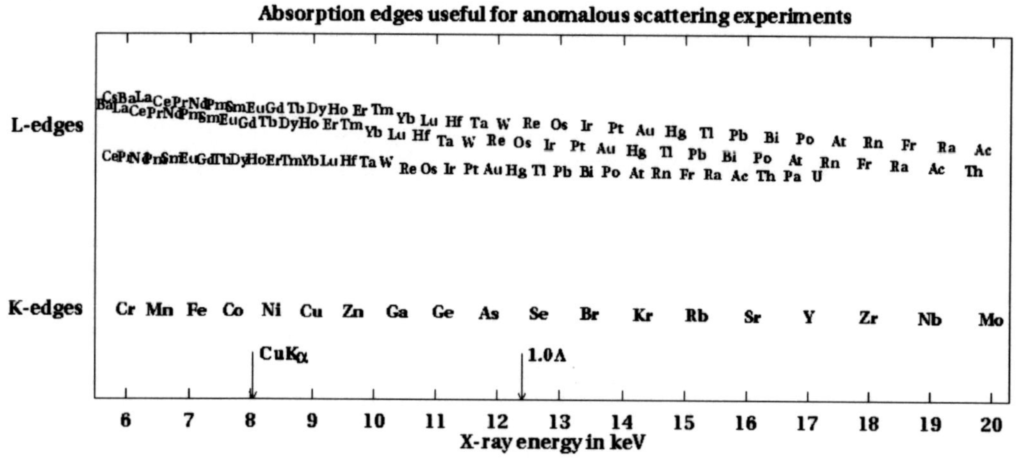

Figure 4.5. Absorption edges that could be targeted in anomalous scattering experiments in MX with L-absorption edges shown in the order L_I, L_{II}, L_{III} from top to bottom of the figure. This figure was generated at http://skuld.bmsc.washington.edu/scatter/

The only common heavy-atom derivatives that could not be easily targeted at $\lambda = 0.933$ Å wavelength are those derived from the from soaking of crystals in either NaBr (Dauter *et al.*, 2000, 2001) or RbBr (Korolev *et al.*, 2001). In these cases an X-ray energy of 13.3 keV is on the low energy side of the K-absorption edges of these atoms and higher energy (shorter wavelength) X-rays would be required (Figure 4.5).

4.3. Sulphur-SAD

Taken to its logical extreme, remote SAD is suitable for much more challenging, and potentially much more rewarding experiments than those shown in Figure 4.4. These target anomalous signals resulting from the naturally occurring presence of sulphur and phosphorous atoms in proteins and nucleic acids respectively and could allow the solution of the crystal structures of almost all *native* proteins – *i.e.* there should no longer be any need to produce heavy atom derivatives to enable structure determination.

The K-absorption edges of S & P are at *very* long wavelengths ($\lambda = 5.01$ Å and 5.78 Å respectively, Figure 4.6). Single-crystal X-ray diffraction experiments at such energies/wavelengths, while in principle feasible [Stuhrmann *et al.,* 1995, 1997], are clearly never going to be routine due to problems with absorption of such soft X-rays by both air and the sample itself, diffraction geometry and the production of X-rays of such a wavelength. However, f'' though small, is still significant at wavelengths that are more accessible on synchrotron X-ray beam-lines designed for MX. Indeed, even at $\lambda = 1.54$ Å, – *i.e.* the Cu-K_α radiation most commonly produced by in-house laboratory X-ray sources - f'' for S is ~0.5 e^- (Figure 4.6). Hendrickson & Teeter (1981) and Wang (1985) have shown that using the SAD technique, it is possible to exploit such small f'' to solve the crystal structures of biological macromolecules. Dauter and colleagues [Dauter *et al.,* 1999] revisited this work in order to assess the potential usefulness of the so-called S-SAD (sulphur-SAD) technique. This work and complementary investigations at longer wavelengths using beam-lines at synchrotron sources [Weiss *et al.,* 2001; Gordon *et al.,* 2001; Cianci *et al.,* 2001; Micossi *et al.,* 2002; Ramagopal *et al.,* 2003a, b; Wang *et al.,* 2006] demonstrated that it is possible to successfully exploit anomalous signals as low as 0.5% for SAD phasing. This corresponds to the anomalous signal that, at λ = 1.8 Å, would be generated by the presence in the sequence of a protein of 1 sulphur-containing amino acid [*i.e.* cysteine or methionine] every 100 residues. Thus, S-SAD could, in principle, be used to solve the crystal structure of almost every protein.

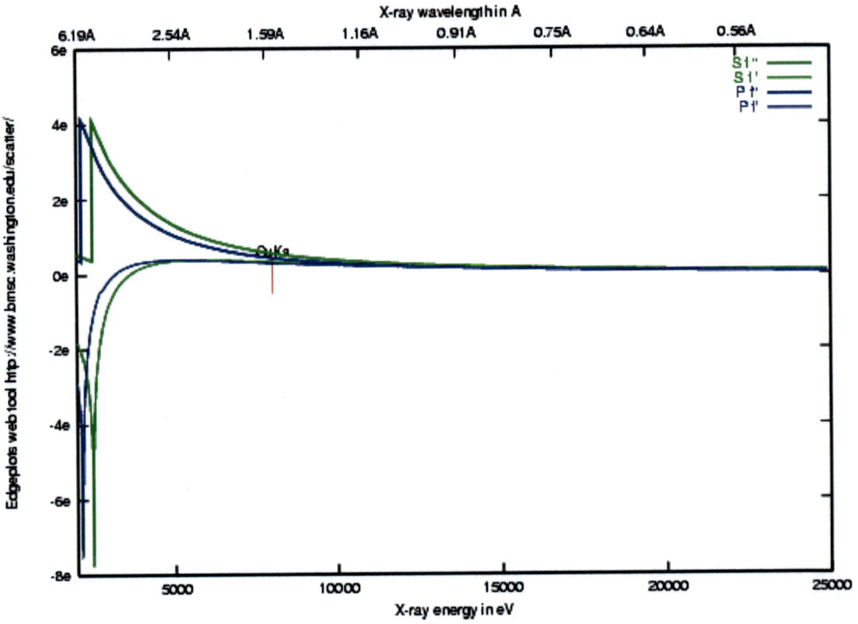

Figure 4.6. Plots of f' and f'', as a function of X-ray energy/wavelength, as calculated from theory [Brennan & Cowan, 1992] for S (green lines) and P (blue lines). The energy/wavelength of Cu-K_α radiation is also marked. This figure was generated at http://skuld.bmsc.washington.edu/scatter/.

It should be obvious from Figure 4.6, that the anomalous signals resulting from the presence of sulphur-containing amino acid residues in crystals of macromolecules can be increased by using longer wavelengths to collect diffraction data. For example, while f'' for S is ~0.7 e^- at λ = 1.8 Å, this rises to ~1.0 e^- at λ =2.1 Å. The potential benefits of using longer wavelengths in phasing macromolecular crystal structures has led to many investigations regarding the usefulness of long wavelength X-rays in macromolecular structure determination.

4.4. The Use of Long Wavelength X-rays in Macromolecular Crystallography

Before synchrotron radiation became readily available for protein crystallography, Cu K_α radiation (E = 8 keV; λ = 1.54 Å) was predominantly used to solve the 3-dimensional structures of biological macromolecules. With the routine use of synchrotron radiation data collection is carried out over a wide range of energies, usually between ~20.0 keV (λ = 0.62

Å) and 6.5 keV (λ = 1.9 Å). Such an energy range allows AD experiments to target anomalous scattering effects resulting from the K-absorption edges of the 1st row transition metals found in metalloproteins and the *L*-edges the of the 3rd row transition metals and lanthanides commonly used to prepare heavy atom derivatives of protein crystals. Crucially, this energy range also allows experiments to be carried out around the *K*-edge of selenium.

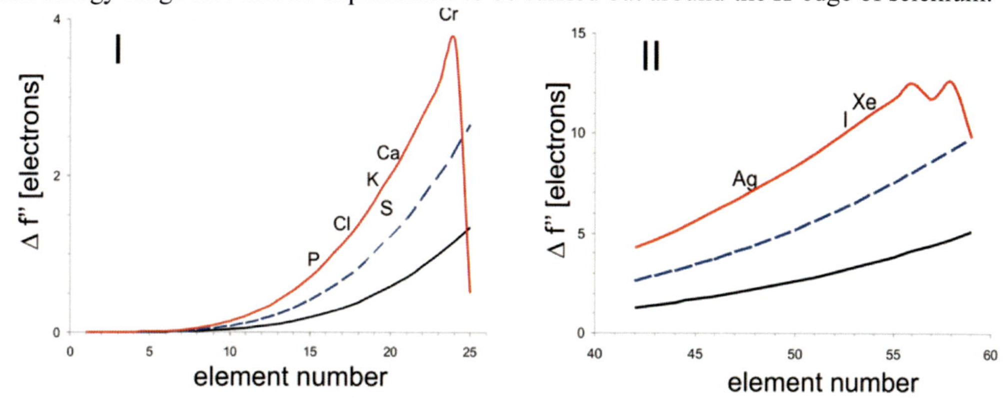

Figure 4.7. The f'' inducible as a function of atomic number using X-ray energies of 12.4 keV (black line), 8.3 keV (blue dashed line) and 6.2 keV (red line).

AD experiments have been carried out at higher energies (>25 keV) than the range normally associated with MX [Helliwell *et al.*, 1993; Takeda *et al.*, 2004; Jakoncic *et al.*, 2006] targeting the *K*-absorption edges of I, Xe and Ho whilst experiments at low energies (~2 keV) have tried to exploit the anomalous signal at the *K*-absorption edges of S or P atoms intrinsic to biological macromolecules [Stuhrmann *et al.*, 1995, 1997]. Both approaches are unlikely to become routine due to the immense technical effort required and the experimental difficulties presented by working at such extreme (for MX) X-ray energies. As a result of these difficulties, the use of X-rays with energies between 8 and 6 keV (λ ~1.5 Å- 2.1 Å) has received more and more attention as the technical and physical difficulties (*e.g.* absorption) can be dealt with readily, at least in part. At such X-ray energies the f'' inducible for some light (P, S, Cl, Ca *etc.*) and medium weight atoms such as I and Xe make their presence in crystals of macromolecules useable from a structure solution point of view (Figure 4.7).

Despite the massive potential of the S-SAD technique, the method is not yet a routine tool in MX and, more often than not, is seen as a method of 'last resort'. The reasons for this are that the anomalous signals available are rather small ($\langle \Delta F/F \rangle \leq 1.5$ %, see equation (4.2)) and the difficulties associated with working with longer wavelength X-rays (increased absorption from both air and sample) mean that measuring such small signals accurately is not trivial. These difficulties tend to be exacerbated at SR sources, where the high intensity of the X-rays produced, coupled with the need to collect highly redundant data [Weiss *et al.*, 2001], mean that radiation damage is also a problem. Indeed, radiation damage problems are probably the reason that a substantial proportion of macromolecular crystal structures solved using S-SAD are undertaken with data collected on in-house X-ray sources using either Cu or Cr anodes.[*c.f.* Sarma & Karplus, 2006; Watanabe, 2006]. However, S-SAD data collections on in-house X-ray sources, despite recent innovations, can take days, if not weeks, and if S-SAD is to become a routine tool in MX, it is essential that more and more S-SAD structure solutions are successfully carried out on beam-lines at modern SR sources.

Thus, to enhance the chances of successful S-SAD experiments at SR beam-lines it is necessary to improve the quality of the data collected and use the spectral range of synchrotron radiation to optimize the anomalous signals available from the molecule under study. Simple measures that can be taken include providing a helium path between crystal and detector. This reduces both attenuation of the diffracted beams (Table 4.1) and the background noise generated by air scatter (Figure 4.8). This means that the signal/noise ratios required to extract small anomalous signals from diffraction data should be obtainable more quickly, thus reducing the chances of radiation damage.

Table 4.1. Attenuation length (air vs. helium) for longer wavelength X-rays.

Wavelength	Air (mm)	He (mm)
1.50 Å	965	333328
1.70 Å	650	333328
1.90 Å	467	249995
2.10 Å	347	224995
2.25 Å	282	199995

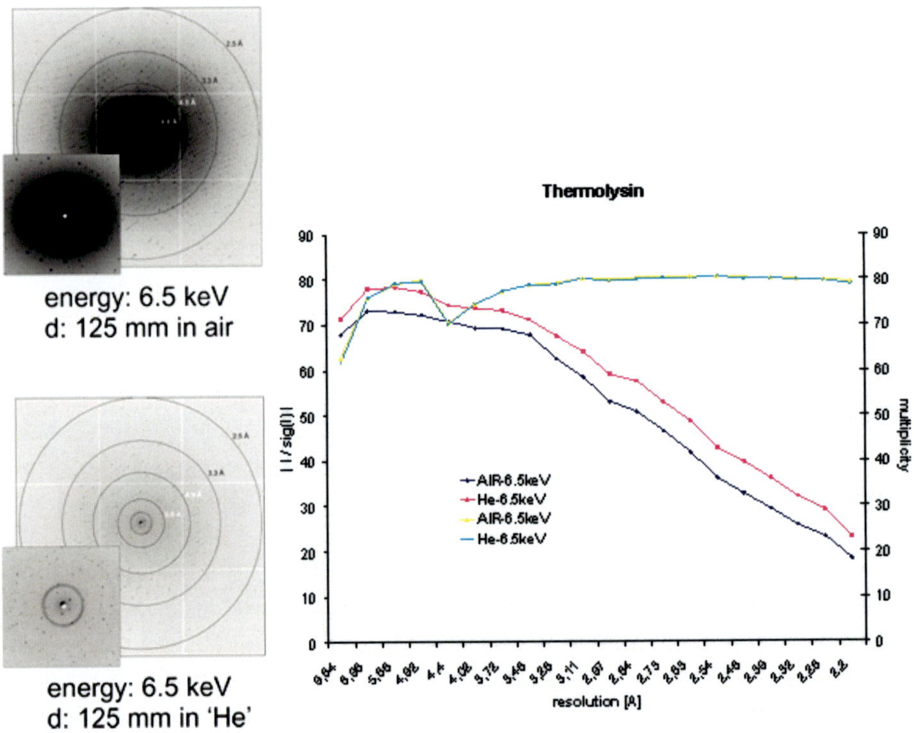

Figure 4.8. Examples of decreased background (left) and increased signal/noise ratio (right) resulting from the use of a He-filled path between crystal and detector for data collected from crystals of macromolecules using X-rays of energy 6.5 keV (λ = 1.9 Å)

A less straightforward measure that would improve the quality of data collected at long wavelengths on SR beam-lines would be the elimination of (or correction for) the contamination of X-ray beam by wavelengths resulting from harmonic reflections from monochromators. Channel-cut Si(111) monochromators produce a contaminant with wavelength λ/3 and diffraction spots due to this wavelength will be superimposed on the low resolution reflections due to the main component of the beam. At long wavelengths, such contamination of the X-ray beam cannot, under normal operating conditions, be eliminated by the Pt-, Rh- or Pd- coated mirrors generally used for focusing X-rays and data quality suffers commensurately.

The quality of long-wavelength diffraction data will also be improved by (as is currently the case) a move away from single-axis to multi-circle goniometry. One obvious advantage of this is that the ability to align crystals so that a mirror plane in the diffraction pattern is perpendicular to the rotation axis will allow the reflections comprising anomalous pairs to be measured on the same oscillation image. The reflections involved will thus have received the same X-ray dose and differences in their intensities should be more reliable. Multi-circle goniometry will also mean that diffraction data can be collected with the crystal in more than one orientation. As has been pointed out [Debreczeni *et al.,* 2003], scaling such data sets together provides an efficient correction for absorption which is, in some cases, crucial to the success of S-SAD experiments.

These and other measures (scaling long wavelength data against reference data collected at much higher energy [Mueller-Dieckmann *et al.*, 2004], improving detector efficiency at wavelengths of interest, timely use of modern anomalous scatterer substructure determination, phasing and model building software combined with experiments aimed at determining the optimum wavelength for the collection of S-SAD data [*i.e.* Mueller-Dieckmann *et al.*, 2004] should help to make S-SAD routine in MX and allow the MX community to benefit from the vast potential of the technique.

Table 4.2. The f'' inducible for Xe and I as a function of X-ray energy/wavelength

Energy (keV)/Wavelength (Å)	f''_{Xe} (e^-)	f''_{I} (e^-)
14.2/0.87	2.8	2.6
12.4/1.00	3.6	3.3
10.0/1.24	5.1	4.8
8.0/1.55	7.4	6.9
7.0/1.78	9.2	8.6
6.5/1.91	10.3	9.7
6.0/2.07	11.6	10.9
5.5/2.25	13.2	12.3

As should be clear from Figure 4.7, the use of long wavelengths in MX should not be confined to the S-SAD technique. Using long wavelengths allows researchers to increase [Table 4.2] the f'' inducible from elements such as Xe and I now commonly used in the derivatisation of protein crystals [Schiltz *et al.,* 1994; Sauer *et al.,* 1997; Evans & Bricogne, 2002] and, in the future, M-absorption edges and the huge f'' available from these, may also become routine targets for AD experiments [Liu *et al.,* 2001].

4.5. Long Wavelength and Molecular Replacement

Another example of the usefulness of diffraction data collected at lower energies is its employment during structure solution using the molecular replacement (MR) method. Following the placement of a molecule in the unit cell using MR, the positions of sulphur and/or other light atoms could, even for a poor MR solution, be located using anomalous difference Fourier maps (ΔF_{ano}, α_{MR} – 90°; Figure 4.9). The position of the atoms thus located, can be used to validate the MR solution obtained or, in cases where the search model does not include the entire polypeptide chain, be used as markers in polypeptide chain tracing.

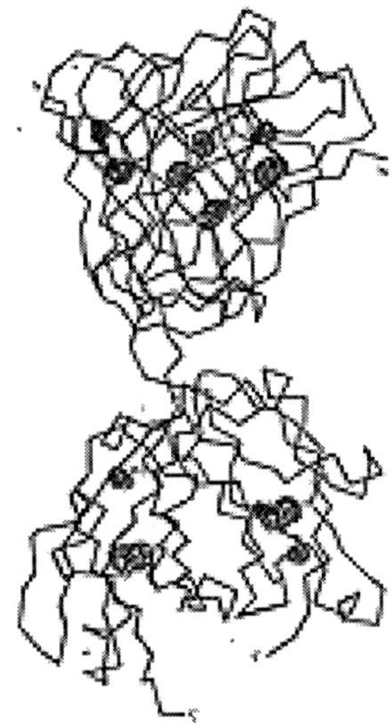

Figure 4.9. An anomalous difference Fourier map (ΔF_{ano}, α_{MR} – 90°) calculated using phases from a MR solution and anomalous differences measured at λ = 1.54 Å. The peaks in the map represent the sulphur atoms from the 7 methionine and 8 cysteine residues in the asymmetric unit. Reproduced with permission of the International Union of Crystallography from Schuermann & Tanner, 2003.

Long wavelength experiments exploiting anomalous scattering can also be used to circumvent the model bias resulting from a conventional MR phasing protocol. Here, MR-phased anomalous difference Fourier syntheses can be used to elucidate the anomalous scattering substructure, which can then be used to initiate SAD phasing (Schuermann & Tanner, 2003) to produce unbiased electron density maps for model building.

4.6. The Accurate Identification of Ions in the Crystal Structures of Macromolecules

Long wavelength X-rays are also an ideal tool for the identification of ions that may be important to the structure and/or function of macromolecules. A large percentage of biological macromolecules are crystallized from solutions which contain light atoms in the salts (i.e. NaCl, KCl, $CaCl_2$), buffers (e.g. Tris-HCl, Na/K-phosphate) or precipitants (e.g. $(NH_4)_2SO_4$). Consequently, some of these atoms may also be found, bound to the biological macromolecule, particularly if they are important to the function of the protein crystallized.

However, a comparison of crystallographic models of biological macromolecules deposited in the protein databank [PDB; Berman *et al.*, 2000] with their crystallisation conditions as deposited in the Biological Macromolecule Crystallization Database [BMCD; Gilliland *et al.*, 1994] shows a striking discrepancy (Figure 4.10) – only a very small percentage of the former contain atoms assigned as moieties/ions present in the crystallisation solution.

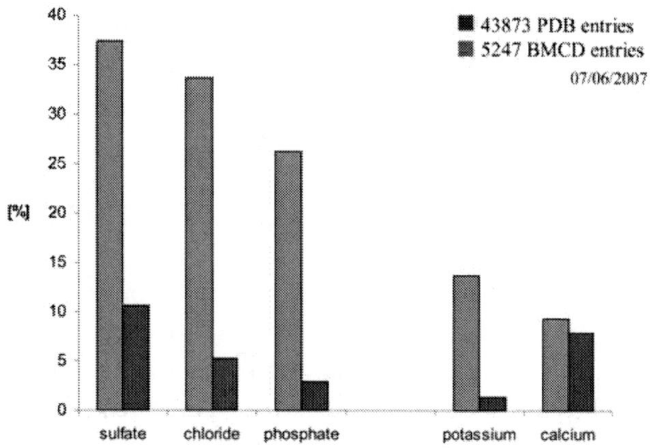

Figure 4.10. Comparison of the number of times certain moieties/ions are present during crystallisation and the number of times these are assigned in the relevant crystallographic model.

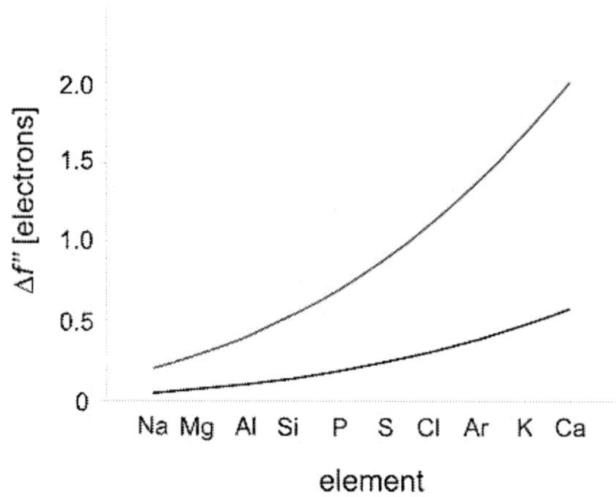

Figure 4.11. The *f''* inducible for light atoms at X-ray energies of 12.4 keV (black line) and 6.2 keV (red line).

This is not too surprising as macromolecular crystal structures are usually refined against high resolution data sets collected at an energy around 12.4 keV (λ ~1 Å) where the anomalous scattering contribution of light elements – particularly for Na and Mg - is generally too low to be detectable (Figure 4.11). However, a careful analysis and determination of the nature of the ions/moieties bound to a biological macromolecule in a crystal structure is required as (a) it provides a more complete and correct crystallographic model and (b) this may provide additional biochemical information including the identification of active site clefts/residues (particularly the case for sulphate/phosphate ions) and highlighting the presence of structural and active site metal ions. The correct assignment of the nature of the atoms in crystal structures, which can be facilitated by using longer wavelengths for data collection to induce larger *f″* (Figure 4.11), may also provide additional proof for the presence of alternative side chain conformations for cysteine and methionine residues and may serve to detect the existence of radiation damage (through the reduction of disulfide bridges) [Mueller-Dieckmann *et al.*, 2007].

4.7. The Use of Radiation Damage to Solve Atomic 3-D structures

That crystals of macromolecules suffer radiation damage when exposed to X-rays was first noticed in the early 1960's [Blake & Philips, 1962] when it was concluded that the damage is proportional to the absorbed X-ray dose. The advent of cryo-cooling techniques seemed to have reduced the problem of radiation damage such that it could be more or less ignored as at 100 K the crystal life time in the X-ray beam is increased by a factor of ~70 [Nave & Hill, 2005]. However the increased photon fluxes and densities available at 3rd generation synchrotron sources mean that radiation damage - which recent studies [Owen *et al.*, 2006] have show to be slightly dependant on dose rate - to crystals of biomolecules is once again a cause for concern.

The most obvious manifestation of radiation damage is a gradual reduction in the diffraction limit of the crystals under study – a process that can be followed by an increase in individual oscillation image temperature factors during scaling procedures. However, long before radiation damage is evident, the photo-electron cascade produced as a result of the interaction between X-rays and the protein and solvent molecules in a crystal [Nave and Hill, 2005; Ravelli and Garman, 2006] causes *site-specific* damage to protein molecules [Burmeister, 2000; Ravelli & McSweeney, 2000; Weik *et al.*, 2000; Nave and Garman, 2005] including the reduction of metal centres [Yano *et al.*, 2005; Beitlich *et al.*, 2007], the breakage of disulfide bridges, the decarboxylation of glutamate or aspartate residues, the loss of hydroxyl groups from serine and tyrosine residues (Figure 4.12) and the loss of occupancy of atoms targeted in AD experiments (Figure 4.2). It should also be noted that radiation damage to protein crystals results in a gradual change – usually an increase - in the unit cell volume and thus a loss of isomorphism [Burmeister, 2000; Ravelli & McSweeney, 2000].

For the reasons mentioned above radiation damage is seen as a nuisance in macromolecular crystallography. However, Ravelli and colleagues have shown how, in certain cases, radiation damage can be turned to the advantage of macromolecular crystallographers by using the isomorphous differences induced as a result of site-specific radiation damage as means to solve the phase problem in MX [Ravelli *et al.*, 2003; Nanao *et al.*, 2005], Figure 4.13). They named the technique, Radiation-damage Induced Phasing (RIP).

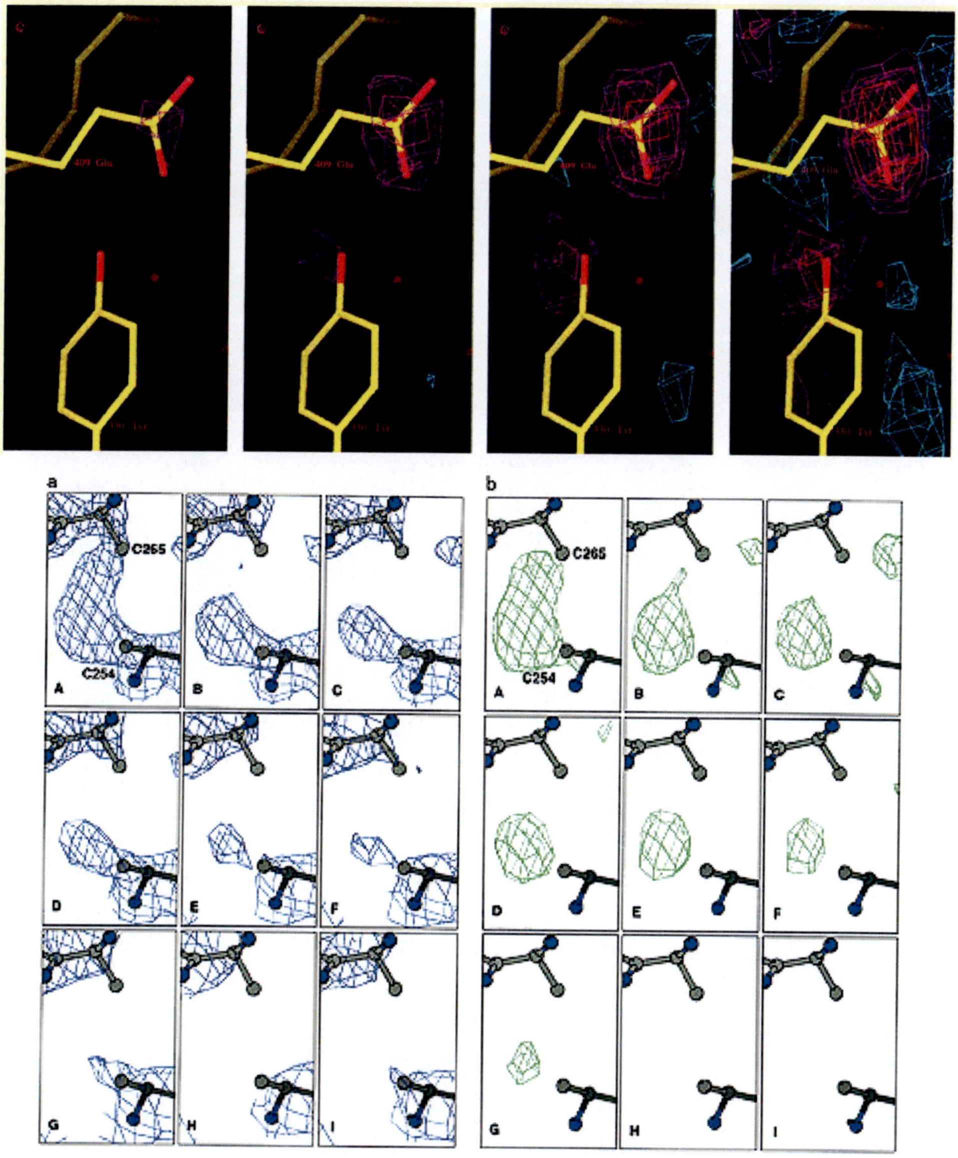

Figure 4.12. Examples of the (top) decarboxylation and loss of hydroxyl groups (reproduced with permission of the International Union of Crystallography from Burmeister, 2000) and (bottom) the breakage of disulphide bonds that occurs as a result of radiation damage to crystals of bio-macromolecules (Reprinted from Weik *et al.,* 2000. Copyright 2001 National Academy of Sciences, USA).

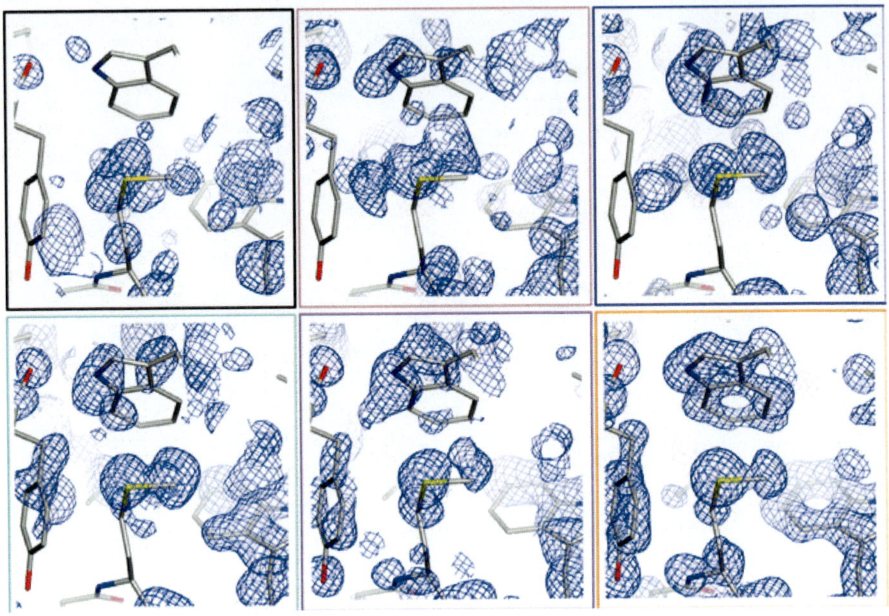

Figure 4.13. The improvement in electron density resulting from RIP as the model of radiation damage is iteratively improved. Reproduced with permission of the International Union of Crystallography from Nanao *et al.*, 2005.

RIP phasing is carried out in three steps. A first data set is collected using an attenuated beam in order to reduce radiation damage. This is followed by a 'burn' during which the crystal is exposed to X-rays but no data taken, in order to induce dramatic changes in the sample structure (proteins or nucleic acids). The accumulated dose during the burn should be around 10 times that accumulated during the initial data collection. Finally a second data set is collected using the same accumulated X-ray dose as for the pre-burn data set. For all three steps carried out, the total accumulated X-ray dose should not exceed the value of 3.0×10^7 Gγ that usually results in a loss of diffracted intensity by a factor of $I_{ln2} / I_0 = \ln 2$ [Owen *et al.*, 2006]. Following a careful scaling of the pre- and post-burn data sets, differences in intensity, ΔF_{RIP}, are used to find initial sites of radiation damage and phases then calculated based on these sites. The radiation damage model is then completed by several iterations of density modification, difference Fourier syntheses and phase calculation.

In a similar manner to S-SAD, the RIP method is still in its infancy. However, and - also like S-SAD – it is sure to benefit from recent software developments such as the programs RADDOSE [Murray *et al.*, 2004] and BEST [Bourenkov & Popov, 2006].

4.8. Using Ultra-violet Light to Help Solve the Phase Problem in MX

RIP or RIPAS [Radiation damage Induced Phasing with Anomalous Scattering; Banumathi et al., 2004] are purely SR based-techniques. Unlike S-SAD, RIP requires X-ray doses and dose rates far in excess of those available from in-house X-ray sources. However, radiation damage to macromolecules can also be induced with energies of light other than those in the X-ray region.

UV-induced damage to DNA is a particularly well studied case since it is involved in the processes of ageing. However, UV-induced damage also occurs in proteins [Wien *et al.*, 2005] with disulphide bonds being sensitive especially when their environment contains aromatic residues [Neves-Petersen *et al.*, 2002]. Using this knowledge, Nanao and Ravelli (2006) developed an experimental phasing protocol, very similar to that one implemented for RIP, based on UV-induced radiation damage. In these experiments, the 'burn' stage described above consists of exposure of the crystal to light from a UV laser at 266nm. Advantages of UV-induced RIP are that the main structural damages induced by UV are located on disulfide bonds, resulting in better signal to noise ratios of peaks in $|\Delta F_{RIP\text{-}UV}|^2$ Patterson functions thus allowing an easier interpretation of structural differences that can be used to initiate phasing procedures.

5. Limitations

Single crystal X-ray diffraction techniques clearly provide important templates for many problems in modern molecular biology and have enabled significant advances in our understanding of the relationships between structure and function, but it is worth remembering at this point that there are a number of limitations. Firstly, the technique requires single crystals of a sufficient quality to enable the collection of useable X-ray diffraction data. Insoluble proteins present a real problem in this regard and for many integral membrane or membrane associated proteins, crystallization is still very difficult. The use of cryo-electron microscopy to define low resolution models combined with more detailed models of components elucidated by X-ray crystallography offers a sensible approach for many protein complexes. A second general problem faced by X-ray techniques is that the end result represents the spatial average of many thousands of molecules that make up the crystal; it is not possible to study just one molecule at a time. If the crystal contains molecules with the same overall conformation (normally a prerequisite for crystallization), but with active centres in different conformations, it is normally difficult to define precise models which account for the different conformations and their relative proportions. X-ray techniques only give an indication of the spatial average, although if the resolution and quality of the X-ray data are sufficiently high then a careful analysis of the electron density and site occupancies can enable different models to be proposed. In addition, the X-ray technique alone is not usually capable of yielding unequivocal information regarding the redox state of the metal centres, unless the resolution and data quality are high and permit a good definition of the local geometry of the metal site. Radiation damage may also play a role, particularly if a very brilliant synchrotron X-ray beam is used to collect the intensity data. It is well established that radiation damage can cause reduction of metal centres and care must be taken to eliminate or minimise this effect, preferably by spectroscopic monitoring of the centres before and after data collection. Figure 5.1 shows the online microspectrophotometer as implemented at the ESRF MX beam lines.

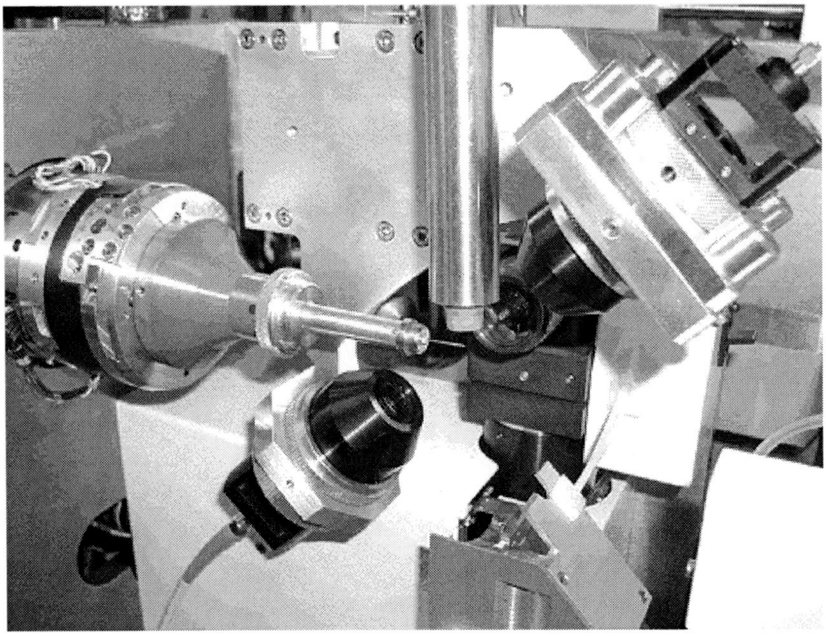

Figure 5.1. The online micro-spectrophotometer as implemented at the ESRF MX beam lines.

A second major limitation of the X-ray technique is that any studies involving transient or short lived intermediates are virtually impossible unless the intermediate can be stabilized in some chemical manner and/or by cryo-trapping. Considerable advances have been made in X-ray structural time-resolved studies with the most detailed crystallographic study being the interaction of dioxygen with myoglobin [Schotte *et al.*, 2003], although in this case the related molecule carbon monoxide was used because of the relative ease with which the Fe-CO bond can be broken photosynthetically. A comprehensive "movie" is now available that shows how the diatomic molecule leaves the heme iron and eventually diffuses into the external environment. These studies are technically very demanding and push the art of time-resolved crystallography using synchrotron radiation to the current limits. For example, the timing of the arrival of the laser pulses required to dissociate the CO from the heme iron and the X-ray pulses required to record the concomitant changes in the diffraction patterns that enable the dissociation process to be followed structurally, has to be very precise. It is unreasonable to expect that such techniques will be generally applicable, but restricted to particular biological systems.

A further limitation is introduced by the resolution of the diffraction data obtainable. Crystals of proteins and enzymes contain a high percentage of solvent, typically between 35% - 80%. As a consequence, they are not normally as well ordered as small molecule structures and regions such as surface loops and side chains tend to be far more flexible than the interior of the molecule. In addition, they contain a large number of diffracting atoms. Although cryofreezing can considerably assist the disorder and flexibility issues, the overall diffracting power is weak and even using modern synchrotron radiation sources this is reflected in the quality of the diffraction data, particularly at high resolution. Indeed, it is usually very difficult to collect diffraction data to atomic resolution, the norm for small molecule structures, and the resolution is often limited to 2.0Å or worse. Limited resolution has a profound effect not only the estimation of coordinate errors, but often in the precise definition of weakly diffracting moieties such as dioxygen and related species.

The increasing industrialization and automation of MX brings with it the danger of uncontrolled data and structure production. Check points and quality control are absolutely required at all stages of structure determination (analogous to the use of the R_{free} parameter during structure refinement [Brunger, 1992]). Increasingly, diagnostics are being installed on synchrotron beam lines with standardized protocols (along the lines of ISO9000-style procedures) being introduced to assure optimal and correct beam line performance. Synchrotron beam lines are seeing mounting numbers of users unfamiliar with a diffraction experiment coming to use the facilities. This puts a corresponding heavier weight on the quality control and on the local experts in ensuring that the users leave with the best data possible. One option being actively considered, and indeed operational at some synchrotrons, is the use of service mode work for academic groups [*i.e.* at Brookhaven National Laboratory, Robinson *et al.*, 2006]. Such services already exist widely for paying industrial customers but are much rarer for academic groups. One future path might be to increase the prevalence of such working styles but this will require a massive change on the part of the user communities to accept the synchrotron-based experts controlling and making decisions on their behalf. A half-way house could well prove to be the operation of beam lines via remote control; samples sent to the synchrotron, placed in the robot changers and then control handed to the users at home in their laboratories.

A critical point is still training. Scientists using a technique must have sufficient training in that method to allow sensible interpretation of results. With high levels of automation and "black-box" software, MX could become vulnerable to non-experts incorrectly or over-interpreting experimental results. The automation allows higher throughput, with more consistent and better quality, but the experimenter must still be able to critically evaluate the outcome.

As indicated above, protein samples are fragile – sustaining damage even when cryo-cooled at 100 K. The very intense radiation of a modern MX beam line, can lead to considerable changes not only with respect to metal centres, but to a protein and its structure. This has been increasingly recognised in the last few years, with radiation damage research on crystalline protein samples becoming an extremely active area of study in its own right [Garman and McSweeney, 2007]. A considerable danger is the over-interpretation or incorrect interpretation of structures, for example for catalytic mechanisms, where observed changes are in fact due to radiation damage effects rather than mechanistic ones. A number of amino acid side chains and ligands can be easily modified by ionising radiation: notably carboxylate groups, disulphide bridges and metal centres reduced by significant exposure to radiation. Although limits on the maximum exposure a cryo-cooled protein sample could sustain have been calculated [Henderson, 1990] and experimentally determined [Owen *et al.*, 2006], only recently have such exposures become common-place on synchrotron beam lines. Work is therefore underway to actively monitor samples by using spectroscopies (UV-Vis and Raman) whilst on the beam line [Bourgeois & Royant, 2005; Pearson *et al.*, 2004; McGeehan *et al.*, 2007] and absorbed dose calculations can now be made based on the X-ray wavelength and crystal contents to allow optimal data collection strategies [Murray *et al.*, 2004].

6. Perspectives

The future for synchrotron-based MX is exciting. The opening of so many new, high quality third-generation radiation sources around the world, and especially in Europe is an opportunity to be seized. The quantity of beam time available will allow the more challenging projects the access to those beam lines and the beam time required for their success, as well as highly specialized beam lines to be constructed to serve the projects requiring specific facilities (for example; micro- and nanofocus, ultra-low and -high resolution data collection, windowless beam lines for low energy operation). The new stations will also allow the deployment of multimodal or multi-technique experiments. As mentioned above, these are already developing with the use in dedicated beam time slots of experiments using on-line microspectrophotometers [Adam *et al.*, 2004; Dubnovitsky *et al.*, 2006] to add an additional dimension to the data collection. Future progress is likely to include the addition of these techniques to the standard MX set-up and also the return of miniaturized kappa-axis geometry diffractometers (*e.g.* the devices developed by Rosenbaum-Rock and the team at EMBL Grenoble Outstation) to allow crystals to be aligned for most effective data collection. At DIAMOND beam lines are being constructed compatible with a P3 biohazard level safety environment. These environments of multimodal-experiments and/or P2 or P3 safety levels must be compatible with sample changers, fluorescence detectors (for edge scanning), sample visualisation etc. The trend to compact, well-designed experiment environments must therefore continue.

The synchrotron light source will increasingly be more than just a source of X-rays. Already most synchrotrons have on-site laboratory and sample preparation facilities. This trend will increase with considerable auxiliary and complementary facilities provided (together with the necessary expertise) to allow users to make the most benefit from their X-ray beam time. Users in the future will fall into two classes; those accessing the beamlines by remote access, requiring "standard"-style facilities and those physically travelling to the synchrotron to use the full panoply of facilities from the beamlines to preparation to complementary facilities and local expertise. This latter set of users may spend some time on the site (weeks or months) for the most demanding experiments. Scientific collaborations such as the Partnership for Structural Biology (see: http://psb.esrf.fr) at Grenoble bring together diverse techniques and knowledge (molecular biology to NMR, electron microscopy, neutron- and X-ray-based techniques) to biological problems.

Detector improvements will play a critical role in the future; currents experiments still throw away valuable signals as today's detectors are relatively inefficient. The current generation of CCD-based devices is expected to shortly give way to pixel detectors based upon the direct detection of X-ray photons rather than conversion through phosphor and a fibre-optic taper to a CCD-chip. The Swiss Light Source has pioneered work on pixel detectors and with the latest incarnation having 2463 by 2457 pixels (total active area of 424 x 435 mm^2), a high dynamic range of 2^{20} with 5 millisecond read-out speed (see http://pilatus.web.psi.ch/pilatus.htm). This detector, and others like it, may allow MX beam lines to function in shutterless mode with continuous rotation of the sample [Broennimann *et al.*, 2006], giving faster data collection and removal of a source of error with the shutter and spindle synchronisation. The speed, high dynamic range and sensitivity of these new-style detectors will greatly enhance experiments: a revolution of the same scale as the arrival of CCD-based detectors is quite possible.

Complementary to macromolecular crystallography, the direct high-resolution imaging of biological particles using coherent diffraction imaging with soft X-rays is being developed [Shapiro *et al.*, 2005]. This technique uses the coherence of light produced by a synchrotron source to collect the Fourier transform of a non-repeating object. The image is reconstructed using phase retrieval by oversampling [Miao *et al.*, 1999; Marchesini *et al.*, 2003]. Eventually the technique of diffraction imaging could bridge the resolution gap between atomic resolution crystal structures and electron microscopy work whilst also allowing thicker samples to be studied than electron microscopy would permit.

Yet the third-generation of synchrotron light sources is not the end; fourth-generation hard X-ray sources are on the drawing board and several have funding approved. The three hard X-ray free electron lasers (XFELs) [see Shintake, 2007] being constructed in Stanford (http://www-ssrl.slac.stanford.edu/lcls/), Hyogo (http://www.riken.jp/XFEL), and Hamburg (http://xfel.desy.de/) are expected to be completed from 2009, 2011, and 2014 respectively. These machines will produce ultra-brilliant pulses of highly coherent X-rays some ten orders of magnitude more brilliant than today's third-generation sources in pulses lasting femtoseconds with repeat rates of tens of Hz. The XFELs are complementary to the storage ring machines, with their quasi-continuous X-ray beam, giving access to shorter time and much higher peak intensity regimes not possible with the third generation. Their application to biological material will be targeted at fast time resolution and coherent diffraction imaging [Neutze *et al.*, 2000; Huldt *et al.*, 2003]. Test experiments on coherent diffraction from artificial samples have been conducted [Chapman *et al.*, 2006] at the Hamburg VUV FEL FLASH (Free-Electron Laser in Hamburg) with highly encouraging results. Unlike the case with storage ring synchrotrons, the samples are completely destroyed after a single femtosecond pulse of radiation but the diffraction pattern may nonetheless be recorded from the sample before it is destroyed. These pilot studies are an important milestone on the road to using the future hard XFELs to provide high-resolution ultra-fast biological imaging of non-repeating objects.

6. Acknowledgements

The authors gratefully acknowledge the information and support that has been provided by numerous scientists, computer experts and engineers working at synchrotron sources throughout the world. Acknowledgement is also given to the macromolecular crystallographic community who are constantly seeking to push the technology to its limits resulting in a very dynamic and exciting environment in which to undertake scientific research.

7. References

Abergel, C., (2004). Spectacular improvement of X-ray diffraction through fast desiccation of protein crystals, *Acta Crystallogr,* **D60**, 1413-1416.

Abola E., Kuhn P., Earnest T. & Stevens R.C. (2000). Automation of X-ray crystallography. *Nature Structural Biology* **7**, 973-977.

Adam, V., Royant, A., Niviere, V., Molina-Heredia, F.P. & Bourgeois, D. (2004). Structure of Superoxide Reductase Bound to Ferrocyanide and Active Site Expansion upon X-Ray-Induced Photo-Reduction. *Structure*, **12**, 1729-1740.

Allan R., Diakun G., Guest M., Keegan R., Nave C., Papiz M., Winn M., Winter G., Diprose J., Esnouf R., Mayo C., Stuart D., Launer L., Walsh M., Fillon J., Henrick K., Pajon A., Cowtan K., Young P., Read R. and Rana O. (2003). An e-science resource for high throughput protein crystallography. *Proceedings of the UK e-science All Hands Meeting, EPSRC,* September 2–4, Nottingham, UK, pp. 230–234.

Amemiya Y., Matsushita T., Nakagawa A., Satow Y., Miyahara J. and Chikawa J.-C. (1988). Design and performance of an imaging plate system for X-ray diffraction study. *Nucl. Instrum. Methods A* **266**, 645-653.

Arndt U.W., Champness J.N., Phizackerley R.P. and Wonacott A.J. (1973). A Single-Crystal Oscillation Camera for Large Unit Cells. *J. Appl. Crystallogr.* **6**, 457-463.

Arzt, S., Beteva, A., Cipriani, F., Delageniere, S., Felisaz, F., Förstner, G., Gordon, E., Launer, L., Lavault, B., Leonard, G., Mairs, T., McCarthy, A., McCarthy, J., McSweeney, S., Meyer, J., Mitchell, E., Monaco, S., Nurizzo, D., Ravelli, R., Rey, V., Shepard, W., Spruce, D., Svensson, O., & Theveneau, P. (2005). Automation of macromolecular crystallography beam lines. *Prog. Biophys. Mol. Bio.l* **89**, 124–152.

Ataka, M. (1993) Protein crystal growth: an approach based on phase diagram determination, *Phase Transitions*, **45**, 205-219.

Ataka, M., Wakayama, N. I., (2002), Effects of a magnetic field and magnetization force on protein crystals growth. Why does a magnet improve the quality of some crystals? *Acta Crystallogr.,* **D58**, 1708-1710.

Banumathi S., Zwart P. H., Ramagopal U. A., Dauter M. and Dauter Z. (2004). Structural effects of radiation damage and its potential for phasing *Acta Crystallogr.* **D60**, 1085-1093.

Barends, T. R. M., Dijkstra, B. W., (2003), Oils used in microbatch crystallisation do not remove a detergent from the drops they cover, *Acta Crystallogr.,* **D59**, 2345-2347.

Baron, A.Q.R., Kohmura, Y., Krishnamurthy, V.V., Shvyd'ko, Y.V. and Ishikawa, T. (1999). Beryllium and aluminium refractive collimators for synchrotron radiation. J. *Synchrotron Rad.* **6**, 953-956.

Barry, B.A., Cooper, I.B., de Riso, A., Brewer, S.H., Vu, D.M. and Dyer, R.B. (2006). Time-resolved vibrational spectroscopy detects protein-based intermediates in the photosynthetic oxygen-evolving cycle. *Proc. Natl. Acad. Sci. USA* **103**, 7288-7291.

Beitlich T., Kühnel K., Schulze-Briese C., Shoeman R. L. and Schlichting I. (2007). Cryoradiolytic reduction of crystalline heme proteins: analysis by UV-Vis spectroscopy and X-ray crystallography. *J. Synchrotron Rad.* **14**, 11-23.

Bergfors, T. M., (1999), Dynamic Light Scattering, *In Proetin Crystallisation: Techniques, Strategies and Tips*, Edited by Bergfors T. M., La Jolla: International University Line, 29-38.

Bergfors, T., (2003), Seeds to crystals, *J. Struct. Biol.*, **142**, 66-76.

Berman, H.M., Westbrook, J., Feng, Z., Gilliland, G., Bhat, T.N., Weissig, H., Shindyalov, I.N. and Bourne, P.E. (2000). The Protein Data Bank. *Nucleic Acids Research* **28**, 235-242.

Berry, I. M., Dym, O., Esnouf, R.M., Harlo, K., Meged, R., Perrakis, A., Sussman, J.L., Walter, T.S, Wilson, J., Messerschmidt, A., (2006), SPINE high-throughput crystallization, crystal imaging and rercognition techniques: current state, performance analysis, new technologies and futures aspects, *Acta Crystallogr*, **D62**, 1137-1149.

Beteva, A., Cipriani, F., Cusack, S., Delageniere, S., Gabadinho, J., Gordon, E., Guijarro, M., Hall, D., Larsen, S., Launer, L., Lavault, B., Leonard, G., Mairs, T., McCarthy, A., McCarthy, J., Meyer, J., Mitchell, E., Monaco, S., Nurizzo, D., Pernot, P., Pieritz, R., Ravelli, R., Rey, V., Shepard, W., Spruce, D., Stuart, D., Svensson, O., Theveneau, P., Thibault, X., Turkenberg, J., Walsh, M. amd McSweeney, S. (2006). High throughput sample handling and data collection at synchrotrons: embedding the ESRF into the high throughput gene to structure pipeline. *Acta Crystallogr* **D62**, 1162-1169.

Biasci, J.-C., Plan, B. and Zhang, L. (2002). Design and performance of ESRF highpower undulator front-end components. *J. Synchrotron Rad.* **9**, 44-46.

Bilderback, D.H., Thiel, D.J., Pahl, R. and Brister, K.E. (1994). X-ray Applications with Glass-Capillary Optics. *J. Synchrotron Rad.* **1**, 37-42.

Bilderback, D.H., Freund, A.K., Knapp, G.S. and Mills D.M. (2000). The historical development of cryogenically cooled monochromators for third-generation synchrotron radiation sources. *J. Synchrotron Rad.* **7,** 53-60.

Biou, V., Leonard, G., Stojanoff, V., Labouré, S., Mattenet, M., Helliwell, J., Felisaz, F., Claustre, L., Lapeyre, F, Brown K. & Thompson, A. (1997). *ESRF Newsletter April* 1997, 21-25.

Blake C. C. F. & Phillips D. C. (1962). *Biological Effects of Ionizing Radiation at the Molecular Level*, pp. 183-191, *International Atomic Energy Agency Symposium*, Brno, Czechoslovakia, 2-6 July 1962. Vienna: IAEA.

Blow, D. M. (2003). How Bijvoet made the difference: the growing power of anomalous scattering . *Methods Enzymol.* **374**, 3-22.

Blow, D. M., Chayen, N. E., Lloyd, L. F., Saridakis, E., (1994), Control of nucleation of protein crystals, *Protein Sci*, **3**, 1638- 1643.

Blundell, T. L. & Johnson, L. N. (1976). *Protein Crystallography*, Academic Press. London, U.K.

Bourenkov G. P. & Popov A. N. (2006). A quantitative approach to data-collection strategies. *Acta Crystallogr.* **D62**, 58-64.

Bourgeois, D. & Royant, A. (2005). Advances in kinetic protein crystallography. *Curr Opin Struct Biol.* **15**, 538-547.

Brennan, S. & Cowan, P.L. (1992). A suite of programs for calculating x-ray absorption, reflection and diffraction performance for a variety of materials at arbitrary wavelengths. *Rev. Sci. Instrum.* **63**, 850-853.

Bricogne G., Vonrhein, C., Flensburg, C., Schiltz, M. & Paciorek, W. (2003). Generation, representation and flow of phase information in structure determination: recent developments in and around SHARP 2.0. *Acta Crystallogr.* **D59**, 2023-2030.

Broennimann, Ch., Eikenberry, E.F., Henrich, B., Horisberger, R., Huelsen, G., Pohl, E., Schmitt, B., Schulze-Briese, C., Suzuki, M., Tomizaki, T., Toyokawa H. and Wagner, A. (2006). The PILATUS 1M detector. *J. Synchrotron Rad.* **13**, 120-130.

Brodersen, D.E., de La Fortelle, E., Vonrhein, C., Bricogne, G., Nyborg, J. &. Kjeldgaard. M. (2000). Applications of single-wavelength anomalous dispersion at high and atomic resolution. *Acta Crystallogr.* **D56**, 431-441.

Brunger, A.T. (1992). The free *R* value: a novel statistical quantity for assessing the accuracy of crystal structures. *Nature* **355**, 472-474.

Bunick, C., North, A., Stubbs, C., (2000), Evaporative microdialysis: an effective improvement in an established method of protein crystallization, *Acta Crystallogr.,* **D56**, 1430-1431.

Burla, M. C., Carrozzini, B., Cascarano, G. L., Giacovazzo, C. , Moustiakimov, M., Polidori, G. & Siliqi, D. (2004). MAD phasing: choosing the most informative wavelength combination. *Acta Crystallogr.* **D60**, 1683-1686.

Burmeister W. P. (2000). Structural changes in a cryo-cooled protein crystals owing to radiation damage, *Acta Crystallogr.* **D56**, 328–341.

Chapman, H.N., Barty, A., Bogan, M.J., Boutet, S., Frank, M., Hau-Riege, S.P., Marchesini, S., Woods, B.W., Bajt, S., Benner, W.H., London, R.A., Plönjes, E., Kuhlmann, M., Treusch, R., Düsterer, S., Tschentscher, T., Schneider, J.R., Spiller, E., Möller, T., Bostedt, C., Hoener, M., Shapiro, D.A., Hodgson, K.O., van der Spoel, D., Burmeister, F., Bergh, M., Caleman, C., Huldt, G., Seibert, M.M., Maia, F.R.N.C., Lee, R.W., Szöke, A., Timneanu, N. and Hajdu, J. (2006). Femtosecond diffractive imaging with a soft-X-ray free-electron laser. *Nature Physics* **2**, 839-843.

Chayen, N.E., (1997), The role of oil in macromolecular crystallisation, *Structure*, **5**, 1269-1274.

Chayen, N.E., Saridakis, E., (2002), Protein Crystallisation for genomics: towards high-throughput optimization techniques, *Acta Crystallogr.,* **D58**, 921-927.

Chayen, N. E., (2003), Crystallisation of membrane proteins in oils, *In Methods and Results in Crystallisation of Membrane Proteins*, edited by Iwata S., La Jolla: International University Line, 131-139.

Chayen , N. E., (2004) Turning Protein Crystallisation from an Art into a Science. *Current Opinion in Structural Biology*, **14**, 577-583.

Cianci, M., Rizkallah, P.J., Olczak, A., Raftery, J., Chayen, N.E., Zagalsky, P.F. & Helliwell, J. R. (2001). Structure of lobster apocrustacyanin A1 using softer X-rays. *Acta Crystallogr.* **D57**, 1219-1229.

Cipriani F., Felisaz F., Launer L., Aksoy J.-S., Caserotto H., Cusack S., Dallery M., di-Chiaro F., Guijarro M., Huet J., Larsen S., Lentini M., McCarthy J., McSweeney S., Ravelli R., Renier M., Taffut C., Thompson A., Leonard G.A. and Walsh M.A. (2006). Automation of sample mounting for macromolecular crystallography. *Acta. Crystallogr.* **D62**, 1251-1259.

Cohen A. E., Ellis P. J., Miller M. D., Deacon A. M. and Phizackerly R. P. (2002). An automated system to mount cryo-cooled protein crystals on a synchrotron beamline, using compact sample cassettes and a small-scale robot. *J. Appl. Crystallogr.* **35**, 720-726.

Cramer, P. & Müller, C.W. (1997). Engineering of diffraction-quality crystals of the NF-kB p52 homodimer:DNA complex. *FEBS letters* **405**, 373-377.

Cusack S, Belrhali H, Bram A, Burghammer M, Perrakis A, Riekel C (1998). Small is beautiful: protein micro-crystallography. *Nat Struct Biol.* **5**, 634-637.

D'Arcy, A., (1994), Crystallizing proteins- a rational approach? *Acta Crystallogr.,* **D50**, 469-471.

Dauter, Z. (2006). Current state and prospects of macromolecular crystallography. *Acta Crystallogr.* **D62**, 1-11.

Dauter, Z., Dauter, M. & Dodson, E. (2002). Jolly SAD. *Acta Crystallogr.* D58, 494-506.

Dauter, Z., Dauter, M., de la Fortelle, E., Bricogne, G. & Sheldrick, G. M. (1999). Can anomalous signal of sulfur become a tool for solving protein crystal structures? *J. Mol. Biol.* **289**, 83-92.

Dauter Z., Dauter M. and Rajashankar K. R. (2000). Novel approach to phasing proteins: derivatization by short cryo-soaking with halides. *Acta Crystollogr.* **D56**, 232-237

Dauter, Z., Li, M. & Wlodawer, A. (2001). Practical experience with the use of halides for phasing macromolecular structures: a powerful tool for structural genomics. *Acta Crystallogr.* **D57**, 239-249.

Deacon, A.M., Cohen, A.E., Ellis, P.J., McPhillips, S., McPhillips, T.M., Miller, M.D., Phizackerley, R.P., Song, J., Soltis, S.M., Tsyba, I., van den Bedem, H., Wolf G. and Zhang, Z. (2004). Automated crystal screening for high-throughput X-ray crystallography at Stanford Synchrotron Radiation Laboratory. *Acta Crystallogr.* **A60**, s12.

Debreczeni, J. É., Bunkóczi, G. Ma, Q., Blaser, H. & G. M. Sheldrick, G.M. (2003). In-house measurement of the sulfur anomalous signal and its use for phasing. *Acta Crystallogr.* **D59**, 688-696.

Doublié, S. (1997). Preparation of selenomethionyl proteins for phase determination. *Methods Enzymol.* **276**, 523-530.

Dubnovitsky, A.P., Ravelli, R.B.G., Popov, A.N. and Papageorgiou, A.C. (2005). Strain relief at the active site of phosphoserine aminotransferase induced by radiation damage. *Protein Science* **14**, 1-10.

Ducruix, A. & Giegé, R. (1999) *Crystallisation of Nucleic acids and Proteins. A Practical Approach,* edn. 2, Oxford: Oxford University Press.

Duke E.M.H., Kehoe R.C., Rizkallah P.J., Clarke J.A. and Nave C. (1998). Beamline 14: a new multipole wiggler beamline for protein crystallography on the SRS. *J. Synchrotron Rad.* **5**, 497-499.

Earnest T.E., Snell G., Cork C., Meigs G., Nordmeyer R., Cornell E., Yegian D. and Jin J. (2002). Automated crystal mounting and alignment system at the advanced light source. *Acta Crystallogrt.* **A 58**., C56.

Eikenberry, E.F., Tate, M.W., Bilderback, D.H. and Gruner, S.M. (1992). X-ray detectors: comparison of film, storage phosphors and CCD detectors. *Inst. Phys. Conf. Ser.* **121**, 273-280.

Eisenstein, M., (2007), The shape of things, *Nature Methods*, **4**, 95-102.

Engström P., Fiedler S., and Riekel C. (1995). Microdiffraction instrumentation and experiments on the microfocus beamline at the ESRF. *Rev. Sci. Instrum.* **2**, 1348-1350.

Evans, G. & Bricogne, G. (2002). Triiodide derivatization and combinatorial counter-ion replacement: two methods for enhancing phasing signal using laboratory Cu Kα X-ray equipment. *Acta Crystallogr* .**D58**, 976-991.

Evans G. & Pettifer R.F. (2000). CHOOCH: a program for deriving anomalous scattering factors from X-ray fluorescence spectra. *J. Appl. Crystallogr.* **34**, 82-86.

Fischetti, R.F., Yoder, D.W., Xu, S., Stepanov, S., Makarov, O., Benn, R., Corcoran, S., Diete, W., Schwoerer-Boehing, M., Signorato, R., Schroeder, L., Berman, L., Viccaro, P.J. and Smith, J.L. (2007). Optical Performance of the GM/CA-CAT canted undulator beamlines for protein crystallography. In: *Synchrotron Radiation Instrumentation*, J.-Y. Choi, S. Rah, eds., American Inst. of Phys., p.754-757.

Flot, D., Gordon, E.J., Hall, D.R., Leonard, G.A., McCarthy, A., McCarthy, J., McSweeney, S., Mitchell, E., Nurizzo, D., Ravelli, R.G.B. & Shepard, W. (2006). The care and nurture of undulator data sets. *Acta Crystallogr.* **D62**, 65-71.

Freund, A. (2003). Recent progress in X-ray optics at the ESRF. *J. Phys. IV France* **104**, 165-170.

Gaponov, Y., Igarashi, N., Hiraki, M., Sasajima, K., Matsugaki, N., Suzuki, M., Kosuge, T. & Wakatsuki, S. (2004). Secure UNIX socket-based controlling system for high-throughput protein crystallography experiments. *J. Synchrotron Rad.* **11**, 17-20.

García-Ruíz, J.M., Gonzalez-Ramirez, L.A., Gavira, J.A., Otálora, F., (2002), *Acta Crystallogr.,* **D58**, 1638-1642.

Garman E. F. & Schneider T. R. (1997). Macromolecular Cryocrystallography. *J. Appl. Crystallogr.* **30**, 211-237.

Garman, E. (2003). 'Cool' crystals: macromolecular cryocrystallography and radiation damage. *Curr. Opin. Struct. Biol.* **13**, 545–551.

Garman, E.F. and McSweeney, S.M. (2007). Progress in research into radiation damage in cryo-cooled macromolecular crystals. *J. Synchrotron Rad.* **14**, 1-3.

Gilliland, G.L., Tung, M., Blakeslee, D.M. and Ladner, J. (1994). The Biological Macromolecule Crystallization Database, Version 3.0: New Features, Data, and the NASA Archive for Protein Crystal Growth Data. *Acta Crystallogr.* **D50**, 408-413.

Girard, E., Legrand, P., Roudenko, O., Roussier, L., Gourhant, P., Gibelin, J., Dalle, D., Ounsy, M., Thompson, A.W., Svensson, O., Cordier, M.-O., Robin, S., Quiniou, R. and Steyer, J.-P. (2006). Instrumentation for synchrotron-radiation macromolecular crystallography. *Acta Crystallogr.* **D62**, 12-18.

González, A., Pédelacq, J.-D., Solà, M., Gomis-Rüth, F. X., Coll, M., Samama J.-P. & Benini, S. (1999). Two-wavelength MAD phasing: in search of the optimal choice of wavelengths. *Acta Crystallogr.* **D55**, 1449-1458.

González, A. (2003). Faster data-collection strategies for structure determination using anomalous dispersion. *Acta Crystallogr.* **D59**, 315-322.

Gordon, E.J., Leonard, G.A., McSweeney, S. & Zagalsky, P.F. (2001). The C1 subunit of α-crustacyanin: the de novo phasing of the crystal structure of a 40 kDa homodimeric protein using the anomalous scattering from S atoms combined with direct methods. *Acta Crystallogr.* **D57**, 1230-1237.

Grübel, G., Als-Nielsen, J. and Freund, A.K. (1994). The troika beamline at the ESRF. *J.Phys.* **IV 4**, C9-C27.

Hara, T., Tanaka, T., Tanabe, T., Marechal, X., Kitamura, H., Elleaume, P., Morrison, B., Chavanne, J., Vaerenburgh, P. and Schmidt, D. (1998). SPring-8 in-vacuum undulator beam test at the ESRF. *J. Synchrotron Rad.* **5**, 406-408.

Harmsen, A., Leberman, R. and Schulz, G.E. (1976). Comparison of protein crystal diffraction patterns and absolute intensities from synchrotron and conventional X-ray sources. *J. Mol. Biol.* **104**, 311-314.

Harp, J. M., Timm, D. E., Bunick, G. J., (1998), Macromolecular Crystal Annealing: Overcoming Increased Mosaicity Associated with Cryocrystallography, *Acta Crystallogr.,* **D54**, 622-628.

Harp, J. M., Leif Hanson, B., Timm, D. E., Bunick, G. J., (1999), Macromolecular crystal annealing: evaluation of techniques and variables, *Acta Crystallogr,* **D55**, 1329-1334.

Hart, H.E., Scrandis, J.B., Mark, R. and Hatcher, R.D. (1966). *J. Opt. Soc. Am.* **56**, 1018-1023.

Helliwell, J.R., Greenhough, T.J., Carr, P.D., Rule, S.A., Moore, P.R., Thompson, A.W. and Worgan J.S. (1982). Central data collection facility for protein crystallography, small angle diffraction and scattering at the Daresbury Laboratory Synchrotron Radiation Source (SRS), *England. J. Phys. E* **15**, 1363-1372.

Helliwell, J.R. and Thompson, A.W. (1983). A remote-control direct-beam exposure device for the Arndt-Wonacott rotation camera. *J. Appl. Crystallogr.* **16**, 579.

Helliwell, J.R., Papiz, M.Z., Glover, I.D., Habash, J., Thompson, A.W., Moore, P.R., Harris, N., Croft, D. and Pantos, E. (1986). The wiggler protein crystallography workstation at the Daresbury SRS: Progress and results. *Nucl. Instrum. Methods A* **246**, 617-623.

Helliwell, J.R., Ealick, S., Doing, P., Irving T. and Szebenyi M. (1993). Towards the measurement of ideal data for macromolecular crystallography using synchrotron sources. *Acta Crystallogr.* **D49**, 120-128.

Helliwell, J. (2001). Synchrotron-radiation instrumentation, methods and scientific utilization in International Tables for Crystallography Volume F (eds. Rossmann, M.G. and Arnold, E.), Kluwer Academic Publishers, London, pp.155-166.

Henderson, R. (1990). Cryoprotection of protein crystals against radiation damage in electron and X-ray diffraction. *Proc. R. Soc. Lond.* Ser. *B* **241**, 6-8.

Hendrickson, W. A. & Teeter, M. M. (1981). Structure of the hydrophobic protein crambin determined directly from the anomalous scattering of sulphur. *Nature* **290**, 107-113.

Hendrickson, W. A. (1991). Determination of macromolecular structures from anomalous diffraction of synchrotron radiation. *Science* **254**, 51-58.

Hendrickson, W. A., Horton, J. R. & LeMaster, D. M. (1990). Selenomethionyl proteins produced for analysis by multiwavelength anomalous diffraction (MAD): a vehicle for direct determination of three-dimensional structure. *EMBO Journal* **9**, 1665-1672.

Hendrickson, W. A. & Ogata, C. M. (1997). Phase determination from multiwavelength anomalous diffraction measurements. *Methods Enzymol.* **276**, 494-522.

Hendrickson W.A. (1999). Maturation of MAD phasing for the determination of macromolecular structures. *J. Synchrotron Rad.* **6**, 845-851.

Heras, B., Edeling, M. A., Byriel, K. A., Jones, A., Raina, S., Martin, J. L., (2003), Dehydration converts DsbG crystal diffraction from low to high resolution, *Structure*, **11**, 139-145.

Heras, B. & Martin, J. L., (2005), Post-crystallisation treatments for improving diffraction quality of protein crystals, *Acta Crystallogr.,* **D61**, 1173- 1180.

Hey A. & Trefethen A. (2003). Data deluge an e-science perspective. In: F. Berman, A. Hey and G. Fox, Editors, Grid Computing Making the Global Infrastructure a Reality, Wiley, New York, pp. 809-824.

Holyoak, T., Fenn, T. D., Wilson, M. A., Moulin, A. G., Ringe, D., Petsko, G. A., (2003), Malonate: a versatile cryoprotectant and stabilizing solution for salt-grown macromolecular crystals, *Acta Crystallogr.,* **D59**, 2356-2358.

Huang, Z. & Ruth, R.D. (1998). Laser-Electron Storage Ring. *Phys. Rev. Lett.* **80**, 976-979.

Huldt, G., Szöke, A. and Hajdu, J. (2003). Diffraction imaging for single particles and biomolecules. *J. Struct. Biol.* **144**, 219-227.

Hülsen, G., Broennimann, C., Eikenberry, E.F. and Wagner, A. (2006). Protein crystallography with a novel large-area pixel detector. *J. Appl. Crystallogr.* **39**, 550-557.

Ihee, H., Rajagopal, S., Vrajer, V., Pahl, R., Anderson, S., Schmidt, M., Schotte, F., Anfinrud, P.A., Wulff, M. and Moffat, K. (2005). Visualizing Reaction Pathways in Photoactive Yellow Protein from Nanoseconds to Seconds. *Proc. Natl. Acad. Sci. USA.* **102**, 7145-7150.

Isoyama, G., Yamamoto, S., Shioya, T., Ohkuma, H., Sasaki, S., Mitsuhashi, T., Yamakawa, T. and Kitamura, H. (1989). Construction of a multiundulator, Revolver No. 19, at the Photon Factory. *Rev. Sci. Instr.* **60**, 1863-1866.

Jacquamet, L., Ohana, J., Joly, J., Borel, F., Pirochi, M., Charrault, P., Bertoni, A., Israel-Gouy, P., Carpentier, P., Kozielski, Blot, D. and Ferrer, J. (2004), Automated Analysis of Vapour Diffusion Crystallisation Drops with an X-Ray Beam, *Structure*, **12**, 1219-1225.

Jakoncic, J., Di Michiel, M., Zhong, Z., Honkimaki, V., Jouanneau, Y. and Stojanoff, V. (2006). Anomalous diffraction at ultra-high energy for protein crystallography. *J. Appl. Crystallogr.* **39**, 831-841.

Jiang, J, & Sweet, R.M. (2004). Protein data bank depositions from synchrotron sources. *J. Synchrotron Rad.* **11**, 319-327.

Kahn R., Fourme R., Bosshard R., Caudron B., Santiard J.C. and Charpak G. (1982). An X-ray diffractometer for macromolecular crystallography based on a spherical drift chamber – hardware, software and multiwavelength data acquisition with synchrotron radiation. *Nucl. Instrum. Methods* **201**, 203-208.

Kahn R., Fourme R., Bosshard R., Chiadmi M., Risler J.L., Dideberg O. and Wery J.P. (1985). Crystal structure study of *Opsanus tau* parvalbumin by multiwavelength anomalous diffraction. *FEBS Lett.* **179**, 133-137.

Karain, W.I., Bourenkov, G.P., Blume, H., and Bartunik, H.D. (2002). Automated mounting, centering and screening of crystals for high-throughput protein crystallography. *Acta Crystallogr. D* **58**, 1519-1522.

Kirkpatrick, P. & Baez, A.V. (1948). *J. Opt. Soc. Amer.* **38**, 766.

Kriminski, S., Caylor, C. L., Nonato, M. C., Finkelstein, K. D., Thorne, R. E., (2002), Flash-cooling and annealing of protein crystals, *Acta Crystallogr,* **D58**, 459-471.

Korczynska, J., Hu, T., Smith, D.K., Jenkins, J., Lewis, R., Edwards, T. and Brzozowski, A. M., (2007), Microscale vapour diffusion for protein crystallisation, *Acta Crystallogr.,* **D63**, 1009-1015.

Korolev, S., Dementieva, I., Sanishvili, R., Minor, W., Otwinowski, Z. and Joachimiak, A. (2001). Using surface-bound rubidium ions for protein phasing. *Acta Crystallogr.* **D57**, 1008-1012.

Kuhn, P. & Soltis, S M. (2001). Macromolecular structure determination in the post-genome era. *Nucl. Instrum. Methods A* **467**, 1363-1366.

Kumasaka, T., Yamamoto, Y., Yamashita, E., Moriyama, H. and Ueki, T. (2002). Thrichromatic Concept Optimizes MAD experiments In Synchrotron X-Ray Crystallography. *Stucture* **10**, 1205-1210.

Lamzin, V.S., Perrakis, A. & Wilson, K.S. (2001) The ARP/wARP suite for automated construction and refinement of protein models. In *International Tables for Crystallography. Volume F: Crystallography of biological macromolecules* (Rossmann, M.G. & Arnold, E. eds.), Dordrecht, Kluwer Academic Publishers, The Netherlands, pp. 720-722.

Lamzin V.S. & Perrakis A. (2000). Current state of automated crystallographic data analysis. *Nature Struct. Biol.* **7**, 978–981.

Lavault B., Ravelli R.B.G. and Cipriani F. (2006). C3D: a program for the automated centring of cryocooled crystals. *Acta Crystallogr. D* **62**, 1348-1357.

Le Bolloc'h, D., Livet, F., Bley, F., Schulli, T., Veron, M. and Metzger, T.H. (2002). X-ray diffraction from rectangular slits. *J. Synchrotron Rad.* **9**, 258-265.

Lengeler, B., Tümmler, J., Snigirev, A., Snigireva, I. and Raven, C. (1998). *J. Appl. Phys.* **84**, 5855–5861.

Leonard, G.A., Sainz, G., de Backer, M.M.E. & McSweeney, S. (2005). Automatic structure determination based on the single-wavelength anomalous diffraction technique away from an absorption edge. *Acta Crystallogr.* **D61**, 388-396.

Leslie A.G., Powell H.R., Winter G., Svensson O., Spruce D., McSweeney S., Love D., Kinder S., Duke E. and Nave C. (2002). Automation of the collection and processing of X-ray diffraction data - a generic approach. *Acta Crystallogr.* **D 58**, 1924-1928.

Lindley, P. (1999). Macromolecular crystallography with a third-generation synchrotron source. *Acta Crystallogr.* **D55**, 1654-1662.

Liu, Y., Ogata, C.M. & Hendrickson, W.A. (2001). Multiwavelength anomalous diffraction analysis at the M absorption edges of uranium. *Proc Natl Acad Sci USA* **98**, 10648-10653.

Loll, P. J., Tretiakova, A., Soderblom, E., Compatibility of detergents with microbatch-under-oil crystallization method, (2003), *Acta Crystallogr.,* **D58**, 1114-1116.

Luecke, H., Schobert, B., Richter, H.T., Cartailler, J.P., and Lanyi, J.K. (1999). Structural changes in bacteriorhodopsin during ion transport at 2 Angstrom resolution *Science* **8**, 255-260.

Marchesini, S., He, H., Chapman, H.N., Hau-Riege, S.P., Noy, A., Howells, M.R., Weierstall, U. and Spence, J.C.H. (2003). X-ray image reconstruction from a diffraction pattern alone. *Phys. Rev.* **B 68**, 140101.

Mayo C.J., Diprose J.M., Walter T.S., Berry I.M., Wilson J., Owens R.J., Jones E.Y., Harlos K., Stuart D.I. and Esnouf R.M. (2005). Benefits of automated crystallisation plate tracking, imaging and analysis. *Structure* 13, 175-182.

McGeehan, J. E., Carpentier, P., Royant, A., Bourgeois, D. & Ravelli, R. B. G. (2007). X-ray radiation-induced damage in DNA monitored by online Raman. *J. Synchrotron Rad.* **14**, 99-108.

McPherson, A. (1999) Crystallisation of Biological Macromolecules, Cold Spring Harbor: Cold Spring Harbor Laboratory Press.

McPherson, A. & Cudney, B., (2006), Searching for silver bullets: An alternative strategy for crystallising macromolecules, *J. Struct. Biol.*, **156**, 387-406.

McPhillips, T.M., McPhillips, S.E., Chiu, H.-J., Cohen, A.E., Deacon, A.M., Ellis, P.J., Garman, E., Gonzalez, A., Sauter, N.K., Phizackerley, R.P., Soltis, S.M. and Kuhn, P. (2002). Blu-Ice and the Distributed Control System: software for data acquisition and instrument control at macromolecular crystallography beamlines. *J. Synchrotron Rad.* **9**, 401-406.

Miao, J., Charalambous, P., Kirz, J. and Sayre D. (1999). Extending the methodology of X-ray crystallography to allow imaging of micrometre-sized non-crystalline specimens. *Nature* **400**, 342-344.

Micossi, E., Hunter, W. N. & Leonard, G. A. (2002). ***De novo*** phasing of two crystal forms of tryparedoxin II using the anomalous scattering from S atoms: a combination of small signal and medium resolution reveals this to be a general tool for solving protein crystal structures. *Acta Crystallogr.* **D58**, 21-28.

Minor W.V. and Otwinowski Z. (1996). Advances in accuracy and automation of data collection and processing. In: Bourne P.E., Watenpaugh K. (Eds.), *Crystallographic Computing* **7**, Proceedings from the Macromolecular Crystallography Computing School.

Mueller-Dieckmann, C., Polentarutti, M., Djinovic Carugo, K., Panjikar, S., Tucker, P. A. & Weiss, M. S. (2004). On the routine use of soft X-rays in macromolecular crystallography. Part II. Data-collection wavelength and scaling models. *Acta Crystallogr.* **D60**, 28-38.

Mueller-Dieckmann, C., Panjikar, S., Schmidt, A., Mueller, S., Kuper, J., Geerlof, A., Wilmanns, M., Singh, R.K., Tucker, P.A. and Weiss, M.S. (2007). On the routine use of soft X-rays in macromolecular crystallography, Part IV – Efficient determination of anomalous substructures in bio-macromolecules using longer wavelenghts. *Acta Crystallogr.* **D63**, 366-380.

Munro, I.H. (1997). Synchrotron radiation research in the UK. *J. Synchrotron Rad.* **4**, 344–358.

Murray, J. W., Garman, E. F. and Ravelli, R. B. G. (2004). X-ray absorption by macromolecular crystals: the effects of wavelength and crystal composition on absorbed dose. *J. Appl. Crystallogr.* **37**, 513-522.

Nanao M. H., Sheldrick G. M. and Ravelli B. G. (2005). Improving radiation damage substructures for RIP. *Acta Crystallogr.* **D61**, 1227-1237.

Nanao M. H. and Ravelli B. G. (2006). Phasing macromolecular structures with UV-induced structural changes. *Structure* **14**, 791-800.

Nave C. & Garman E. F. (2005). Towards an understanding of radiation damage in cryocooled macromolecular crystals. *J. Synchrotron Rad.* **12**, 257-260.

Nave C. & Hill M. A. (2005). Will reduced radiation damage occur with very small crystals? *J. Synchrotron Rad.* **12**, 299-303.

Neves-Petersen M. T., Gryczynski Z., Lakowicz J., Fojan P., Pedersen S., Petersen E. and Petersen S. B. (2002). High probability of disrupting a disulphide bridge mediated by an endogenous excited tryptophan residue. *Protein Sci.* **11**, 588 - 600.

Neutze, R., Wouts, R., van der Spoel, D., Weckert, E. and Hajdu, J. (2000). Potential for biomolecular imaging with femtosecond X-ray pulses. *Nature* **406**, 752-757.

Normile D. & Jianxiang Y. (2001). Japan and China gear-up for "Postgenome" research. *Science* **294**, 84

Nurizzo, D., Mairs, T., Guijarro, M., Rey, V., Meyer, J., Fajardo, P., Chavanne, P., Biasci, J.-C., McSweeney S. & E. Mitchell (2006). The ID23-1 structural biology beamline at the ESRF. *J. Synchrotron Rad.* **13**, 227-238.

Ogata, C.M. (1998). MAD phasing grows up. *Nat.Struct. Biol.,* **5**, 638-640.

Ohana J., Jacquamet L., Joly J., Bertoni A., Taunier P., Michel L., Charrault P., Pirocchi M., Carpentier P., Borel F., Kahn R. and Ferrer J. L. (2004). CATS: a Cryogenic Automated Transfer System installed on the beamline FIP at ESRF. *J. Appl. Crystallogr.* **37**, 72-77.

Owen R. L., Rudiño-Piñera E. and Garman E. F. (2006). Experimental determination of the radiation dose limit for cryocooled protein crystals. *Proc Natl Acad Sci USA* **103**, 4912-4917.

Pajon A., Ionides J., Diprose J., Fillon J., Fogh R., Ashton A.W., Berman H., Boucher W., Cygler M., Deleury E., Esnouf R., Janin J., Kim R., Krimm I., Lawson C.L., Oeuillet E.,

Poupon A., Raymond S., Stevens T., van Tilbeurgh H., Westbrook J., Wood P., Ulrich E., Vranken W., Xueli L., Laue E., Stuart D.I. and K. (2005). Design of a data model for developing laboratory information management and analysis systems for protein production. *Proteins* **58**, 278-284.

Pearson, A.R., Mozzarelli, A. and Rossi, G.L. (2004). Microspectrophotometry for structural enzymology. *Curr. Opin. Struct. Biol.* **14**, 656-662.

Pebay-Peroula, E., Rummel, G., Rosenbush, J.P. and Landau, E.M. (1997). Bacteriodhopsin. *Science* **277**, 1676-1681.

Perrakis, A., Cipriani, F., Castagna, J.C., Claustre, L., Burghammer, M., Riekel, C. and Cusack, S. (1999). Protein microcrystals and the design of a diffractometer: current experience and plans at EMBL and ESRF/ID13. *Acta Crystallogr* **D55**, 1765-1770.

Peterson, M. R., Harrop, S. J., McSweeney, S. M., Leonard, G. A., Thompson, A. W., Hunter, W. N. & Helliwell, J. R. (1996). MAD Phasing Strategies Explored with a Brominated Oligonucleotide Crystal at 1.65Å Resolution. *J. Synchrotron Rad.* **3**, 24-34.

Phillips, J.C., Wlodawer, A., Yevitz, M.M. and Hodgson, K.O. (1976). Applications of Synchrotron Radiation to Protein Crystallography: Preliminary Results. *Proc. Natl. Acad. Sci. USA* **73**, 128-132.

Phizackerley R.P., Cork C.W. and Merritt E.A. (1986). An area detector data acquisition system for protein crystallography using multiple-energy anomalous dispersion techniques. *Nucl. Instrum. Methods* **A 246**, 579-595.

Pohl, E, González, A., Hermes, C. & van Silfhout, R.G. (2001). Overview of the tunable beamlines for protein crystallography at the EMBL Hamburg Outstation; an analysis of current and future usage and developments. *J. Synchrotron Rad.* **8**, 1113-1120.

Pohl E., Ristau U., Gehrmann T., Jahn D., Robrahn B., Malthan D., Dobler H. and Hermes C. (2004). Automation of the EMBL Hamburg protein crystallography beamline BW7B. *J. Synchrotron Rad.* **11**, 372-377.

Popov, A.N. and Bourenkov, G.P. (2003). Choice of data-collection parameters based on statistical modelling. *Acta Crystallogr.* **D59**, 1145-1153.

Rajasekar, A. Wan, M., Moore, R., Schroeder, W., Kremenek, G., Jagatheesan, A., Cowart, C., Zhu, B., Chen, S.-Y., and Olschanowsky, R. (2003). Storage resource broker - managing distributed data in a grid. *Computer Society of India Journal, Special Issue on SAN* **33**, 42-54.

Ramagopal, U.A., Dauter, M. & Dauter, Z. (2003a). SAD manganese in two crystal forms of glucose isomerase. *Acta Crystallogr.* **D59**, 868-875.

Ramagopal, U.A., Dauter, M. & Dauter, Z. (2003b). Phasing on anomalous signal of sulfurs: what is the limit? *Acta Crystallogr.* **D59**, 1020-1027.

Ravelli R. B. G. and Garman E. F. (2006). Radiation damage in macromolecular cryocrystallography. *Curr Opin Struct Biol.* **16**, 624-629.

Ravelli R. B. G. and McSweeney S. M., (2000). The 'fingerprint' that X-rays can leave on structures. *Structure Fold Des.* **8**, 315–328.

Ravelli, R. B. G., Nanao, M. H., Lovering, A., White, S. & McSweeney, S. (2005). Phasing in the presence of radiation damage. *J. Synchrotron Rad.***12**, 276-284.

Ravelli R. B. G., Schrøder Leiros H.-K. , Pan B., Caffrey M. and McSweeney S. (2003). Specific radiation damage can be used to solve macromolecular crystal structures. *Structure*, **11**, 217-224

Rice, L. M., Earnest, T. N. & Brunger, A. T. (2000). Single-wavelength anomalous diffraction phasing revisited. *Acta Crystallogr.* **D56**, 1413-1420.

Riekel, C., Burghammer, M. and Schertler, G. (2005). Protein crystallography microdiffraction. *Curr. Opin. Struct. Biol.* **15**, 556-562.

Robinson H., Soares, A.S., Becker, M., Sweet, R. and Héroux, A. (2006). Mail-in crystallography program at Brookhaven National Laboratory's National Synchrotron Light Source. *Acta Crystallogr.* **D62**, 1336-1339.

Rodgers D.W. (1994). Cryocrystallography. *Structure* **2**, 1135-1140.

Rosenbaum, G., Holmes, K.C. and Witz, J. (1971). Synchrotron Radiation as a Source for X-ray Diffraction. *Nature* **230**, 434-437.

Roth M., Carpentier P., Kaikati O., Joly J., Charrault P., Pirocchi M., Kahn R., Fanchon E., Jacquamet L., Borel F., Bertoni A., Israel-Gouy P. and Ferrer J. L. (2002). FIP: a highly automated beamline for multiwavelength anomalous diffraction experiments. *Acta Crystallogr.* **D 58**, 805-814.

Rupp B., Segelke B., Krupka H. I., Lekin T. P., Schäfer J., Zemla A., Toppani D., Snell G. and Earnest T. (2002). The TB structural genomics consortium crystallization facility: towards automation from protein to electron density. *Acta Crystallogr.* D58, 1514-1518.

Sarma, G.N & Karplus, P.N. (2006). In-house sulfur SAD phasing: a case study of the effects of data quality and resolution cutoffs. *Acta Crystallogr.* **D62**, 707-716.

Sauer O., Schmidt A. and Kratky C. (1997). Freeze-Trapping Isomorphous Xenon Derivatives of Protein Crystals. *J. Appl. Crystallogr.* **30**, 476-486

Schmidt, T., Ingold, G., Imhof, A., Patterson, B., Patthey, L., Quitmann, C., Schulze-Briese, C. and Abela, R. (2001). Insertion devices at the Swiss Light Source (phase I). *Nucl. Intrum. Meth.* **A467-468**, 126-129.

Schiltz M., Prangé T. and Fourme R. (1994). On the preparation and X-ray data collection of isomorphous xenon derivatives. *J. Appl. Crystallogr.* **27**, 950-960.

Schneider, T. R. & Sheldrick, G. M. (2002). Substructure solution with SHELXD. *Acta Crystallogr.* **D58**, 1772-1779

Schotte, F., Lim, M., Jackson, T.A., Smirnov, A., Soman, J., Olson, J.S., Phillips, G.N., Wulff, M. and Anfinrud, P.A. (2003). Watching a Protein as it Functions: 150 ps Time-resolved Structures of a Myoglobin Mutant at 1.7 Å Resolution. *Science*, **300**, 1944-1947.

Schuermann, J.P. and Tanner, J.J. (2003). MRSAD: using anomalous dispersion from S atoms collected at Cu *K*□ wavelength in molecular-replacement structure determination. *Acta Crystallogr.* **D59**, 1731-1736.

Shapiro, D., Thibault, P., Beetz, T., Elser, V., Howells, M., Jacobsen, C., Kirz, J., Lima, E., Miao, H., Neiman, A.M. and David Sayre' D. (2005). Biological imaging by soft x-ray diffraction microscopy. *Proc. Natl. Acad. Sci. USA* **102**, 15343-15346.

Shintake, T. (2007). Review of the worldwide SASE FEL development. *Proceedings of PAC07*, Albuquerque, New Mexico, USA. pp89-93.

Skinner J.M. and Sweet R.M. (1998). Integrated software for a macromolecular crystallography synchrotron beamline. *Acta Crystallogr.* **D54**, 718–725.

Skinner, J.M., Cowan, M., Buono, R., Nolan, W., Bosshard, H., Robinson, H.H., Héroux, A., Soares, A.S., Schneider, D.K. and Sweet, R.M. (2006). Integrated software for macromolecular crystallography synchrotron beamlines II: revision, robots and a database. *Acta Crystallogr.* **D62**, 1340-1347.

Smith, J.L. (1991) Determination of three-dimensional structure by multiwavelength anomalous diffraction. *Curr. Opin. Struct. Biol.* **1**, 1002-1011.

Snigirev, A., Kohn, V., Snigireva, I. and Lengeler, B. (1996). A compound refractive lens for focusing high-energy X-rays. *Nature* **384**, 49-51.

Stevens R.C. & Wilson I.A. (2001). Industrialising structural biology. *Science* **293**, 519-520.

Stuhrmann, S., Hutsch, M., Trame, C., Thomas, J. and Stuhrmann, H.B. (1995). Anomalous Dispersion with Edges in the Soft X-ray Region: First Results of Diffraction from Single Crystals of Ribosomes Near the ***K***-Absorption Edge of Phosphorus. *J. Synchrotron Rad.* **2**, 83-86.

Stuhrmann, S., Bartels, K. S., Braunwarth, W., Doose, R., Dauvergne, F., Gabriel, A., Knöchel, A., Marmotti, M., Stuhrmann, H. B., Trame, C. and Lehmann, M. S. (1997). Anomalous Dispersion with Edges in the Soft X-ray Region: First Results of Diffraction from Single Crystals of Trypsin Near the ***K***-Absorption Edge of Sulfur. *J. Synchrotron Rad.* **4**, 298-310.

Stura, E. A., (1999), Seeding, *In Protein Crystallisation: Techniques, Strategies, and Tips*, Edited by Berfors T. M.. La Jolla:International University line, 141-153.

Takeda, K., Miyatake, H., Park, S-Y., Kawamoto, M., Kamiya, N. & Miki, K. (2004). Multi-wavelength anomalous diffraction method for I and Xe atoms using ultra-high-energy X-rays from SPring-8. *J. Appl. Crystallogr.* **37**, 925-933.

Tate, M.R., Eikenbeery, E.F., Barna, S.L., Wall, M.E., Lawrence, J.L. and Gruner, S.M. (1995). A Large-Format High-Resolution Area X-ray Detector Based on a Fiber-Optically Bonded Charge-Coupled Device (CCD). *J. Appl. Crystallogr.* **28**,196-205.

Teng, T.-Y. (1990). Mounting of crystals for macromolecular crystallography in a free-standing thin film. *J. Appl. Crystallogr.* **23**, 387–391.

Terwilliger, T. C. (2000). Maximum likelihood density modification. *Acta Crystallogr.* **D56**, 965-972.

Terwilliger, T. C. (2003) Automated main-chain model building by template matching and iterative fragment extension. *Acta Crystallogr.* **D59**, 38-44.

Terwilliger, T.C. & Berendzen. J. (1999). Automated MAD and MIR structure solution. *Acta Crystallogr.* **D55**, 849-861.

Thompson, A.W., Habash, J., Harrop, S., Helliwell, J.R., Nave, C., Atkinson, P., Hasnain, S. S., Glover, I.D., Moore, P.R., Harris, N., Kinder, S. and Buffey, S. (1992). A new macromolecular crystallography Station (9.5) on the SRS wiggler beamline for very rapid Laue and rapidly tunable monochromatic measurements: Commissioning and first results. *Rev. Sci. Instrum.* **63**, 1062-1064.

Ueno, G., Hirose, R., Ida, K., Kumasaka, T & Yamamoto, M. (2004). Sample management system for a vast amount of frozen crystals at SPring8. *J. Synchrotron Rad.* 37**,** 867-873.

Ueno, G., Kanda, H., Kumasaka, T. and Yamamoto, M. (2005). Beamline Scheduling Software: administration software for automatic operation of the RIKEN structural genomics beamlines at SPring-8. *J. Synchrotron. Rad.* **12**, 380-384.

van den Bedem H., Miller M.D. and Wolf G. (2003). Towards automated data collection at the Stanford Synchrotron Radiation Laboratory. *Synchrotron Rad. News* **16**, 15-19.

Von Dreele, R.B., Stephens, P.W., Smith, G.D. and Blessing, R.H. (2000). *Acta Crystallogr.* **D56**, 1549-1553

Wakatsuki, S., Belrhali, H., Mitchell, E.P., Burmeister, W.P., McSweeney, S.M., Kahn, R., Bourgeois, D., Yao, M., Tomizaki, T. and Theveneau, P. (1998). ID14 'Quadriga', a beamline for Protein Crystallography at the ESRF. *J. Synchrotron Rad.* **5**, 215-221.

Walsh, M. A., Evans, G., Sanishvili, R., Dementieva I. & Joachimiak A. (1999a). Taking MAD to the extreme: ultrafast protein structure determination. *Acta Crystallogr.* **D55**, 1168-1173.

Walsh, M. A., Evans, G., Sanishvili, R., Dementieva I. & Joachimiak A. (1999b). MAD data collection - current trends. *Acta Crystallogr.* **D55**, 1726-1732.

Wang, B. C. (1985). Resolution of phase ambiguity in macromolecular crystallography. *Methods Enzymol.* **115**, 90-112.

Wang, J., Dauter, M. & Z. Dauter. Z. (2006). What can be done with a good crystal and an accurate beamline? *Acta Crystallogr.* **D62**, 1475-1483.

Watanabe, N. (2006). From phasing to structure refinement in-house: Cr/Cu dual-wavelength system and a loopless free crystal-mounting method. *Acta Crystallogr.* **D62**, 891-896.

Weik M., Ravelli R.B.G., Kryger G., McSweeney S., Raves M.L., Harel M., Gros P., Silman I., Kroon J. & Sussman J.L. (2000). Specific chemical and structural damage to proteins produced by synchrotron radiation. *Proc Natl Acad Sci USA* **97**, 623–628.

Wien F., Miles A. J., Lees J. G., Vrønning Hoffmann S. and Wallace B.A. (2005). VUV irradiation effects on proteins in high-flux synchrotron radiation circular dichroism spectroscopy. *J. Synchrotron Rad.* **12**, 517-523

Weiss, M. S., Sicker, T. & Hilgenfeld, R. (2001). Soft X-rays, high redundancy, and proper scaling: a new procedure for automated protein structure determination via SAS. *Structure* **9**, 771-777.

Weiss, M. S., Mander, G., Hedderich, R., Diederichs, K., Ermler, U. & Warkentin, E. (2004). Determination of a novel structure by a combination of long-wavelength sulfur phasing and radiation-damage-induced phasing. *Acta Crystallogr.* **D60**, 686-695.

Wilson. W. W., (2003), Light scattering as a diagnostic for protein crystal growth - a practical approach, *J. Struct. Biol.*, **142**, 56-65.

Woolfson, M. & Fan H-F. (1995). *Physical and Non-Physical Methods of Solving Crystal Structures.* Cambridge University Press, Cambridge, U.K.

Yadav, M. K., Gerdts, C. J., Sanishvili, R., Smith, W. W., Roach, L. S., Ismagilov, R. F., Kuhn P. and Stevens, R. C., (2005), In situ data collection and structure refinement from microcapillary protein crystallization. *J. Appl. Crystallogr.*, **38**, 900-905.

Yano J., Kern J., Irrgang K.-D., Latimer M. J., Bergmann U., Glatzel P., Pushkar Y., Biesiadka J., Loll B., Sauer K., Messinger J., Zouni A. and. Yachandra V. K (2005). X-ray damage to the Mn4Ca complex in single crystals of photosystem II: A case study for metalloprotein crystallography. *Proc Natl Acad Sci USA* **102**, 12047-12052.

Yeh, J.I. & Hol, W. G., (1998), A flash-annealing technique to improve diffraction limits and lower mosaicity in crystals of glycerol kinase, *Acta Crystallogr.,* **D54**, 479-480.

Zhang, Z., Sauter, N.K., van den Bedem, H., Snell, G. and Deacon, A.M. (2006). Automated diffraction image analysis and spot searching for high-throughput crystal screening. *J. Appl. Crystallogr.* **39**, 112-119.

Zhu, D. W., Garneau, A., Mazumdar, M., Zhou, M., Xu, G-J., Lin, S.-X., (2006), Attempts to rationalise protein crystallisation using relative crystallisability, *J. Struct. Biol.*, **154**, 297-302.

Zwart, P. H., Banumathi, S., Dauter, M. & Dauter, Z. (2004). Radiation-damage-induced phasing with anomalous scattering: substructure solution and phasing. *Acta Crystallogr.* **D60**, 1958-1963.

In: Astrophysics and Condensed Matter
Editor: Thomas G. Hardwell, pp. 71-104
ISBN: 978-1-60456-332-0

Chapter 2

ON STUDY OF UNIVERSAL DIAMAGNETISM OF SPINLESS BOSONS: FROM N FINITE PARTICLES TO FIELD THEORY

Debnarayan Jana*
Dept. of Physics, University College of Science and Technology
92 A. P. C. Road, Kolkata -700 009, W.B., India

Abstract

The phenomenon of diamagnetism in which a negative susceptibility emerges as the response of a system under an external magnetic field is intriguing and fascinating from various aspects of physical science. We demonstrate explicitly that diamagnetism is a robust property of charged spinless bose systems at finite temperature regardless of their interactions. We also point out the connection of this diamagnetism with other dissipative systems and problems in condensed matter physics. Starting from a single particle, the discussion will go on to N(finite) particles system and finally, end on to an interacting charged scalar field theory in an arbitrary spatial dimension d. A novel regularization scheme particularly suitable for magnetic field problem has been utilized to regularize the zero point divergent energy. Finally, as an illustration of this theory, we take up two classic examples of charged bosonic type of excitations in condensed matter physics.

Keywords: Diamagnetism, Spinless Bosons, Renormalization

PACS numbers: 05.30.J, 11.10, 75.20, 75.20.H, 72.80.Ng, 72.90.$+y$

1. Introduction

Diamagnetism is the material property [1] when it is subjected to an external magnetic field. The magnetic field is responsible for the cyclotron motion of each charged particle in the system. In such a diamagnetic material, an induced dipole moment is developed whose direction is opposite to the applied magnetic field. Hence, the orbital magnetic susceptibility is characterized by a negative value as opposed to conventional paramagnetic or

*E-mail address: djphy@caluniv.ac.in

ferromagnetic material originated entirely from spin. In this respect, it is to be remembered that particles comprising the system must be charged otherwise the magnetic field will not couple them. Within this many body framework, it is interesting to note that there is no *classical* diamagnetism. Moreover, this diamagnetism has been regarded one of the surprises in theoretical physics [2, 3]. In the above approach, the spin and statistics does not enter essentially. However, in most practical situation, the system is basically a quantum many body system held at finite non-zero temperature. Therefore, a detailed understanding of diamagnetism involves quantum mechanics as well as statistical mechanics. In many text books, there exists classical theory of magnetism. Langevin's theory of paramagnetism and diamagnetism is not consistent in this respect because it postulates the existence of stationary orbits as well as the magnetic moments of atoms and molecules. These two assumptions cannot be satisfied classically. A careful look will reveal that in the final expression of diamagnetic susceptibility the ensemble averaged radius of the electrons has to be computed quantum mechanically. There is also some interesting issues regarding this approach [4]. Landau [5] first pointed out the remarkable fact that a collection of quantum free electrons possess a non-zero diamagnetism. Classical theory predicts that the motion of electrons in plane perpendicular to the magnetic field follows a closed path and is periodic in nature. In going over to quantum theory, this periodic motion is quantized and subsequently, when the magnetic field is switched on, the electronic energy changes resulting a non-zero diamagnetic susceptibility.

In dealing with many particles which form an essential ingredient in condensed matter physics, the spin and statistics automatically come close together. All the elementary particles in nature can be divided into two category - one is boson while the other is fermion. We show explicitly that diamagnetism is intimately connected with the spinless boson system. Elementary excitations [6, 7, 8] in condensed matter physics play an important role in discussing the transport and thermodynamics properties of many body system at low temperature. These elementary excitations are quite helpful in mapping the strongly interacting many body system to a weakly non-interacting one.
Elementary excitations can be classified as fermionic and bosonic nature. The simplest example of fermionic excitations are the excitations in metals near the Fermi surface. In case of strongly correlated electron systems, it is possible to think in terms of elementary excitations very close to the Fermi surface.

The second category of elementary excitations originates from electrons are the collective one which are bosonic in nature. It is amazing to note that though the elementary particles in the system are fermions in nature but the collective excitations could be bosonic. The simplest example is the quantum of lattice vibrations - phonons. In fact, our theory can help in distinguishing the nature of elementary excitations. The appropriate field theory to consider the behavior in an external magnetic field is the charged scalar field.

The progress of the theory of diamagnetism has its own way in condensed matter physics. The magnetic anisotropy of an organic crystal (e.g. napthalene, anthacene, hexamethyl benzene etc.) can be probed by the diamagnetic susceptibilty [10]. In this context, the magnetic measurement can be regarded as a most powerful method to crystal analysis even compared to X-ray method. Experimentally, graphite was found to possess higher value (almost eight times) of diamagnetic susceptibility compared to its allotropes charcoal and diamond. The diamagnetism of two dimensional graphite [11] has been contrasted to the large value of carbon nanotube [12, 13], the one dimensional form of graphite. Although protein, like all organic molecules, is diamagnetic in nature, but the atomic as well as electronic structure of the active site can be obtained from measuring the paramagnetic response [14]. The diamagnetic susceptibility for two helium atoms has been estimated using the Hartree Fock and perturbation theory as a function of atomic separation [15]. It also demonstrates the basic physics in magnetic system and opened automatically the path in the applications of many body physics. For example, the Larmor diamagnetism and van Vleck Paramagnetism has been discussed in relativistic quantum theory from the Gordon decomposition approach [16]. The diamagnetism of charged oscillators in an arbitrary magnetic field [17] has been calculated *exactly* at finite temperature. The response of dirty metals [18] have also been grouped in Landau diamagnetism. A non-perturbative solutions for Bloch electrons have been recently found in the literature [19] to illustrate some of the basic features in magnetic field system. The intimate relation of Landau diamagnetism and magnetization of interacting disorder conductors have been demonstrated recently by Montambaux [20]. The statistical mechanics of pair quantum gases in a magnetic field has been reviewed [21]. The most useful applications of this diamagnetism are in superconductivity since the charged elementary excitations are approximately bosons. The fluctuations induced diamagnetic susceptibility has been discussed in the literature [22] from late seventy and has been observed in experiment [23] with superconductors above the transition temperature. The high field diamagnetic fluctuation in Niobium [24] and MgB_2 [25] also has been a subject of current interest. More recently, cuprate system [26, 27] and quantum hall effect of spinless bosons [28] have been discussed in the literature. Diamagnetic orbital response of the mesoscopic system [29, 30, 31]is also an interesting emerging area of condensed matter physics. Even in relativistic platform, diamagnetism concept has been reformulated recently in the literature [32]. In the high energy region, the concept has been reutilized in discussing the response of quark matter in a strong magnetic field [33]. In cross-disciplinary subject, there has been attempt in the literature to connect chaotic dynamics to experimentally accessible macroscopic quantum observable quantity. More specifically, it has been established that the chaotic dynamics in non-integrable elliptic billiards system in a uniform magnetic field can induce substantial reduction and large fluctuations of the diamagnetic susceptibility [34]. On the other hand, the integrable circular billiard case yields results very close to the standard Landau diamagnetism.

Our aim is to make a general prediction of the magnetic response of the charged spinless bosonic system under an external magnetic field. We demonstrate clearly through our calculation that the charged scalar fields are always diamagnetic at any finite temperature in any homogeneous or inhomogeneous static magnetic field.

In this chapter, the various sections are arranged in the following order. After discussing the famous van-Leeuwen theorem in the next section, we move on to the contrasting quantum situation for a single particle as well as N(finite) particles in section III. In section IV, we point out the intriguing connection of this diamagnetism to other systems in condensed matter. In section V, we calculate the partition function of charged scalar field theory in presence of magnetic field. In section VI, we illustrate two simple examples from condensed matter systems at low temperature. Finally, in section VII we give the conclusions.

2. Classical Case

The response of a collection of classical point charged particle in presence of an external magnetic field in thermal equilibrium is non-magnetic [35]. This theorem is known as Bohr-van Leeuwen theorem. This is based on the fact that classically, the magnetic field is a no-work force. Later Peierls [2] gave a very simple elegant argument based on partition function to conclude that the partition function is unaffected by the application of an external magnetic field. A pictorial proof [2] in terms of closed orbits of the inner electrons and the orbits at the boundary of the systems also shows the cancellation of two types of currents resulting zero magnetization in the classical approximation. An exact analytical proof [36] for free electrons also supports this feature. Below we present a simple proof [37] of classical *interacting* N particle system in term of partition function.

The N-particle Hamiltonian is given by

$$H_N = \sum_{i=1}^{N} \frac{1}{2m_i} \left(\vec{p_i} - \frac{e_i \vec{A}(r_i)}{c} \right)^2 + \sum_{i=1}^{N} V(r_i) + \sum_{i<j} W(|\vec{r_i} - \vec{r_j}|) \tag{1}$$

Here $\vec{A}$ is the vector potential, $V(r_i)$ is the one body potential and $W(|\vec{r_i} - \vec{r_j}|)$ is the two body pairwise interaction. The partition function of this system can be written as

$$Z_N(\vec{A}) = \frac{1}{N!} \int_{-\infty}^{\infty} \int_{-\infty}^{\infty} \prod_{i=1}^{N} d^3p_i\, d^3r_i\, \exp(-\beta H_N) \tag{2}$$

If we introduce the new integration variables $\vec{p_i'} = \vec{p_i} - e\vec{A}(\vec{r_i})/c$, the phase space volume $d\Gamma = \prod_{i=1}^{N} d^3p_i'\, d^3r_i = \prod_{i=1}^{N} d^3p_i\, d^3r_i$, since the jacobian of the transformation is equal to unity. This is seen from the fact that the vector potential itself depends only on the coordinates. Consequently, the vector potential is not involved in the partition function as seen explicitly below.

$$\begin{aligned} Z_N(\vec{A}) &= \frac{1}{N!} \int_{-\infty}^{\infty} \int_{-\infty}^{\infty} \prod_{i=1}^{N} m_i^3\, d^3v_i\, d^3r_i\, \exp\left[-\beta \left(\sum_i \frac{1}{2} m_i v_i^2 + V(r_i) \right) \right] \\ &\times \exp\left(-\beta \sum_{i<j} W(|\vec{r_i} - \vec{r_j}|) \right) \\ &= Z_N(0) \end{aligned} \tag{3}$$

Thus, the free energy is unaffected by the application of an external magnetic field. As a result, all derivatives of free energy $F = -k_B T \ln Z$ with respect to magnetic field (magnetization, susceptibility) are equal to zero. Notice in particular the important role played by the lower as well as the upper limits of the two integrations. Instead of infinity, any finite value, however, will disprove the theorem. Regarding the applicability of this theorem one has to keep in mind that the statistical equilibrium can be achieved only if a gas is confined in a closed volume. Otherwise, it is meaningless to calculate the partition function for a non-equilibrium state. The impossibility of explaining the magnetic properties of a material in classical terms is one of the catastrophes of classical physics. In other words, it points out that the effect of magnetic field in thermal equilibrium must be sought in quantum mechanics. All expressions describing the magnetic properties of media are in one way or another dependent on Planck's constant.

3. Quantum Aspects: Single Particle Case

The physics of a system in presence of an external magnetic field is interesting from the theoretical point of view because of the invalidity of perturbation technique [38]. Before we go on to N particle systems, we briefly summarize the single particle picture. Without going into explicit calculation, the discreteness of the energy levels can be understood simply from the quantization condition. Classically, the electron can have any magnitude of velocity with no restriction on the radius. However, quantum mechanically, this is not allowed as the circumference must be an integral number of de Broglie wavelength. This implies that $2\pi r = nh/mv$ with $r = mvc/eB$. Naturally, it automatically indicates that the kinetic energy can have only certain discrete values given by $E_n = \frac{1}{2}mv^2 = \frac{n\hbar\omega_c}{2}$. This value agrees with the exact calculation, if one replaces n by $n + 1/2$. The quantum version of single particle in two dimension is given by the Hamiltonian

$$H = \frac{1}{2m}\left(\vec{p} - \frac{e\vec{A}}{c}\right) = \frac{\vec{\Pi}^2}{2m} \tag{4}$$

The non-vanishing commutation relation between the Π operators form the central part to this problem. In fact, $[\Pi_x, \Pi_y] = -i\hbar m\omega$ with $\omega = \frac{eB}{mc}$ remains valid even for an inhomogeneous magnetic field. By suitably defining the raising and lowering operators, the Hamiltonian can be re-casted to one dimensional Harmonic oscillators. From this, one can easily obtain the Landau spectrum $E_n = (n+1/2)\hbar\omega$ with $n = 0, 1, 2..$ and the degeneracy of each level is $\frac{\phi}{\phi_0}$. Here, ϕ is the typical flux enclosed by the given area and ϕ_0 is the flux quantum ($\frac{hc}{e}$). It is interesting to note that a charged particle in a magnetic field naturally builds up the time scale given by the inverse of cyclotron frequency ω. Physically speaking, if one is interested in computing any measured quantity, then it is required to be averaged over long times compared to this time scale. Another length scale emerges automatically in this magnetic field problem as shown below. Apart from energy, there exists another constant of motion L_z as can be seen directly by computing the commutator $[L_z, H]$. Introducing the complex coordinates $z = x + iy$, we find the form of the simultaneous eigen states of L_z and H as

$$\psi(z, \bar{z}) = z^m \exp\left(-\frac{|z|^2}{4l^2}\right) \tag{5}$$

Here, $l = \sqrt{\frac{\hbar c}{eB}}$ is known as the magnetic length which controls the spreading of the wave packet in the plane. This form of the eigen function actually forms an important basis in constructing the celebrated Laughlin wave function [39] in Fractional Quantum Hall Effect [40, 41]. An alternative way of viewing this length scale is as follows. The length scale emerges by equating the harmonic oscillator potential energy with the ground state energy ($\frac{1}{2}m\omega^2 l^2 = \frac{1}{2}\hbar\omega$). This length scale is independent mass of the charged particle. A careful look in to the Hamiltonian of single particle in a spherical symmetric potential

$V(r)$ interacting with a uniform external magnetic field reveals that there are two competing terms given by

$$\begin{aligned} H &= \frac{p^2}{2m} + V(r) - \frac{e}{2mc}\left(\vec{p}\cdot\vec{A} + \vec{A}\cdot\vec{p}\right) + \frac{e^2}{2mc^2}(\vec{A}\cdot\vec{A}) \\ &= H_0 + H_1 + H_2 \end{aligned} \tag{6}$$

A quick look at the above equation points out that the second term H_1 is responsible for paramagnetism while the third term H_2 is for diamagnetism. Note carefully the change of sign before the two terms. In most atomic situations, the term A^2 is unimportant because of the fact that only at fields of 10^5 Tesla or higher (this field is much larger than the fields produced in the laboratory), the two terms are comparable. If we chose the vector potential as $\vec{A} = \frac{1}{2}\vec{r}\times\vec{B}$, then for spherically symmetric Hamiltonian, the ground state has the essential property $\vec{L}|0>= 0$. Again since, $H_1|0>= 0$ the first order Zeeman effect vanishes exactly. We get the standard negative diamagnetic susceptibility from non-zero and positive H_2. In addition, this susceptibility has almost same order of magnitude for all atomic systems. However, we can even chose a different vector potential which can produce the same magnetic field. These two vector potentials are related by a gauge transformation $\vec{A} = \vec{A'} + \vec{\nabla}\Lambda_1$. In such a situation, an application of time independent perturbation theory in second order along with the invariance of the physical quantity under gauge transformation indicates an important sum rule [42] given by

$$\sum_n (E_n - E_0)| < n|\Lambda_1|0 > |^2 = \frac{\hbar^2}{2m}(< 0|(\nabla\Lambda_1)^2|0 >) \tag{7}$$

The famous Thomas-Riche-Kuhn sum rule in introductory quantum mechanics relating the electric dipole transition matrix elements can be derived as a special case of this sum rule for $\Lambda_1 = x$. For $\Lambda_1 = xy$, we obtain another sum rule involving the electric quadrupole transition matrix elements given by

$$\sum_n (E_n - E_0)| < n|xy|0 > |^2 = \frac{\hbar^2}{3m}(< 0|\vec{r}^2|0 >) \tag{8}$$

Because of the arbitrary nature of the function Λ, in general, many important sum rules can be derived. However, for non-zero angular momentum of the ground state of many particle system, it is not a priori clear which term dominates over the other to decide the response of the system to an external uniform magnetic field. But, for any quantum system, in-spite of its very low magnitude, the existence of its diamagnetism is definite resulting a universal phenomena.

It was Landau [5] who first showed that the classical partition function is replaced by the sum because of the existence of discrete energy levels and found that the diamagnetic susceptibility is non-zero. There is a simple way of deriving the non-oscillatory part of the Landau Diamagnetism at zero temperature [43] and the ground state energy goes as the square of the magnetic field. In this single particle picture, statistics do not come into the picture. But the effect of magnetic field on the systems is two fold - namely the orbital motion of the particle and the spin part. We know that diamagnetic effect comes solely from the orbital part while paramagnetism from the spin part. A simple statistical mechanical calculation shows that the free energy for the orbital part is given by

$$F_{orbit}(B,\beta) = F_{orbit}(0) + \beta^{-1}\log\left(\frac{sinh(\frac{\beta\hbar\omega}{2})}{\frac{\beta\hbar\omega}{2}}\right) \tag{9}$$

while for that of spin part is simply

$$F_{spin}(B,\beta) = -\beta^{-1}\log\left(2\cosh(\beta\hbar\omega/2)\right) \tag{10}$$

where $\omega = \frac{eB}{mc}$. An interesting comment can be given here to elucidate about the variation of the ratio of the orbital partition function and computed free energy difference as a function of magnetic field. This is however not emphasized often in our courses. This is important from the follow up discussion of the interacting case.

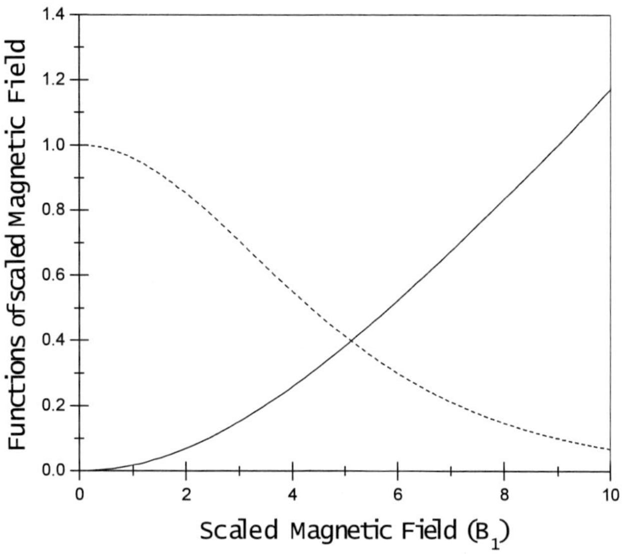

Figure 1. The variation of partition functions and the free energy for single charged particle in a uniform magnetic field. The dashed curve represents the variation of $Z(B)/Z(0)$ while the full line correspond to $F(B) - F(0)$ with scaled magnetic field B_1.

In figure 1, we show the variation of ratio of the partition function $Z(B)/Z(0)$ and $F(B) - F(0)$ as a function of scaled magnetic field at some particular temperature. The graph indicates that the ratio of partition function in orbital case is *always* less than unity and the difference of the free energy is always *positive*. On the other hand, for the spin part, the reverse is true. This fact is embedded in the actual finite temperature calculation of the susceptibility

$$\chi_{orbit}(B,\beta) < 0, \quad \chi_{spin}(B,\beta) > 0 \tag{11}$$

All atoms, ions or molecules having an odd number of electrons are paramagnetic in nature. The exact expression of the diamagnetic susceptibility in terms of dimensionless parameter $\beta\mu_B B$(μ_B is the Bohr magneton) is given by

$$\frac{\chi k_B T}{\mu_B^2} = -\left[1 - \coth^2(\beta\mu_B B) + \left(\frac{1}{\beta\mu_B B}\right)^2\right] \tag{12}$$

This expression indicates that the diamagnetic susceptibility is largest in the vicinity of zero magnetic field and decreases monotonically with increasing B. It is worthy to mention that Peierls [44] investigated the oscillatory effects in the Fermi gas in the limiting situation when the magnetic energy is higher than the thermal energy ($\beta \mu_B B \gg 1$). He showed that in a real three dimensional system, since the energy spectrum is continuous, there should not be any discontinuities in the magnetization. However, in two dimensions, the magnetization undergoes discontinuities at some specific field values. In normal measurements of magnetic susceptibility of a metal, one actually measures the sum of paramagnetic and diamagnetic susceptibility of electrons and ion cores. In such a situation, the paramagnetic component can be separated by magnetic resonance method.

3.1. Sign of Electric Susceptibility vs Magnetic Susceptibility

In this section, we would like to discuss the sign of the response of the electric susceptibility in contrast to magnetic susceptibility. We also indicate a possibility of reversing the sign of the electric susceptibility. The static susceptibility of any material is always positive in contrast to the two possibilities in magnetic systems. The original arguments of this behavior is due to no other than Landau and Lifshitz [45, 46]. We follow here closely more transparent arguments [47, 48].

The argument presented below is based on finite temperature thermodynamic perturbation theory. We show below explicitly that a term *linear* in field variable at thermal equilibrium with *positive* temperature yields always positive susceptibility. The starting point is the usual definition of canonical partition function Z and the free energy F given by

$$Z = \sum_n \exp(-\beta \epsilon_n), \quad F = -(1/\beta)\, log\, Z \tag{13}$$

In the expression of partition function, we would like to substitute the n-th order energy values taking into the corrections up to second order in perturbation theory. This leads to the following expression of the free energy

$$F = F_0 + \left\langle \epsilon_n^{(1)} \right\rangle + \left\langle \epsilon_n^{(2)} \right\rangle - (\beta/2)\left[\left\langle (\epsilon_n^{(1)})^2 \right\rangle - (\left\langle \epsilon_n^{(1)} \right\rangle)^2\right] \tag{14}$$

Here, $\epsilon_n^{(1)}$ and $\epsilon_n^{(2)}$ are the first order and second order correction to the n-th level energy of the system. The angular brackets in the above expression signify the thermal average given by

$$\langle X_n \rangle = \frac{1}{Z_0} \sum_n X_n \exp(-\beta \epsilon_n^{(0)}) = \sum_n w_n X_n \tag{15}$$

$\epsilon_n^{(0)}$ represents the unperturbed energy levels and $Z_0 = \sum_n \exp(-\beta \epsilon_n^{(0)})$. Now, with the help of formal perturbation theory, we obtain the free energy as

$$F = F_0 + \langle H'_{nn} \rangle - (\beta/2)\left[\left\langle (H'_{nn})^2 \right\rangle - (\langle H'_{nn} \rangle)^2\right] + \sum_n \sum_m |H'_{nm}|^2 \frac{w_n}{\epsilon_n^{(0)} - \epsilon_m^{(0)}} \tag{16}$$

With a little simplification in last term, we can rewrite the above equation of the free energy as

$$F = F_0 + \langle H'_{nn} \rangle - (\beta/2)\left[\left\langle (H'_{nn})^2 \right\rangle - (\langle H'_{nn} \rangle)^2\right] + \frac{1}{2}\sum_n \sum_m |H'_{nm}|^2 \frac{w_n - w_m}{\epsilon_n^{(0)} - \epsilon_m^{(0)}} \tag{17}$$

A close inspection to equation (17) reveals that all the second order corrections are negative since $(\epsilon_n^{(0)} - \epsilon_m^{(0)})$ and $(w_n - w_m)$ must have opposite sign for *positive* temperature. This immediately fixes the positive sign of the electric susceptibility. The appearance of the positivity lies in the fact that the perturbation due to the electric field appears as a linear term in the Hamiltonian. For diamagnetism, a *quadratic* term appears in the Hamiltonian, so that the above arguments do not hold good. In fact, a positive coefficient of B^2 appears in the free energy which fixes the negative sign of the diamagnetic susceptibility. Negative susceptibility may result when the populations are inverted [47] in a maser or laser. In such a case, the sign of the *imaginary* part of the susceptibility is reversed at the operating frequency. A negative sign of the static dielectric susceptibility implies the *emission* lines. This is supported by Krammers-Kronig relation which connects the real and imaginary part of the susceptibility.

The argument given above can be further applied to relativistic system [49]. The Lagrangian of a typical relativistic charged scalar field in presence of uniform electric field is given by

$$L = (D_0\Phi)^*(D_0\Phi) - (\vec{\nabla}\Phi^*) \cdot (\vec{\nabla}\Phi) - m^2(\Phi^*\Phi) - V(\Phi^*\Phi) \tag{18}$$

where $D_0 = \partial_0 - ieA_0$. The effective potential $V(\Phi^*\Phi)$ is written in a gauge invariant form and most of the situation in field theory, it is taken as a smooth polynomial function of $(\Phi^*\Phi)$. In addition, one also require that $V(\Phi^*\Phi)$ must be renormalizable. This, however, restricts the power of the polynomial. The highest power of the polynomial can be easily counted in any arbitrary dimension [49]. However, in condensed matter physics, because of the existence of natural cutoff, higher powers of the polynomial may be allowed. Since we are interested in the equilibrium statistical mechanics of this system, the corresponding Hamiltonian must be independent of time. Therefore, it follows that the Hamiltonian is given by

$$H = \Pi^*\Pi + (\vec{\nabla}\Phi^*) \cdot (\vec{\nabla}\Phi) + m^2(\Phi^*\Phi) + V(\Phi^*\Phi) - ie[\Pi^*(A_0\Phi) - \Pi(A_0\Phi)^*] \tag{19}$$

In this Hamiltonian, like the non-relativistic case, the electric field appears as the linear vector potential term A_0. Now, one can easily use the previous second order perturbation theory to show that the ground state energy always decreases with the electric field. This implies automatically the response of this relativistic one as always positive instead of negative as found in the magnetic case below.

In case of magnetic field, there is both linear as well as the quadratic term of the vector potential with opposite sign. In such a situation, thus the net effect of a magnetic field is not clear at first. However, we will show below how to analyze the situation.

3.2. N(finite) Particle - Discrete Models

The starting point is the standard 1d Hubbard model of fermion in an external magnetic field or flux

$$H_{fermion} = -\sum_{i=1}^{L} (c_i^\dagger c_{i+1} \exp(i\phi_f/L) + h.c.) + U \sum_i n_{i\uparrow} n_{i\downarrow} \tag{20}$$

Here, $c_i^\dagger$ and c_i are the fermionic creation and annihilation operator and $c_{L+1} = c_1$, where L is the length of the one dimensional lattice. A uniform vector potential $A_x\vec{x}$ modifies the hopping integral by the usual Peierls substitution $t \rightarrow t\exp(\pm i\phi/L)$, where $\phi = L(A_x e/\hbar c)$. We choose the lattice constant $a_0 = 1$. Using Jordan-wigner transformation and an appropriate unitary transformation, the above fermionic Hamiltonian can be reduced to a simple hard core boson problem in one dimension [51] as given by

$$H_{boson} = -\sum_{i=1}^{L} (\sigma_i^+ \sigma_{i+1}^- \exp(i\phi/L) + h.c.) + U\sum_i n_{i\uparrow}n_{i\downarrow} \tag{21}$$

Here, σ's are the Pauli spin operators and The periodic boundary condition is imposed by identifying σ_{L+1} to σ_1. The standard use of Rayleigh-Ritz variational principle along with the fact that $|\cos(\theta)| \leq 1$ suggests [51] that

$$E_b^0(\phi) \geq E_b^0(0) \tag{22}$$

where $E_b^0(\phi)$ is the ground state energy in presence of the flux ϕ. In other words, the bosons always prefer to have zero flux. This result can be generalized to any arbitrary dimension and on any lattice. It will be interesting to have a straightforward extension of this nice result to finite temperature.

The electronic diamagnetism in two dimensional lattice in a periodic potential has been addressed by Hasegawa et. al [52]. They calculated the density of states of tight-binding non-interacting spinless electrons numerically for magnetic flux $\phi = \frac{p}{q}$ in a plaquette. They also presented strong numerical evidence of the fact that for fixed electron density ν $(0 < \nu < 1)$, the total energy of these spinless electrons have an absolute minimum for $\phi = \nu$. Moreover, later on, it was also shown [53] that the nature of these minima are cusp like. Even this argument has been generalized to three dimensional lattices [54]. Here, in three dimensional lattices, the energy is not necessarily minimum for $\phi = \nu$. Moreover, the results of these three dimensional lattices depend strongly on the choice of the two dimensional lattices. These conclusions are in sharp contrast with the basic fact that the kinetic energy of such electrons in an external magnetic field always increases and thus violates the usual theory of Landau diamagnetism. In all these calculations, the Peierls substitution

$$\begin{aligned} \epsilon(\vec{k}) &\rightarrow \epsilon\left(-i\hbar\vec{\nabla} - \frac{e\vec{A}}{c}\right) \\ t(m-m') &\rightarrow t(m-m')\exp\left[\frac{ie}{\hbar c}\int_m^{m'} \vec{A}(r)\cdot d\vec{r}\right] \end{aligned} \tag{23}$$

have been used which is normally taken into consideration for a slowly varying magnetic field. To understand this paradox, one has to be careful about the Peierls substitution. It was remarked long ago by Kohn [55] that this substitution as an uncontrolled approximation which might lead to questions regarding the reliability of the result. In fact, Alexandrov and Capellmann [56] first pointed out that the above substitution leads to quantitative as well as qualitative errors in computing the energy differences with and without the magnetic field for nearly free electrons and tight-binding electrons. For nearly free electrons, they established that the substitution was an asymptotic approximation leading to errors on the sign of the energy difference at large ν. In the tight-binding situation, they pointed out using harmonic oscillator as pseudo-atomic potential that the the hopping integral was reduced in strong magnetic field. This implies that the kinetic energy of the spinless electrons in a magnetic field is higher than without it. In this way, the usual theory was reestablished for the above two cases. However, the negative value of the energy difference might occur in a very small narrow window of the filling factor ν. This however does not mean the violation of the Landau diamagnetism.

All these might imply that there is no possibility of orbital paramagnetism in two dimensional lattice. In fact, if the Fermi energy is in the saddle point of the band structure or is in the narrow indirect band gap, then the orbital paramagnetism can be proved for two dimensional model in a weak magnetic field [57].

However, when one takes into account the many electrons with spins, then the Pauli exclusion principle can cause the lowering of energy. For a half-filled band (density $= 1/2$), even with an on site interaction among the electrons, Lieb [58] proved that the typical magnetic flux required for the lowest energy of the systems is 2π times the electron density.

3.3. N(finite) Particle - Continuum Model

B. Simon [50] in 1976 first demonstrated through Kato's inequality that N spinless Bosons in an external magnetic field show diamagnetism. For the sake of completeness, we reproduce briefly B. Simon's arguments below for the ground state of the quantum system. The starting Hamiltonian is

$$H_N = -\sum_{i=1}^{N} \frac{1}{2m_i}\left(\hbar\vec{\nabla}_i - \frac{ie_i\vec{A}(r_i)}{c}\right)^2 + \sum_{i=1}^{N} V(\vec{r_i}) + \sum_{i<j} W(|\vec{r_i} - \vec{r_j}|) \tag{24}$$

Let $\Psi(\vec{r}_1, \vec{r}_2, ...\vec{r}_N)$ be the wave function. Since $|\Psi|^2 = (\Psi^*\Psi)$, we note that using 3N dimensional gradients

$$\begin{aligned} (|\Psi|\hbar\vec{\nabla}|\Psi|) &= |Re(\Psi^*\hbar\vec{\nabla}\Psi)| \\ &= |Re[\Psi^*(\hbar\vec{\nabla} - ie\vec{A}/c)\Psi]| \\ &\leq |\Psi||(\hbar\vec{\nabla} - ie\vec{A}/c)\Psi| \end{aligned} \tag{25}$$

Therefore, with x denoting the N vectors $(\vec{r}_1, \vec{r}_2, ...\vec{r}_N)$, we get

$$|\hbar\vec{\nabla}|\Psi||^2(x) \leq |(\hbar\vec{\nabla} - ie\vec{A}/c)\Psi|^2(x) \tag{26}$$

It follows automatically then that

$$\begin{aligned} \int dx \left[(\hbar\vec{\nabla}|\Psi|)^2 + V(x)|\Psi|^2 + W(x)|\Psi|^2\right] \leq \int dx \Big[&|(\hbar\vec{\nabla} - ie\vec{A}/c)\Psi|^2 \\ &+ V(x)|\Psi|^2 + W(x)|\Psi|^2\Big] \end{aligned} \tag{27}$$

Now if one chooses Ψ to be the ground state wave function for H_N, then the right hand side equal to $E(A)$, whereas by the variational principle for the ground state energies, the left hand side is greater than $E(0)$. Hence, we get

$$E(0) \leq E(A) \tag{28}$$

This is an *exact* and *non-perturbative* result. Few important remarks are here. The above argument is strictly valid for the ground state only. Because the transformation of $\Psi \longrightarrow |\Psi|$ in fact destroys the orthogonality condition between the eigen states [50]. Moreover, we notice that the above proof fails for fermions. The replacement of Ψ by $|\Psi|$ used above is not allowed by Fermi statistics. Even spinless fermions do not obey this inequality. For a counterexample, we choose a spherically symmetric potential V and concentrate on a particular eigenstate $n = 1$. Then, energies of $l \neq 0$ states decrease in lowest order perturbation theory for a suitable choice of $\vec{A}$. This shows that the above inequality fails for fermions. Note that $|\psi|$ is not differentiable, hence to prove the above inequality, one can borrow the arguments from Kato [59] by writing $\psi_\epsilon = \sqrt{|\psi|^2 + \epsilon^2}$ and then taking $\epsilon \longrightarrow 0$ limit. The intriguing aspect of this argument lies in the fact that without solving the energy spectra, one can prove *exactly* that the free energy in presence of magnetic field is higher than that of zero magnetic field. The above proof does not depend on the nature of applied external magnetic field and is also independent of the nature of interaction between the particles. In this sense, the simple proof bears a remarkable significant character.

The spinless character of bosons is also important here because in such a situation no Zeeman term is generated in the Hamiltonian. The presence of Zeeman term however does not make clear analysis of the nature of variation of the energy with magnetic field.

The diamagnetism of magnetic Schrödinger operator and paramagnetism of Pauli operators have been rigorously proven for *inhomogeneous* magnetic field [60]. The paramagnetism for non-relativistic electrons in an external potential [61] has also been shown for a non-uniform magnetic field (such as polynomial in $(x^2 + y^2)^N$).

The diamagnetism of a model disordered system was demonstrated long back ago by Luttinger [62] using a new variational scheme at low temperature and it was shown that for not large field, the susceptibility per electron was quite significant varying with temperature as $T^{-2/5}$.

4. Relation of Diamagnetism to Other Systems

4.1. Dissipative Systems

The Landau diamagnetism has been one of the play ground of theoretical physics for verifying various concepts relating to dissipation. For example, the dynamics of this charged particle in a magnetic field connecting to ohmic bath can be computed exactly [63]. This approach points out important issues of confined geometry and the approach towards the equilibrium of a quantum dissipation system. Besides, this approach highlights the significance the Bohr-van-Leeuwen theorem in this dissipative system as well as general coherence and de-coherence aspects in condensed matter physics [64].

The Feynman-Vernon Hamiltonian [65] for the system is written as

$$H = \frac{1}{2m}\left(\vec{p} - \frac{e\vec{A}}{c}\right) + \sum_j \left[\frac{1}{2m_j}p_j^2 + \frac{1}{2}m_j\omega_j^2(\vec{q}_j - \vec{r})^2\right] \tag{29}$$

Here, $\vec{p}, \vec{r}$ are the corresponding momentum and position operator of the particle while $\vec{p}_j, \vec{q}_j$ are for the reservoir particle. In this ohmic dissipation model, the typical distribution of oscillators taken as

$$J(\omega) = \frac{\pi}{2}\sum_j m_j\omega_j^3\delta(\omega - \omega_j) = \gamma\omega \tag{30}$$

The quantum Langevin equation reads as

$$m\ddot{\vec{r}} + m\gamma\dot{\vec{r}} - \frac{e}{c}(\dot{\vec{r}} \times \vec{B}) = \vec{f}(t) \tag{31}$$

where $\vec{f}(t)$ is the operator form of the noise and γ is the friction constant. The magnetization in this dissipative environment is a time dependent quantity given by

$$M_z(t) = \frac{|e|}{2c}\langle(x\dot{y} - y\dot{x})\rangle = \frac{|e|}{4c}Im\left\langle(\dot{z}z^\dagger + z^\dagger\dot{z})\right\rangle \tag{32}$$

In the limit of zero damping ($\gamma = 0$) and $t \to \infty$, it is found that the equilibrium value of the magnetization [63] is given by the following expression

$$M_z^0 = -\frac{|e|\hbar}{2mc}\coth\left(\frac{\hbar\omega}{2k_BT}\right) \tag{33}$$

A careful look into the equation (33) reveals that this value corresponds to the contribution from the boundary electrons only [1]. In order to get the correct expression to restore the Bohr-van-Leeuwen theorem, we have to incorporate a confining potential $\frac{1}{2}k(x^2 + y^2)$ in the Hamiltonian and then calculate the trace for the magnetization. At the end, we have to take the $k \to 0$ limit to obtain the final result. The general expression obtained [63] in this way

$$\begin{aligned} M_z &= \frac{|e|\hbar}{2mc}\left[\sum_{n=1}^{\infty}\frac{4n\pi\eta\nu_c}{(\nu_c^2 + \eta^2 - n^2\pi^2)^2 + 4n^2\pi^2\nu_c^2}\right. \\ &\quad \left. + \frac{\nu_c}{\eta^2 + \nu_c^2} - \frac{1}{2}\frac{\sinh(2\nu_c)}{sinh^2(\nu_c) + \sin^2(\eta)}\right] \end{aligned} \tag{34}$$

Here, the dimensionless parameters ν_c and η are $\frac{\hbar\omega}{2k_BT}$ and $\frac{\hbar\gamma}{2k_BT}$ respectively. In the limit of zero damping, we find the celebrated Landau result

$$M_z^0 = \frac{|e|\hbar}{2mc}\left[\frac{2k_BT}{\hbar\omega} - \coth\left(\frac{\hbar\omega}{2k_BT}\right)\right] \tag{35}$$

This non-equilibrium methodology illustrates that the path to the equilibrium from a non-equilibrium is not unique. In other words, it essentially demonstrates the remarkable difference whether the system is equilibrated in presence of a boundary or free space. Moreover, it also emphasizes that the limits $k \to 0$ and $t \to \infty$ are *not* interchangeable. To get the correct expression, one has to equilibrate the particles first ($t \to \infty$) and then slowly the confining potential is switched off ($k \to 0$) or equivalently thermodynamic limit. The reverse order eventually violates the Bohr-van-Leeuwen theorem. This is also in accord with the fact that although the number of boundary electrons are fewer; however, they make quite significant contribution in the final expression of magnetization. When the dissipation η is strong enough, M_z vanishes and thus restores Bohr-van-Leeuwen theorem. For a mesoscopic system, the surface effects being significant in compared to its bulk counterpart, this study may be of relevance in understanding some physical properties of the system.

Generally speaking, the equilibrium quantity computed in this fashion should not contain the coupling constant γ which characterizes the system-bath interaction. Here, however, it is difficult to separate the system and bath. Moreover, since the friction coefficient γ characterizes the transport property of such a system, the magnetization can be re-casted to a form involving effective Drude resistivity and effective Hall resistivity [66]. In this sense, Landau diamagnetism of such a quantum system is interestingly connected to the transport properties of dissipative system and provides an important missing link between equilibrium and non-equilibrium statistical mechanics. This has been recently demonstrated thoroughly by a path integral approach [67]. In this work, the thermodynamical properties of the system has been derived from the free energy. However, although transport properties is treated mostly in kinetic theory but it is related to the expectation value of the operator $(\vec{r} \times \vec{v})$. Because of the presence of this velocity $\vec{v}$, the damping factor γ appears in the expression of magnetization. Moreover, Landau diamagnetism is of quantum origin with no classical counterpart. Even Jayannavar and Kumar [68] have pointed out the non-existence of dissipative classical diamagnetism. The diamagnetic susceptibility of two dimensional charged particle system has been computed in the dissipative region [69] which may be relevant for cuprate superconductors in the normal state.

4.2. Random Walk and Polymer

We would like to discuss an important connection of the partition function of bosons in magnetic field with Brownian motion of the particles. This is due to famous Feynman's path integral approach. In discussing the partition function of N interacting bosons, he observed that the quantum nature of the bosons can be transformed to a classical 'ring polymer' problem [70] with extra degrees of freedom. He showed explicitly that in this formulation the interatomic potential does not alter the existence of Bose Einstein condensation in He^4. A polymer is a macromolecule consisting of large number of monomers joined into a chain. Every molecule has a basic chain group known as links. Each polymer is characterized by a definite number of links, N, known as the degree of polymerization. In fact, this number is directly proportional to the molecular mass of the chain. The statistical mechanics of such a long chain molecule can be described by the random walk and self-avoiding walk [71, 72]. In the subsequent section, we follow the path integral approach [73, 74] to the problem to connect with the magnetic field problem.

The end to end distribution function of a planar (two dimensional) random walk of N steps each of length l_0 is given by

$$G(\vec{r}, N) = \int_{r(0)=0}^{r(N)=r} \mathcal{D}r(t)\ \exp\left\{-\frac{1}{l_0}\int_0^N dt\ \left(\frac{dr}{dt}\right)^2\right\} \tag{36}$$

Obviously, N has the dimension of length. If the walker has further restriction of enclosing a given area A_1, then the probability

$$\begin{aligned} P(A_1, N) &= \frac{1}{G(0,N)}\int_{r(0)=0}^{r(N)=r} \mathcal{D}r(t)\ \delta\left(A_1 - \frac{1}{2}\left|\int_0^N d\tau\ \left(x\frac{\partial y}{\partial\tau} - y\frac{\partial x}{\partial\tau}\right)\right|\right) \\ &\times \exp\left\{-\frac{1}{l_0}\int_0^N dt\ \left(\frac{dr}{dt}\right)^2\right\} \end{aligned} \tag{37}$$

Notice that the area can have both positive and negative sign. Interestingly, the Fourier transform of $P(A_1, N)$ has a nice interpretation as given below. We can write

$$P(A_1, N) = \frac{1}{2\pi}\int dk\ P(k, N)\ \exp(ikA_1) \tag{38}$$

where

$$P(k, N) = \frac{1}{G(0,N)}\int_{r(0)=0}^{r(N)=r} \mathcal{D}r(t) \exp\left\{-\int_0^N d\tau\ \mathcal{L}(r,\tau)\right\} \tag{39}$$

The Lagrangian $\mathcal{L}$ is defined as

$$\mathcal{L}(r,\tau) = \frac{1}{l_0}\left(\frac{dr}{d\tau}\right)^2 + \frac{ik}{2}\left|x\frac{dy}{d\tau} - y\frac{dx}{d\tau}\right| \tag{40}$$

We note that this Lagrangian has a remarkable similarity with the Lagrangian for the charged particle in a constant magnetic field with symmetric vector potential $\vec{A}$ given by

$$\mathcal{L}_B = \frac{m}{2}\left(\frac{dr}{d\tau}\right)^2 + ie\left(\frac{d\vec{r}}{d\tau}\right)\cdot\vec{A} \tag{41}$$

A close inspection reveals the following identification of the parameters $\frac{m}{2} \leftrightarrow \frac{1}{l_0}$ and $eB \leftrightarrow k$. This establishes the connection between the polymer having constrained to a given area A_1 and the changed particle in a constant magnetic field.

Even one can write the partition function instead of distribution defined as

$$G(0, N) = \frac{Z(0,N)}{\tilde{S}} \tag{42}$$

where $\tilde{S}$ is the two dimensional area and the partition function Z is given by

$$Z(0, N) = \int_{\tilde{S}} d^2r\ G(\vec{r}_1 = \vec{r}_2 = \vec{r}, N) \tag{43}$$

Then, it is easy to see that

$$P(A_1, N) = \frac{Z(A_1, N)}{Z(0, N)} \tag{44}$$

By definition, the probability $P(A_1, N) \leq 1$ and subsequently, it helps us to write an important inequality in terms of free energy

$$F(A_1, N) \geq F(0, N) \tag{45}$$

Though here we have used the constant magnetic field in symmetric vector potential gauge, this result can be generalized to any arbitrary (inhomogeneous) magnetic field. This approach has been recently utilized to derive an important inequality [75] for the spinless bose system in an external magnetic field. This method based on Wiener measure automatically takes into account the *interacting* case to obtain an important connection [75] between Brownian motion and magnetism. The probability was shown as the ratio of two partition functions

$$P(B) = \frac{Z(B, \beta)}{Z(0, \beta)} \tag{46}$$

Again, by definition $P(B) \leq 1$, $F(B, \beta) \geq F(0, \beta)$. Note that this inequality was obtained without explicitly computing the partition function; only by using the analogy with Brownian motion. Even in this approach, the zero field susceptibility $\chi = -\frac{\partial^2 F}{\partial B^2}|_{B=0}$ of the magnetic system is related to the variance [75] of the distribution of areas in diffusion problem as

$$\chi = \frac{1}{\beta}[\ln(P(B))'']_{B=0} = -\frac{e^2}{\beta}\overline{(A - \bar{A})^2} = -\frac{e^2}{\beta}\, VarA \leq 0 \tag{47}$$

This result has been suitably used to compute the distribution of solid angles for Brownian motion on a sphere [76] and polymer problems [77]. In all these discussions, the magnetic field has been cleverly used as a some kind of counter in measuring the typical closed area in a specific Brownian motion. This diamagnetic susceptibility should be contrasted with the magnetic spin susceptibility which is a measure of the spin fluctuation and hence is positive as shown below.

$$\begin{aligned} \chi &= \frac{\partial m}{\partial B} = \frac{\partial}{\partial B}\frac{Tr\left(\frac{1}{N}\sum_i S_i\, e^{-\beta(H_0 - \sum_j \vec{S}_j \cdot \vec{B})}\right)}{Tr\left(e^{-\beta(H_0 - \sum_j \vec{S}_j \cdot \vec{B})}\right)} \\ &= \frac{\beta}{N}\left[\left\langle\left(\sum_i S_i\right)^2\right\rangle - \left(\sum_i \langle S_i\rangle\right)^2\right] > 0 \end{aligned} \tag{48}$$

The analogy of the polymer problem with magnet can be further extended in the density matrix language [78, 79]. The density matrix $\rho(\vec{r}, \vec{r'}; \beta)$ is defined as

$$\rho(\vec{r}, \vec{r'}; \beta) = \sum_i \psi_i^*(\vec{r'})\, e^{-\beta H}\psi_i(\vec{r}) \tag{49}$$

so that the partition function

$$Z(\beta) = \int \rho(\vec{r}, \vec{r}; \beta)\, d^3r \tag{50}$$

In this language, the equation of motion $-\frac{\partial \rho}{\partial \beta} = H\rho$ can be transformed as

$$\left[\frac{\partial}{\partial \beta} - \frac{\hbar^2}{2m}\nabla^2 - \frac{e\hbar B}{2mic}\left(x\frac{\partial}{\partial y} - y\frac{\partial}{\partial x}\right) + \frac{e^2B^2}{8mc^2}\left(x^2+y^2\right)\right]\rho = 0 \tag{51}$$

The third term is responsible for the angular momentum of the system. It is evident from the above equation that zero magnetic field ($B = 0$) refers to a simple differential equation for conduction of heat flow

$$\frac{\partial \rho}{\partial \beta} = K\nabla^2\rho \tag{52}$$

where K is the thermal diffusivity. Equivalently, this implies that finding the temperature at any point (x, y, z) at time β in an infinite medium of diffusivity $\frac{\hbar^2}{2m}$ is due to an instantaneous point source of heat of unit strength at (x', y', z') at time $\beta = 0$. The solution [80] of the density matrix is given by

$$\rho(B=0) = \left(\frac{m}{2\pi\hbar^2\beta}\right)^{3/2} \exp\left(-\frac{m}{2\hbar^2\beta}(|\vec{r} - \vec{r}'|)^2\right) \tag{53}$$

For a general non-zero magnetic field, the solution was due to Sondheimer and Wilson [81].

The equation (51) in the zero angular momentum state has a remarkable similarity with the end to end distribution $G(\vec{r}, \vec{r}'; N)$ of a polymer chain confined in a tube as given by Doi and Edwards model [82]

$$\left(\frac{\partial}{\partial N} - \frac{l_1^2}{6}\nabla^2 + \frac{\omega^2}{6}(x^2+y^2)\right) G(\vec{r}, \vec{r}'; N) = 0 \tag{54}$$

Here l_1 is the Kuhn length and ω is the strength of the harmonic potential. In this representation, the tube around the chain has been modelled by this harmonic potential. A quick look to the equation (51)reveals the following identification of the two systems as

$$\beta \leftrightarrow N, \quad \frac{\hbar^2}{2m} \leftrightarrow \frac{l_1^2}{6}, \quad \frac{e^2B^2}{8mc^2} \leftrightarrow \frac{\omega^2}{6} \tag{55}$$

However, the quantity of interest in this theory is the average radius $< x^2 + y^2 >= a^2$. In terms of complex coordinates $z = x + iy$ and $\bar{z} = x - iy$, the Hamiltonian H can be re-casted as

$$H = -\frac{2\hbar^2}{m}\partial_z\partial_{\bar{z}} + \frac{e^2B^2}{8mc^2}|z|^2 - \frac{eB\hbar}{2mc}(z\partial_z - \bar{z}\partial_{\bar{z}}) \tag{56}$$

If we are interested in the ground state properties, we can safely neglect the contribution of the angular momentum. The ground state wave function is then simply

$$\psi_0(z, \bar{z}) = \mathcal{N} \exp\left(-\frac{eB}{4\hbar c}|z|^2\right) = \mathcal{N} \exp\left(-\frac{\omega}{2l_1}|z|^2\right) \tag{57}$$

Therefore, the average radius in the ground state can be computed as

$$a^2 = \int d^2z \, |z|^2 \, |\psi_0(z, \bar{z})|^2 \tag{58}$$

Thus, using the analogy of the magnetic field problem, one can calculate the radius of the polymer confined in a tube provided the integral converges. This analogy of the tube problem with that for an electron in a constant magnetic field can be utilized critically to discuss the various physical properties of many interacting tubes.

It was shown in all the above proofs of the spinless bosons systems that the free energy of N particles in an external magnetic field is greater than the free energy without an external magnetic field. Thus, the free energy of such a system always increases with the magnetic field resulting the negative susceptibility. However, it is not clear whether the free energy is a *monotonically* increasing function of the magnetic field B or not.

5. Field Theory Aspects

All the above systems deal with N finite particle resulting the energy to be compared as finite. However, in a continuum field theory, the degrees of freedom being infinite, one has to be very careful about the comparison of the two energies. The energies are divergent in nature (either in presence or absence of magnetic field) and one has to develop a suitable regularization scheme in such a way that the difference of them is finite and cut-off independent. A simple example [83] from mathematical physics illustrates this idea used earlier in Casimir's effect [84]. We have developed a novel regularization scheme [49, 85] particularly suitable for dealing this magnetic field problem.

5.1. Relativistic Bosonic Field Theory

There are several reasons for studying the scalar fields in an external magnetic field at any finite temperature. First of all, the results of N particle system do not always translate in same form to a corresponding field theory [85]. Secondly, in field theory one has to develop a proper regularization scheme to have a finite value of the observable (in this case the energy). Thirdly, to study the elementary excitations in the system in the long wavelength limit, it is better to consider the corresponding field theoretic version rather than the particle picture. More ambitiously, to look into the various symmetries being broken, it is easier to understand from the field theoretic version rather than the N particle picture.

We start with the Lagrangian density of the complex scalar field which is charged and spinless. Below we follow the natural scale ($\hbar = 1, c = 1$) in field theory. In presence of a homogeneous constant magnetic field, the relativistic free Lagrangian of the system is given by

$$\mathcal{L} = (D_\mu \Phi)^*(D_\mu \Phi) - m^2(\Phi^* \Phi) \tag{59}$$

where $\mu = 0, 1, 2$. The complex scalar fields are usually written in terms two real fields Φ_1 and Φ_2 as

$$\Phi = \frac{\Phi_1 + i\Phi_2}{\sqrt{2}}, \Phi^* = \frac{\Phi_1 - i\Phi_2}{\sqrt{2}} \tag{60}$$

The relativistic action of complex charged scalar field in the imaginary time formalism can be written as

$$S = \int\int d^d x \, d\tau \left[(D_\mu \Phi)(D^\mu \Phi)^* + m^2(\Phi\Phi^*) \right] \tag{61}$$

Here $\mu = 0, 1, 2$ and $D_\mu = \partial_\mu - ieA_\mu$. The imaginary time τ has the range from 0 to $\beta = \frac{1}{k_B T}$. The vector potential is $\vec{A}$ and the respective magnetic field $\vec{B} = \vec{\nabla} \times \vec{A}$. The bosonic condition is imposed via the boundary condition $\Phi(x, 0) = \Phi(x, \beta)$.

5.1.1. 2d Free Theory

In a magnetic field problem, the important physics takes place in two space dimension. Below we restrict ourselves to two dimension first to compute the free energy of charged scalar fields in a homogeneous uniform magnetic field. Using the standard finite temperature field theory formulation [86], we can write the partition function for free charged scalar fields as

$$\begin{aligned} Z(B, \beta) &= \int \mathcal{D}[\Pi_1]\mathcal{D}[\Pi_1] \int \mathcal{D}[\Phi_1]\mathcal{D}[\Phi_2] \\ &\times \exp\left[\int d^2x \, d\tau \left(i\Pi_1 \frac{\partial \Phi_1}{\partial \tau} + i\Pi_2 \frac{\partial \Phi_2}{\partial \tau} - H + \mu(\Phi_2 \Pi_1 - \Phi_1 \Pi_2) \right) \right] \end{aligned} \tag{62}$$

Here, the chemical potential in connection with the conserved charged $Q = \int d^2x \, (\Pi_1 \Phi_2 - \Pi_2 \Phi_1)$ with $\Pi_i = \partial_0 \Phi_i$ and H is the corresponding Hamiltonian of the free fields given by

$$\begin{aligned} H &= \int d^2x \left[\frac{1}{2}(\Pi_1^2 + \Pi_2^2) + \frac{1}{2}\left((\nabla\Phi_1)^2 + (\nabla\Phi_2)^2 \right) \right. \\ &\quad \left. + \frac{1}{2}(e^2 A^2 + m^2)(\Phi_1^2 + \Phi_2^2) - \vec{j} \cdot \vec{A} \right] \end{aligned} \tag{63}$$

where $\vec{j}$ is the current density given by

$$\vec{j} = -e \left(\Phi_1 \vec{\nabla} \Phi_2 - \Phi_2 \vec{\nabla} \Phi_1 \right) \tag{64}$$

To obtain the possible modes of the system, we first solve the Klein Gordon equation for the complex scalar field in the gauge $\vec{A} = (-By, 0)$. Therefore, the expansion involves in this system is basically as a set of harmonic oscillator with frequency $\omega_l = \sqrt{m^2 + (2l + 1)eB}$ with $l = 0, 1, ...\infty$. The states are highly degenerate with degeneracy for each l given as $\frac{eBA}{2\pi}$, where A is the area of the given system. It is interesting to note that the degeneracy is independent of l and proportional to magnetic field B and area A. The standard calculation finally yields [49] the free energy as

$$
\begin{aligned}
F(B,\beta) &= 2\pi AeB\sum_{l=0}^{\infty}\left[\omega_l + \frac{1}{\beta}\ln\left(1-\exp(-\beta(\omega_l-\mu))\right)\right. \\
&\left. +\frac{1}{\beta}\ln\left(1-\exp(-\beta(\omega_l+\mu))\right)\right] \qquad (65)
\end{aligned}
$$

It is obvious from the above expression that the first term is due to zero point fluctuations of the vacuum in a magnetic field while the rest are from the finite temperature contribution. Because of the summation of all Landau levels, the zero point energy is divergent. In conventional field theory, this infinite zero point energy is not taken seriously because it can be used for the redefinition of the zero scale of the energy. However, the *change* in zero point energy is observable in contrast to infinite zero point energy. Below, we use Casimir's idea for regularizing the energy as well as the renormalizing the energy difference between with and without the magnetic field. It is also interesting to note that $B \to 0$ limit of the free energy does not yield the result for without magnetic field. This is due to unusual *discrete* spectra in a magnetic field in contrast to *continuous* spectra in the zero magnetic field situation. The zero point energy without the magnetic field can be calculated as simply

$$
F(0,\beta=\infty) = 2\pi A\int d^2p\,\sqrt{p^2+m^2} \qquad (66)
$$

The above integral, of course, diverges because of the upper limit of p. To make the zero point energy finite, One may use regularizing cut-off function $\exp(-\frac{p^2+m^2}{2\Lambda^2})$ with a large Λ where one can keep the upper limit of the integration to infinity [87]. However, below we use a simple novel scheme to regularize this quantity.

5.1.2. Regularization Scheme

The presence of infinite degrees of freedom in any field theory makes the zero point energy to diverge. But the change in zero point energy caused by some external agent is, however, finite and observable. We show below explicitly the difference between the two energies with and without the magnetic field is finite and convergent. This prescription must be supplemented by a regularization scheme. A true renormalized energy difference corresponds to the convergent situation when one takes the cut-off used in the theory to infinity and thus, the difference becomes independent of cut-off in the asymptotic limit.

We use the mode matching regularization scheme to renormalize the energy difference. It has been established [49] clearly as the most proper and efficient way of regularizing the magnetic field problem. This approach has also been justified [85] from the phase space argument as well as from relativistic one.

As a first step, it is easily understood that the finite upper limit of the discrete sum and continuous integral will make the two quantities finite. However, to compare the two energies, we need to have a relation between the two upper limits. Here, lies the importance of the physically motivated mode matching scheme. Fixing the discrete sum of Landau level index up to L, we find the number of modes in presence of magnetic field as $2\pi AeB(L+1)$. However, restricting the momentum cut-off to Λ in continuous spectra renders the total mode to $\pi A\ \Lambda^2$. Equating these two modes, we find an interesting equation involving the two cut-offs as

$$\Lambda^2 = 2eB(L+1) \tag{67}$$

One might wonder that the momentum cut-off depend on the magnetic field or non-integer values of Landau level as given by the relation (67). Though this is an artifact of this theory, it is to be regarded as an intermediate step to the renormalization procedure. Using the above relation (67), we show explicitly the positive energy difference for a particular set of parameters in figure 2.

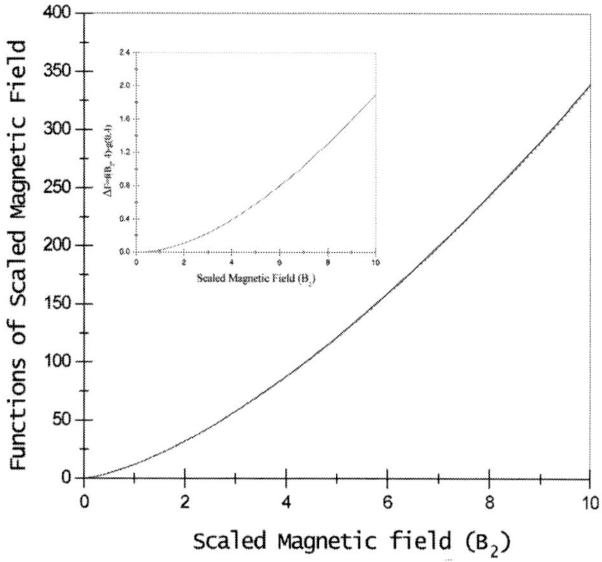

Figure 2. We show the graph of of two energy $f(B,L) = \frac{F(B,L)}{2\pi A}$ and $g(0,\Lambda) = \frac{F(0,\beta=\infty)}{2\pi A}$ as a function of the scaled magnetic field. Note that the two energies are plotted for $m = 1$ and $L = 4$ and they are almost identical to each other. However, there exists a finite positive difference between them as shown in the inset of this graph.

But the positive nature remains valid for any set of parameters. This establishes the diamagnetic nature of the vacuum of charged scalar fields. However, this result is true for a finite cut-off. With the increase of the cut-offs, the energy difference is basically the difference between two infinities. In such a situation, we find the positive finite energy difference. Moreover, we can write [49] the energy difference as

$$\begin{aligned} \Delta f(B,m) &= \sum_{l=0}^{\infty} a_l(B,m) \\ a_l(B,m) &= eB\left[\sqrt{m^2+(2l+1)eB} - \int_0^1 d\alpha\,\sqrt{m^2+2(l+\alpha)eB}\right] \end{aligned} \tag{68}$$

Using the convex nature of $f(z_l) = \sqrt{z_l+\alpha}$, where $z_l = \frac{1+2l\rho}{2\rho}$ with $\rho = \frac{eB}{m^2}$, we can easily rewrite the above expression in terms of $c_l(\rho)$.

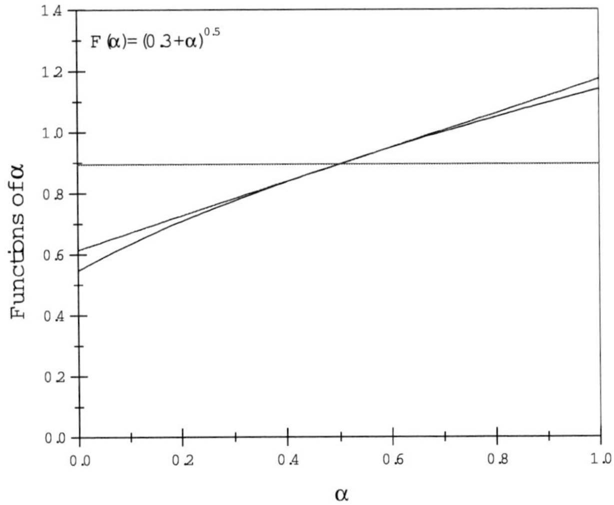

Figure 3. We show here the graphical proof of the positivity of the quantity $a_l(B, m)$. The convexity property of the function $f(z_l)$ plays an important role here. See reference [49] for better clarification of this point.

In figure 3, we show the positivity of $c_l(\rho)$ for each value of l. Moreover, the sum converges [49] to give the renormalized energy difference.

A careful look [49, 88] in this situation however, reveals that

$$\Delta f^R = \Delta f^0 \; + O\left(\frac{1}{g(L)}\right) \tag{69}$$

where $g(L) = \sqrt{L}$. The typical β function in field theory language reads as

$$\Lambda \frac{\partial \Lambda}{\partial L} = eB \tag{70}$$

One might think of another regularization scheme such as

$$E_L = \sqrt{m^2 + (2L+1)eB} = \sqrt{\Lambda^2 + m^2} \tag{71}$$

which connects the two cut-offs as

$$\Lambda^2 = (2L+1)eB \tag{72}$$

Though the above scheme (72) satisfies the β function, however, a detailed analysis [49] shows that this scheme in the asymptotic limit fails to give the convergent energy difference. In particular, we demonstrate clearly in figure 4 that the energy difference following this scheme depends critically on the cut-off.

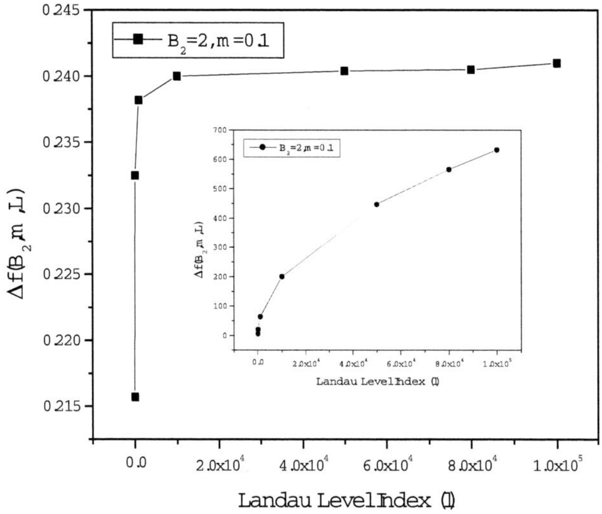

Figure 4. We depict the importance of the mode matching regularization scheme over the another scheme known as energy matching of the highest level. The inset is the difference of the energy Δf^W scheme due to the un-physical energy scheme. It is evident that mode matching scheme gives the asymptotic renormalizable energy difference while other does not. Note in particular the y-axis scale for clarification.

A brief look into the expression of the energy difference[49] in the asymptotic limit gives

$$\Delta f^W = \sqrt{L} + O\left(\frac{1}{\sqrt{L}}\right) \tag{73}$$

Hence, this scheme turns out to be of no use as a proper renormalization scheme. On the other hand, the mode matching scheme has been successfully used for bare QED [85] in $(2+1)$ dimension to renormalize the energy difference and to extract the counterintuitive diamagnetic nature of the vacuum. This formalism can be easily extended to $(3+1)$ dimension without any difficulty. In $(3+1)$ dimension, Nielsen [89] used the Poisson summation technique to regularize the energy difference. However, it turns out that the scheme due to Nielsen fails to provide a cut-off independent result. In this sense, the mode matching regularization scheme illustrates a unique way of renormalizing the energy difference of field theoretical system in a magnetic field.
The QED effective action calculation at finite temperature and density in an external homogenous and static magnetic field demonstrates [90] the expected de Haas-van alphen oscillation and these results have some astrophysical implication [91].
Before we end this section, we would like to point out that the free energy density in two dimension in the *zero mass* limit varies with magnetic field as $B^{3/2}$. This has also been justified from general physical consideration [49]. This variation has been also seen in the calculation of free energy of massless fermions on the walls at zero temperature [92]. In fact the methodology used in that paper to regularize the divergent expression can be directly used in the massless limit. More specifically, the following mathematical formula has been explicitly used to obtain the renormalized expression of the free energy at zero temperature

and finite temperature in the massless limit:

$$\sum_{n=1}^{\infty} f(n) = \int_{\delta}^{\infty} f(x) \sum_{n=-\infty}^{\infty} \delta(x-n)\, dx$$
$$= \sum_{m=-\infty}^{\infty} \int_{\delta}^{\infty} f(x) \exp(2\pi i\, m\, x)\, dx \tag{74}$$

Here, δ has the limits lying between 0 and 1. The above identity is valid for any smooth regular function $f(x)$.

5.1.3. Interacting Theory in an Arbitrary Dimension

Without the interaction, we have evaluated the partition function exactly because the integrand in the action was a gaussian one. However, the presence of polynomial function of $V(\Phi\Phi^*)$ makes the analysis little bit complex. For interacting charged scalar fields defined on a lattice, Brydges et. al. [93] have proved the diamagnetic inequality. In conventional field theoretic approach, this can be solved using the perturbation technique [94]. Below we present a non-perturbative proof of diamagnetic inequality for the interacting charged scalar fields in an arbitrary spatial dimension. The partition function of the interacting charged scalar field theory in presence of an external magnetic field can be written as

$$Z(A) = \int\int \mathcal{D}[\Phi]\mathcal{D}[\Phi^*]\, \exp(-S) \tag{75}$$

$$S = \int\int d^d x\, d\tau \left[(D_\mu \Phi)(D^\mu \Phi)^* + m^2(\Phi\Phi^*) + V(\Phi^*\Phi)\right] \tag{76}$$

In an arbitrary dimension d, for an interacting case, we have followed an elegant approach [49, 95] to write the partition function as

$$\frac{Z(A)}{Z(0)} = \langle\langle exp(-S_{int})\rangle\rangle \leq 1 \tag{77}$$

where the action S has been divided into parts S_0 which does not contain A and S_{int} containing A only. From the above equation (77), it is easy to notice that $F(B) \geq F(0)$. Few remarks are in order to elucidate the above inequality.

- The uniqueness of this proof lies in the redefinition of the measure of the integral through the interaction of self-interaction.
- The above proof is an *exact* and *non-perturbative*. Without solving the modes of the interacting system, we have been able to prove the inequality.
- The proof is valid for an arbitrary finite temperature and finite inhomogeneous magnetic field. In this sense, the proof is quite robust and general.
- The above proof also goes through even if the interaction is mediated by a dynamical electromagnetic field A_μ besides the usual electromagnetic field $\vec{A}_{ext}$.
- The arguments given above are independent of the details of the interaction.

- To calculate the response (susceptibility) as a function of magnetic field and temperature which is most relevant for an experiment, it is suggested to do the perturbative approach. In this formalism, it is even difficult to calculate the corrections due to self-interaction.
- The above proof is strictly valid for *bare* regularized quantities. For example, in three dimension, when the divergent part is subtracted, the response changes to paramagnetic one [96].
- We have not discussed the condensation of this relativistic system in a magnetic field. This has been one of the interesting future direction in this area [97].

5.2. Non-relativistic Bosonic Field Theory

Before we end the discussion of the diamagnetism of scalar field theory, we would like to analyze a typical non-relativistic system. This kind of system appears on the low energy bosonic excitations of strongly interacting electron system in an external magnetic field. A typical example arises in Fractional Quantum Hall Effect [40, 41]. The elementary excitations known as composite bosons [98] arises only in 2d electron system in a strong magnetic field at low temperature. Mathematically speaking, the electrons moving in such a flat land in a strong magnetic field are mathematically equivalent to a collection of composite bosons in a much weaker magnetic field. In two dimensions, a fermion can be thought of as a boson to which an odd number of flux quantum ($\phi_0 = \frac{hc}{e}$) is attached. This magnetic field is introduced via the statistical gauge field. The name statistical is because of the fact that the statistical nature of the particles is changed on introduction of this gauge field. The binding of the statistical flux to the boson is achieved by the Chern-Simons interaction. Therefore, apart from the external magnetic field in this system, we have another magnetic field originated by the statistical vector potential $\vec{a}$. The partition function of this two dimensional system in a strong magnetic field can be written as

$$Z = \int\int\int \mathcal{D}[\Phi]\mathcal{D}[\Phi^*]\mathcal{D}[a_\mu]\, e^{-S} \tag{78}$$

Here, a_μ is the statistical vector potential which is used to generate the statistical gauge field $\vec{b}$. This statistical magnetic field is proportional to the density of electrons and is generated internally. Mathematically, the statistical magnetic field $\vec{b}$ can be written as

$$\vec{b} = \vec{\nabla} \times \vec{a} = k\rho\phi_0 \tag{79}$$

where the density $\rho(\vec{r}) = \sum_i \delta^2(\vec{r}_i - \vec{r})$ and $k = 2n+1, n = 0, 1, 2...$ The effective action S consists of the following parts as given below.

$$\begin{aligned}
S &= i\int_0^\beta d\tau \int d^2r \left[\frac{e}{2k\phi_0}\epsilon^{\mu\nu\rho}a_\mu\partial_\nu a_\rho\right] + S_0(\Phi^*, \Phi, \vec{a}+\vec{A}) + S_{int}, \\
S_0(\Phi^*, \Phi, \vec{a}+\vec{A}) &= \int_0^\beta d\tau \int d^2r\ L_0(\Phi^*, \Phi, \vec{a}+\vec{A}) \\
L_0 &= \Phi^*(\partial_0 - ieA_0)\Phi + \frac{1}{2m^*}\left[|(-i\vec{\nabla} - e(\vec{a}+\vec{A}))\Phi|\right]^2 \\
S_{int} &= \frac{1}{2}\int_0^\beta d^2r \int d^2r'\, |\Phi(r)|^2\, V(|\vec{r}-\vec{r}'|)\, |\Phi(r')|^2
\end{aligned} \tag{80}$$

Note that the statistical gauge field always appears in the combination of $(\vec{a}+\vec{A})$ only. This helps us to write the ratio of the partition function with and without the magnetic field as

$$\frac{Z(B)}{Z(0)} =<< exp(-\tilde{S}_{int} >>\leq 1 \tag{81}$$

The angular bracket signifies the average has been taken on some suitably defined $\exp(-\tilde{S}_0)$ on the specified space of field configurations. Hence, these Chern-Simon bosons also obey the standard diamagnetic inequality [95]. It is to be noted that for $k = 2n$, the chern-Simon boson becomes fermion, hence the above diamagnetic inequality does not hold good. In addition of a Zeeman term like $g\mu_B\ B\Phi^*\sigma_z\Phi$ in the above effective action breaks the inequality.

6. Applications

We here discuss the two systems where elementary excitations are charged and bosonic in nature. One is the Cooper pair in superconductivity and another is the composite bosons in Fractional Quantum Hall effect [40, 41].

6.1. Cooper Pairs in Superconductivity

The Ginzburgh-Landau theory is adopted for conventional superconductors where fluctuations are very small. Since there is no complete theory, dimensional analysis along with physical arguments give us great insight for fluctuation induced diamagnetism above and close to T_c. This calculated susceptibility can be verified with experimental observation. Moreover, since this diamagnetism does not involve the time dependence of the order parameter, it is easier to compute the susceptibility in this formalism. This formalism was originally developed for normal type superconductors [99, 100, 101].

This formalism uses the droplet picture of critical size used in critical phenomena. The typical excitations in superconductors can be thought of as the collection of these *non-interacting* droplets having appropriate unique critical lengths $\xi(T)$. The fluctuations are less significant far away from the critical temperature T_c. The typical energy [49] required to produce the droplet of size $\xi(T)$ above the critical temperature T_c can be estimated as

$$\Delta E \sim \frac{1}{\xi} \times \xi^d \times |\phi|^2 \sim k_B T_c \tag{82}$$

where $|\phi|^2$ is the density of cooper pairs (elementary excitations in superconductors). The above equation implies immediately the relation between the density and the coherence length $\xi(T)$ as

$$|\phi|^2 \sim \frac{k_B T_c}{\xi^{d-1}} \tag{83}$$

Now, using the phenomenological expression of atomic diamagnetism in scaled form, it is easy to see that the fluctuation induced diamagnetism [49] takes the form as

$$\chi_d \sim\ -\frac{|\phi|^2\xi^2}{\xi^{-1}} \sim\ -(T-T_c)^{-(4-d)\nu/2} \tag{84}$$

where we have used the standard relation of variation of the coherent length scale $\xi^2 \sim (T - T_c)^{-\nu}$. For three dimensional system, it is seen that the susceptibility varies with temperature with an exponent $1/2$. This has been verified in an experiment with high temperature superconductor [102]. Actually, it should be remembered here that the susceptibility computed in three dimensional case is $\frac{\chi}{1+4\pi\chi}$. This implies that as $T \to T_c$, $\chi \to -1/4\pi$ pointing out that there is no real divergence at the critical temperature T_c.

In the neighborhood of the critical temperature, however, the fluctuations are quite strong. The above theory can break down within a very narrow region of the critical temperature. The break down occurs because of high values of T_c and λ/ξ ratio (λ and ξ are the corresponding penetration depth and coherent length) of the high temperature superconductors. Actually, Lobb [103] pointed out the fluctuation induced diamagnetism in three dimension should take the form as

$$\chi = -\frac{k_B T}{\phi_0^2}\xi(T) \sim \; -(T - T_c)^{-0.67} \tag{85}$$

Besides, these high T_c superconductors are strongly anisotropic because of the existence of two coherence lengths $\xi_{\parallel}$ and $\xi_{\perp}$. There could be strong fluctuations in the neighborhood of the critical temperature of such high T_c superconductors. In such a situation, using the above droplet picture, it was found [95] that the susceptibility expression in 3d is modified in the form as

$$\chi \sim \; -\frac{(T - T_c)^{-\nu_{\perp}}}{\left(1 + (T - T_c)^{(\nu_{\parallel} - \nu_{\perp})}\right)^{1/2}} \tag{86}$$

We have assumed that $\xi_{\parallel}^2 \sim (T - T_c)^{-\nu_{\parallel}}$ and $\xi_{\perp}^2 \sim (T - T_c)^{-\nu_{\perp}}$. The isotropic result which is valid mostly for normal type of superconductors follows immediately from the above relation in the limit of $\nu_{\parallel} = \nu_{\perp} = \nu$. Finally, we can generalize the above expression for an arbitrary d dimensions in this anisotropic case [95] as

$$\chi_d^{aniso} \sim -\frac{(d-1)^{(4-d)/2}(T - T_c)^{-(\nu_{\parallel}+\nu_{\perp})(4-d)/2}}{[A(d)(T - T_c)^{-\nu_{\parallel}} + B(d)(T - T_c)^{-\nu_{\perp}}]^{(4-d)/2}} \tag{87}$$

here $A(d) = d - 1 - \theta(d-2)$ and $B(d) = \theta(d-2)$. It is assumed that in plane correlation length ($\xi_{\parallel}$) and out of plane correlation length ($\xi_{\perp}$) diverge at the same critical temperature with different exponents $\nu_{\parallel}$ and $\nu_{\perp}$ respectively. The relevant values of $\nu_{\parallel}$ and $\nu_{\perp}$ are to be taken from the appropriate critical systems as indicated in [103] for calculating near the critical region. For finite magnetic field, however, all the relations indicated above need the appropriate corrections. It will be interesting to study the fluctuation conductivity of superconductors in this model.

Below a field of order 1 Oe in the field-cooling mode, a new phenomenon known as paramagnetic Meissner effect [104](PME) is observed in certain Bi HTSC ceramics [105]. PME has also been observed in Nb, and other high temperature superconductors such as YBCO and LCO. Contrary to ordinary Meissner effect, they seem to turn on below T_c without the need of an external magnetic field. This PME is not caused by paramagnetic or ferromagnetic impurities. This is due to the fact that no large PME magnetization is seen to start abruptly at T_c and no Curie-Weiss type susceptibility is seen above T_c. As has been pointed out earlier that the free energy could in principle decreases within a very narrow window of magnetic field retaining the higher value in compared to the value at the origin, one might sort an Φ^4 field theory perturbatively to compute the partition function. This may shed some light on the typical elementary excitations in these materials.

6.2. Composite Bosons in FQHE

An effective 2D degenerate electron system is required to observe the Quantum Hall effect. Basically, in such a system, electrons are trapped in a quantum well formed at the interface of a semiconductor and an insulator or two semiconductors. It is observed that at low temperature and at sufficient high magnetic field, the transverse conductivity $\sigma_{xy} = \nu\frac{e^2}{h}$ (ν being a rational fraction) is quantized along with the longitudinal conductivity $\sigma_{xx} \to 0$, both with extreme high accuracy. The FQHE is observed at strong magnetic field compared to its IQHE counterpart so that the electrons are fully almost polarized. This implies that the spins in this system can be safely ignored. However, there are situations (e.g. in GaAs) where Zeeman energy due to spin cannot be ignored and may be even comparable to the cyclotron energy.

As has been stated clearly in the single particle case, for interacting situation also, the problems related to magnetic field are non-trivial and cannot be understood simply by means of perturbation theory with non-interacting as the starting point. The many body ground state wave function was guessed by Laughlin [39] from the variational approach. This approach can be justified from the generalized Chern-Simon-Landau-Ginzburgh (CSLG) field theoretical approach in which one can systematically calculate the relevant quantities at the mean field level and incorporate the corrections beyond mean field level. In CSLG, the basic link is that in two dimensions, an electron can be considered as a boson to which an odd number of magnetic flux quanta are attached. In this exact mapping, bosons are coupled to a new non-trivial gauge field $\vec{a}$ besides the usual electromagnetic field generated by $\vec{A}$. The bosons dressed with such a non-local gauge field $\vec{a}$ is termed as composite bosons. The aim of this approach was to reduce the strong interacting 2D electron system to a relative weakly interacting composite bosons in a magnetic field where one can do the systematic calculation of the transport quantities. It is interesting to note down the response of such composite bosons in a non-zero external magnetic field. The second quantized version of the Hamiltonian of this composite bosons reads as

$$\begin{aligned} H_{CB} &= \int d^2r\, \Phi^\dagger \left[\frac{1}{2m}\left(-i\hbar\vec{\nabla} - \frac{e}{c}\vec{A} - \frac{e}{c}\vec{a}\right)^2 + eA_0\right]\Phi \\ &+\frac{1}{2}\int\int d^2r\, d^2s\, (\rho(r)-\bar{\rho})V(|r-s|)(\rho(s)-\bar{\rho}) \end{aligned} \tag{88}$$

and the gauge field is given as

$$a^\alpha(r) = -\frac{\phi_o}{2\pi}\frac{\theta}{\pi}\epsilon^{\alpha\beta}\int d^2s \frac{r^\beta - s^\beta}{|r-s|}\rho(s) \tag{89}$$

where $\theta = (2k+1)\pi$. The density $\rho(x) = \Phi^\dagger(x)\Phi(x)$ and the current density $\vec{j}$ given by

$$\vec{j} = \frac{\hbar}{2mi}\left(\Phi^\dagger\vec{\nabla}\Phi - \vec{\nabla}\Phi^\dagger\Phi - \frac{2ie}{\hbar c}(\vec{A}+\vec{a})\rho\right) \tag{90}$$

also satisfies the equation of continuity $\partial_t\rho + \vec{\nabla}\cdot\vec{j} = 0$. At exactly magic filling factors $\nu = \frac{1}{2k+1}$, the ground state is simply a homogeneous bosonic field without any field. In other words $\Phi(r) = \sqrt{\bar{\rho}}, \vec{a} = -\vec{A}, a_0(r) = 0$, Φ is simply a charged bose condensed state not coupled to an external magnetic field. The validity of this mean field theory has been verified [37] which serves as the starting point of the extensive perturbative calculation. Even the qualitative aspects of FQHE can be understood from this CSLG approach. The famous Laughlin wave function can be recovered from this theory by considering the standard Gaussian fluctuations around the mean field theory. But away from the exact magic filling factor, we can think of composite bosons in a residual magnetic field whose response may be interesting.

7. Conclusions and Perspectives

In this chapter, we have demonstrated clearly the response of spinless particles in presence of an external magnetic field starting from the non-relativistic single particle to relativistic scalar fields. In case of field theory, a unique and efficient regularization scheme has been developed to renormalize the energy difference. The non-perturbative proof for an interacting scalar field in an arbitrary magnetic field at any finite temperature has been included to complete the picture. To elucidate the practical importance of this theory, we have described two situations in condensed matter physics - one with Cooper pairs in superconductors and another with composite bosons in fractional quantum Hall effect. In fact, the fluctuation induced diamagnetism of Cooper pairs can be verified in an experiment both above and below the critical temperature.

Acknowledgements

I have been benefitted from critical discussions with various people around the world; Prof. J. Samuel, Prof. C. S. Sukhre and Prof. N. Kumar from Raman Research Institute (Bangalore, India), Prof. Diptiman Sen from Indian Institute of Science (Bangalore, India), Dr. Sudhansu Sekhar Mondal from Indian Association of Cultivation of Science (Kolkata, India), Prof. J. Frölich from ETH (Jurich, Switzerland), Prof. G. Morandi (Bologna, Italy), Prof. P. Elmfors and Prof. D. Persson (Göteberg, Sweden). I would like to acknowledge the facility provided by the Center for Condensed Matter Sciences, National Taiwan University, Taiwan during my stay as a visiting researcher.

References

[1] J. H. Van Vleck, *The Theory of Electric and Magnetic Susceptibilities*, (Oxford University Press, London, 1932).

[2] R. E. Peierls, *Surprises in Theoretical Physics*, (Princeton University Press, Part I, 1992).

[3] N. Kumar, *Curr. Sci.*, **52**, 1108 (1983).

[4] S. L. O'Dell and R. K. P. Zia, *Am. J. Phys.*, **54**, 32 (1986).

[5] L. Landau, *Z. Phys.*, **64**, 629 (1930).

[6] L. D. Landau and E. M. Lifshtitz, *Statistical Physics*, Part II, (Pergamon Press. 1958).

[7] P. Nozieres, *Theory of Interacting Fermi Systems*, (Addison-Wesley Publishing company, 1964).

[8] D. Pines and P. Nozieres, *The Theory of Quantum Liquids*, (Benjamin, 1966).

[9] P. W. Anderson, *Concepts in Solids*, (World Scientific, 1997).

[10] C. V. Raman, *Proc. Phys. Soc.*, **42**, 309 (1930).

[11] M. P. Sharma, L. G. Johnson and J. W. McClure, *Phys. Lett. A*, **44**, 445 (1973).

[12] A. S. Kotosonov, S V. Kuvshinnikov, *Phys. Lett. A*, **229**, 377 (1997).

[13] Alexander A. Ovchinnikov, *Phys. Lett. A*, **195**, 95 (1994).

[14] A. S. Brill, F. G. Fiamingo and B. S. Gerstman, *Am. J. Phys.*, **63**, 1096 (1995).

[15] L. W. Bruch and F. Weinhold, *J. Chem. Phys.*, **113**, 8667 (2000).

[16] R. Szmytkpwski, *Phys. Rev.*, **A65**, 032112 (2002).

[17] J. M. Manoyan, *J. Phys. A: Math. Gen.*, **20**, 3035 (1987).

[18] R. A. Serota and S. Oh, *Phys. Rev.*, **B41**, 10523 (1990).

[19] A. Trellakis, *Phys. Rev. Lett.*, **91**, 056405 (2003).

[20] G. Montambaux, *Phys. Rev. Lett.*, **86**, 4640 (2001).

[21] J. Daicic, N. E. Frankel, R. M. Gailis and V. Kowalenko, *Phys. Rep.*, **237**, 63 (1994).

[22] J. P. Gollub, M. R. Beasley and M. R. Tinkham, *Phys. Rev. Lett.*, **25**, 1646 (1970).

[23] V. Ambegaokar, Bruce R. Patton and J. W. Wilkinson, *Solid State Commun.*, **7**, 1287 (1969).

[24] S. Salem-Sugui, Jr. M. Friesen, A. D. Alvarenga, F. G. Gandra, M. M. Doria and O. F. Schilling, *Phys. Rev.*, **B66**, 134521 (2002).

[25] A. Lascialfari, T. Mishnov, A.Rigamonti, P. Tedesco and A. Varlamov, *Phys. Rev.*, **B65**, 18051 (2002); J. Mosqueira, M. V. Ramallo, S. R. Curras, C. Torron and F. Vidal, *Phys. Rev.*, **B65**, 174522 (2002).

[26] J. L. Gonzalez and E. V. L. de Mello, *Phys. Rev.*, **B69**, 134510 (2004).

[27] A. Sewer and H. Beck, *Phys. Rev.*, **B64**, 014510 (2001); A. K. Ghosh and A. N. Basu, *Phys. Rev.*, **B54**, 1306 (1996).

[28] N. Regnault and Th. Jolicoer, *Phys. Rev.*, **B69**, 235309 (2004).

[29] R. Deblock, R. Bel, B. Reulet, H. Bouchiat and D. Mailly, *Phys. Rev. Lett.*, **89**, 206803 (2002).

[30] D. Herman, H. Mathur and G. Murthy, *Phys. Rev. B*, **69**, 041301 (2004).

[31] B. Shapiro, *Waves in Random Media*, **9**, 271 (1999).

[32] W. Kutzelnigg, *Phys. Rev.*, **A67**, 032109 (2003).

[33] S. Chakrabarty, *Phys. Rev.*, **D54**, 1306 (1996).

[34] K. Nakamura and H. Thomas, *Phys. Rev. Lett.*, **61**, 247 (1988).

[35] J. H. van Leeuwen, *J. Phys. (Paris)*, **2**, 361 (1921).

[36] S. K. Ma, *Statistical Mechanics*, (World Scientific, Singapore 1987).

[37] D. Jana, S. Ghosh and D. Das, *Ind. J. Phys.*, **78(8)**, 713 (2004).

[38] D. Jana, *Phys. Teach.*, **44**, 21 (2003).

[39] R. B. Laughlin, *Phys. Rev. Lett.*, **50**, 1395 (1983).

[40] R. E. Prange and S. M. Girvin, *The Quantum Hall Effect* (Springer, Berlin, 1987).

[41] T. Chakraborty and P. Pietiläinen, *The Fractional Quantum Hall Effect*, Springer-Verlag, Berlin, NewYork, 1988. For a modern view points see S. M. Girvin, in Les Houches Summer School *Topologocal Aspects of Low Dimensional Systems* edited by A. Comtet, T. Jolicoeur, S. Ouvry and F. David, **69**, 53 (1999).

[42] J. L. Friar and S. Fallieros, *Am. J. Phys.*, **49**, 847 (1981).

[43] A. Dupre, *Am. J. Phys.*, **49**, 34 (1981).

[44] R. E. Peierls, *Z. Phys.*, **81**, 186 (1933).

[45] L. D. Landau and E. M. Lifshitz, *Electrodynamics of Continuous Media* (Pergamon, New York, 1960), Sec. 14, p.63.

[46] L. D. Landau and E. M. Lifshitz, *Statistical Physics*, (Pergamon, New York, 1980), 3rd Edition. Sec. 32, p.95.

[47] T. M. Sanders, Jr., *Am. J. Phys.*, **56**, 448 (1988).

[48] D. Jana, *Phys. Edu.*, **19**, 247 (2002).

[49] D. Jana, *Nucl. Phys.*, **B473**[FS], 659 (1996).

[50] B. Simon, *Phys. Rev. Lett.*, **36**, 1083 (1976); B. Simon, *Indiana Univ J. Math. Phys.* **26**, 1067 (1977).

[51] B. S. Shastry, *Mod. Phys. Lett.*, **B6**, 1427 (1992).

[52] Y. Hasegawa, P. Lederer, T. M. Rice and P. B. Wiegmann, *Phys. Rev. Lett.*, **63**, 907 (1989).

[53] G. Montambaux, *Phys. Rev. Lett.*, **63**, 1657 (1989).

[54] D. Peter, D. Mayon and M. Cyrot, *Phys. Rev. Lett.*, **65**, 386 (1990).

[55] W. Kohn, *Phys. Rev.*, **115**, 1460 (1959).

[56] A. S. Alexandrov and H. Capellman, *Phys. Rev. Lett.*, **66**, 365 (1991).

[57] G. Vignale, *Phys. Rev. Lett.*, **67**, 358 (1991).

[58] E. H. Lieb, *Phys. Rev. Lett.*, **73**, 2158 (1994). See also for an interesting conjecture in his homepage http://www.math.princeton.edu.

[59] T. Kato, *Israel J. Math.*, **13**, 135 (1973).

[60] L. Erdös, *J. Math. Phys.*, **38**, 1289 (1997).

[61] J. E. Avron and R. Seiler, *Phys. Rev. Lett.*, **42**, 931 (1979). See also for a counerexample to the paramagnetic conjecture, J. Avron and B. Simon, *Phys. Lett. A*, **75**, 41 (1979).

[62] J. M. Luttinger, *Phys. Rev. Lett.*, **37**, 1221 (1976).

[63] S. Duttagupta and J. Singh, *Phys. Rev. Lett.*, **79**, 961 (1997).

[64] S. Duttagupta, *Pramana*, **59**, 203 (2002).

[65] R. P. Feynman and F. L. Vernon, *Ann. Phys.(NY)*, **24**, 118 (1963).

[66] S. Duttagupta, A. M. Jayanavar and N. Kumar, *Curr. Sci.*, **80**, 861 (2001).

[67] M. Bandopadhyay and S. Duttagupta, *J. Stat. Phys.*, **123**, 1273 (2006); *ibid*, *J. Phys.: Condens. Matter*, **18**, 10029 (2006).

[68] A.M. Jayannavar and N. Kumar, J. Phys. A, **14**, 1399 (1981).

[69] T. M. Hong and J. M. Wheatley, *Phys. Rev. B*, **43**, 5702 (1991).

[70] R. P. Feynman, *Phys. Rev.*, **90**, 1116 (1953).

[71] M. Doi and S. Edwards, *The Theory of Polymer Dynamics*, (Clarendon, Oxford, 1986).

[72] D. Jana, *Phys. Teach.*, **45**, 29 (2003).

[73] H. Kleinert, *Path Integrals in Quantum Mechanics, Statistics and Polymer Physics* , (World Scientific, Singapore, 1990).

[74] A. Kholodenko, *J. Math. Phys.*, **37**, 1287 (1996).

[75] S. Sinha and J. Samuel, *Phys. Rev.* **B50**, 13871 (1994).

[76] M. M. G. Krishna, J. Samuel and S. Sinha, *J. Phys. A: Math. Gen.*, **33**, 5965 (2000).

[77] S. Sinha, *Phys. Rev. E*, **70**, 11801 (2004).

[78] A. L. Kholodenko and T. A. Vilgis, *J. Phys. I France*, **4**, 843 (1994).

[79] A. Kholodenko, *J. Math. Phys.*, **37**, 1314 (1996).

[80] H. S. Carslaw and J. C. Jaeger, *Conduction of Heat in Solids*, (Oxford University Press, 1947), p. 217.

[81] E. H. Sondheimer and A. Wilson, *Proc. Roy. Soc. A*, **210**, 173 (1951).

[82] M. Doi and S. Edwards, *J. Chem. Soc. Faraday Trans II*, **74**, 1789 (1978).

[83] D. Jana, *Phys. Edu.*, **18**, 177 (2001).

[84] H. B. G. Casimir, *Proc. Kon. Ned. Aked. Wet.*, **51**, 793 (1948).

[85] D. Jana, *Fizika (Zagreb)*, **B11**, 63 (2002).

[86] J. I. Kapusta, *Finite Temperature Field Theory*, (Cambridge University Press, Cambridge, 1989).

[87] K. Deitz, *Phys. Scr.*, **T21**, 65 (1988).

[88] D. Li and B. Rosenstein, *Phys. Rev. B*, **60**, 9704 (1999).

[89] N. K. Nielsen, *Am. J. Phys.*, **49**, 1171 (1981).

[90] P. Elmfors, D. Persson and B. Skagerstam, *Phys. Rev. Lett.*, **71**, 480 (1993).

[91] P. Elmfors, D. Persson and B. Skagerstam, *Astropart. Phys.*, **2**, 299 (1994).

[92] M. B. Voloshin, *Phys. Lett. B*, **389**, 475 (1996).

[93] D. Brydges, J. Fröhlich and E. Seiler, *Ann. Phys.(NY)*, **121**, 258 (1979).

[94] U. Kraemmer, A. K. Rebhan and H. Shultz, *Ann. Phys. (NY)*, **238**, 286 (1995).

[95] D. Jana, *Int. Jour. Mod. Phys.*, **B15**, 2811 (2001).

[96] D. Jana, *Nucl. Phys.*, **B485**[FS], 747 (1997).

[97] P. Elmfors, P. Liljenberg, D. Persson and B. Skagerstam, *Phys. Lett. B*, **348**, 462 (1995).

[98] S. C. Zhang, T. H. Hansson and S. A. Kivelson, *Phys. Rev. Lett.*, **62**, 82 (1989). S. C. Zhang, *Int. J. Mod. Phys.*, **B6**, 25 (1992). S. Kivelson , D. -H. Lee and S. C. Zhang, *Scientific American*, **274**, 64 (1996).

[99] A. Schmid, *Phys. Rev. B*, **180**, 527 (1987); H. Schmidt, *Z. Phys.*, **216**, 336 (1968).

[100] L. G. Aslamasov and A. I. Larkin, *Phys. Lett. A*, **26**, 238 (1968).

[101] M. Tinkham, *Introduction to Superconductivity*, (McGraw Hill, New York, 1996).

[102] W. C. Lee, R. A. Klemm and D. C. Johnson, *Phys. Rev. Lett.*, **63**, 1012 (1989).

[103] C. J. Lobb, *Phys. Rev. Lett.*, **36**, 3930 (1987).

[104] M. Sigrist and T. M. Rice, *Rev. Mod. Phys.*, **67**, 503 (1995).

[105] W. Braunisch, N. Knauf, V. Kataev, S. Neuhausen, A. Grutz, A. Kock, B. Roden, D. Khomskii and D. Wohlleben, *Phys. Rev. Lett.*, **68**, 1908 (1992).

In: Astrophysics and Condensed Matter
Editor: Thomas G. Hardwell, pp. 105-135
ISBN: 978-1-60456-332-0

Chapter 3

Magnetism in Pure and Doped Manganese Clusters

***Mukul Kabir*[1],* *Abhijit Mookerjee*[1] *and D.G. Kanhere*[2]**
[1]S. N. Bose National Centre for Basic Sciences, JD Block, Sector III,
Salt Lake City, Kolkata 700 098, India
[2]Department of Physics and Centre for Modeling and Simulation,
University of Pune, Pune - 411 007, India

Abstract

In this chapter, we report electronic and magnetic structure of pure and (As-) doped manganese clusters from density functional theory using generalized gradient approximation for the exchange-correlation energy. Ferromagnetic to ferrimagnetic transition takes place at $n = 5$ for pure manganese clusters, Mn_n, and remarkable lowering of magnetic moment is found for Mn_{13} and Mn_{19} due to their closed icosahedral growth pattern and results show excellent agreement with experiment. On the other hand, for As-doped manganese clusters, Mn_nAs, ferromagnetic coupling is found only in Mn_2As and Mn_4As and inclusion of a single As stabilizes manganese clusters. Exchange coupling in the Mn_nAs clusters are anomalous and behave quite differently from the Ruderman-Kittel-Kasuya-Yosida like predictions. Finally, possible relevance of the observed magnetic behaviour is discussed in the context of Mn-doped GaAs semiconductor ferromagnetism.

Keywords: Cluster, magnetism, semiconductor ferromagnetism

1. Introduction

The terms "clusters" and "microclusters" are usually used to describe aggregates of number of atoms, starting with the diatomic molecule and reaching an upper bound of several hundred thousands of atoms, which are too small to resemble small pieces of crystals. They bridge the gap between the atom and crystals and serve as the dome of basic physics at reduced dimension. The structural, electronic and magnetic properties of these aggregates

*E-mail address: mukul@bose.res.in. (Corresponding author)

differ widely from their corresponding bulk material and these properties can change dramatically with the addition of just one or few atoms to it. Surface rearrangements can take place on crystals with adatoms, but these changes are less drastic than the changes occur when one or few atoms are added to smaller clusters.

Particularly, transition metal (TM) clusters and their magnetic properties are of more interest and are unique. How the magnetic properties behave in the reduced dimension and how it evolve with cluster size to the bulk? — are the great fundamental questions with a potential technological importance. Several unexpected magnetic ordering have already been reported in clusters: (1) *Non-zero magnetic moment in the clusters of nonmagnetic bulk material* — Cox and co-workers found that the bare rhodium clusters display non-zero magnetic moment for less than 60 atoms in it [1], which is an indication of either ferromagnetic or ferrimagnetic ordering even though the bulk rhodium is Pauli paramagnet at all temperature. In Rh_n clusters magnetism show strong size dependence: Magnetic moment per atom decreases as the cluster size increases and become non-magnetic above 60 atoms[1]. (2) *Enhancement of magnetic moment in the clusters which is already ferromagnetic in bulk*. For Fe_n clusters, it was found that the magnetic moment per atom oscillates with the size of the cluster, slowly approaching its bulk value[2], and finally (3) *Finite magnetic moment observed in the clusters which is antiferromagnetic as bulk* – The Cr_n [3]and Mn_n [4, 5] have shown such property. It has been found that Cr_n clusters have magnetic moment 0.5-1 μ_B per atom. Knickelbein observed, Mn_5 - Mn_{99} clusters posses finite magnetic moment, which is otherwise antiferromagnetic as bulk. All these magnetic measurements on free clusters have been done by using Stern-Gerlach molecular deflection experiments at "low" temperatures.

As we will be discussing the pure as well as doped manganese clusters in this chapter, we discuss some few aspects of manganese. Among all the $3d$ transition metal elements, manganese is unique as an atom, clusters or crystals. It has $3d^5\ 4s^2$ electronic configuration and has high (2.14 eV) $3d^5\ 4s^2 \rightarrow 3d^6\ 4s^1$ promotion energy. This large promotion energy reduces the degree of $s-d$ hybridization as atoms are brought together and leads to weaker bonding. For example, Mn_2 dimer is a weakly bound van der Walls dimer with very low binding energy, ranging $0.1 \pm 0.1 - 0.56 \pm 0.26$ eV per atom[6]. If an additional atom is added and the process keeps on going to form different sized clusters, the binding energy increases monotonically but the improvement is not much and remains the lowest among all the $3d$-transition metal clusters. In the solid phase, manganese exists in four allotropic forms exhibiting a complex phase diagram [7]. The stable form is known as $\alpha-$Mn, which has a very complex lattice structure with as much as 58 atoms in the unit cell and has lowest cohesive energy (2.92 eV). For an isolated manganese atom, according to Hund's rule, the half-filled localized $3d$ electrons give rise to a magnetic moment of 5 μ_B.

We discuss electronic and magnetic properties of Mn_n ($n \leq 20$) and Mn_nAs ($n \leq 10$) clusters from the density functional theory. The pure manganese clusters undergo a ferromagnetic $\rightarrow$ ferrimagnetic transition at $n = 5$. In the recent Stern-Gerlach experiments a few interesting observations were made [4, 5]. (1) Although the experimental uncertainty in the measurement decreases with the increase in cluster size, Knickelbein found an extraordinarily large uncertainty in the measured magnetic moment of Mn_7, 0.72 ± 0.42 μ_B/atom[5]. (2) A relative decrease in the magnetic moment was observed at $n = 13$ and 19. We have discussed these issues in a recent paper[8]. Parks *et al.* found that free Mn_n

clusters are nonreactive towards H_2 for small sizes $n \leq 15$. However, this reaction rate show an abrupt increase at $n = 16$ [9] and it was argued to be attributed from a nonmetal-metal transition at $n = 16$. All these issues are discussed in a great detail here.

Next we will move on to see how a single dopant affect the electronic and magnetic properties. Recently we observed [10] that if we dope a single As-atom into a Mn_n cluster that binding energy of the resultant Mn_nAs cluster improves substantially due to their strong $p - d$ hybridization, i.e. As-atom stabilizes the Mn_n clusters. Mn_2As and Mn_4As are the only clusters that we found to have ferromagnetic Mn-Mn coupling and for all other sizes this coupling is turned out to be ferrimagnetic. It was also found that the exchange coupling in these doped-clusters are anomalous and behave differently from the Ruderman-Kittel-Kasuya-Yosida like predictions. All these issues will be discussed in the context of semiconductor ferromagnetism. Before going to the results, we briefly discuss the projector augmented wave formalism of pseudopotential what we have been used throughout.

2. The Projector-Augmented-Wave Formalism

The most widely used electronic structure methods can be divided into two classes. First one is the linear method [11] developed from the augmented-plane-wave method [12, 13] and Koringa-Kohn-Rostocar method [14, 15] and the second one is the norm-conserving pseudopotentials developed by Hamann, Schlüter and Chiang [16]. In that scheme, inside some core radius, the all electron (AE) wave function is replaced by a soft nodeless pseudo (PS) wave function, with the restriction to the PS wave function that within the chosen core radius the norm of the PS wave function have to be the same with the AE wave function and outside the core radius both the wave functions are just identical. However, the charge distribution and moments of AE wave function are well reproduced by the PS wave function only when the core radius is taken around the outer most maxima of AE wave function. Therefore, elements with strongly localized orbitals pseudopotentials require a large plane wave basis set. This was improved by Vanderbilt [17], where the norm-conservation constraint was relaxed and a localized atom centered augmentation charges were introduced to make up the charge deficit. But the success is partly hampered by rather difficult construction of the pseudopotential. Later Blöchl [18] has developed the projector-augmented- wave (PAW) method, which combines the linear augmented plane wave method with the plane wave pseudopotential approach, which finally turns computationally elegant, transferable and accurate method for electronic structure calculation. Further Kresse and Joubert [19] modified this PAW method and implemented in their existing Veina ab-initio pseudopotential package (VASP). Here in this section we will discuss briefly the idea of the method.

2.1. Wave Functions

The exact Kohn-Sham density functional is given by,

$$E = \sum_n f_n \langle \Psi_n | -\frac{1}{2}\nabla^2 | \Psi_n \rangle + E_H[n + n_Z] + E_{xc}[n], \tag{1}$$

where $E_H[n + n_Z]$ is the hartree energy of the electronic charge density n and the point charge densities of the nuclei n_Z, $E_{xc}[n]$ is the electronic exchange-correlation energy and

f_n are the orbital occupation number. $|\Psi_n\rangle$ is the all-electron wave function. This physically relevant wave functions $|\Psi_n\rangle$ in the Hilbert space exhibit strong oscillations and make numerical treatment difficult. Transformation of this wave functions $|\Psi_n\rangle$ into a new pseudo wave functions $|\tilde{\Psi}_n\rangle$ in the PS Hilbert space,

$$|\Psi_n\rangle = \tau|\tilde{\Psi}_n\rangle, \tag{2}$$

within the augmentation region Ω_R, then makes PS wave function $|\tilde{\Psi}_n\rangle$ computationally convenient.

Let us now choose a PS partial wave function $|\tilde{\phi}_i\rangle$, which is identical to the corresponding AE partial waves $|\psi_i\rangle$ outside the augmentation region and form a complete set within the augmentation region Ω_R. Within the augmentation region every PS wave function can be expanded into PS partial waves,

$$|\tilde{\Psi}\rangle = \sum_i c_i|\tilde{\phi}_i\rangle, \tag{3}$$

where the coefficients c_i are scalar products,

$$c_i = \langle\tilde{p}_i|\tilde{\Psi}_n\rangle, \tag{4}$$

with some fixed function $\langle\tilde{p}_i|$ of the PS wave function, which is called the projector function. By using the linear transformation[18],

$$\tau = 1 + \sum_i (|\phi_i\rangle - |\tilde{\phi}_i\rangle)\langle\tilde{p}_i|, \tag{5}$$

the corresponding AE wave function $|\Psi_n\rangle$ in the Eq. 1.2 is then,

$$|\Psi_n\rangle = |\tilde{\Psi}_n\rangle + \sum_i (|\psi_i\rangle - |\tilde{\psi}_i\rangle)\langle\tilde{p}_i|\tilde{\Psi}_n\rangle, \tag{6}$$

where i refers to the atomic site $\mathbf{R}$, the angular momentum quantum numbers $L = l, m$ and an additional index k for the reference energy ϵ_{kl} and, with $\tilde{p}_i$ being the projector functions, which within the augmentation region Ω_R satisfy the condition,

$$\langle\tilde{p}_i|\tilde{\phi}_j\rangle = \delta_{ij}. \tag{7}$$

The AE charge density at a given point $\mathbf{r}$ is the expectation value of the real-space projector operator $|r\rangle\langle r|$ and hence given by,

$$n(\mathbf{r}) = \tilde{n}(\mathbf{r}) + n^1(\mathbf{r}) - \tilde{n}^1(\mathbf{r}), \tag{8}$$

where $\tilde{n}(\mathbf{r})$ is the soft PS charge density calculated from the PS wave function

$$\tilde{n}(\mathbf{r}) = \sum_n f_n\langle\tilde{\Psi}_n|\mathbf{r}\rangle\langle\mathbf{r}|\tilde{\Psi}_n|\mathbf{r}\rangle, \tag{9}$$

and onsite charge densities are defined as,

$$n^1(\mathbf{r}) = \sum_{ij} \rho_{ij}\langle\phi_i|\mathbf{r}\rangle\langle\mathbf{r}|\phi_j\rangle, \tag{10}$$

and,

$$\tilde{n}^1(\mathbf{r}) = \sum_{ij} \rho_{ij} \langle \tilde{\phi}_i | \mathbf{r} \rangle \langle \mathbf{r} | \tilde{\phi}_j \rangle. \tag{11}$$

ρ_{ij} are the occupation of each augmentation channel (i, j) and are calculated from the PS wave functions applying the projector function,

$$\rho_{ij} = \sum_n f_n \langle \tilde{\Psi}_n | \tilde{p}_i \rangle \langle \tilde{p}_j | \tilde{\Psi}_n \rangle. \tag{12}$$

It should be pointed out here that for a complete set of projector functions the charge density $\tilde{n}^1$ is exactly same as $\tilde{n}$ within Ω_R.

2.2. The Total Energy Functional

The total energy can be written as a sum of three terms,

$$E = \tilde{E} + E^1 - \tilde{E}^1, \tag{13}$$

where the first term,

$$\begin{aligned} \tilde{E} &= \sum_n f_n \langle \tilde{\Psi}_n | -\frac{1}{2}\nabla^2 | \tilde{\Psi}_n \rangle + E_{xc}[\tilde{n} + \hat{n} + \tilde{n}_c] + E_H[\tilde{n} + \hat{n}] \\ &+ \int v_H[\tilde{n}_{Zc}][\tilde{n}(\mathbf{r}) + \hat{n}(\mathbf{r})] d\mathbf{r} + U(\mathbf{R}, Z_{ion}). \end{aligned} \tag{14}$$

$n_{Zc}(\tilde{n}_{Zc})$ is the sum of the point charge density of the nuclei $n_Z(\tilde{n}_z)$ and the frozen core charge density $n_c(\tilde{n}_c)$, i.e. $n_{Zc} = n_z + n_c$ and $\tilde{n}_{Zc} = \tilde{n}_z + \tilde{n}_c$ and $\hat{n}$ is the compensation charge, which is added to the soft charge densities $\tilde{n} + \tilde{n}_{Zc}$ and $\tilde{n}^1 + \tilde{n}_{Zc}$ to reproduce the correct multipole moments of the AE charge density $n^1 + n_{Zc}$ located in each augmentation region. As n_{Zc} and $\tilde{n}_{Zc}$ have same monopole $-Z_{ion}$ and vanishing multipoles, the compensation charge $\hat{n}$ is chosen so that $\tilde{n}^1 + \hat{n}$ has the same moments as AE valence charge density n^1 has, within each augmentation region. $U(\mathbf{R}, Z_{ion})$ is the electrostatic interaction potential between the cores.

The second term in the total energy is,

$$\begin{aligned} E^1 &= \sum_{ij} \rho_{ij} \langle \phi_i | -\frac{1}{2}\nabla^2 | \phi_j \rangle + \overline{E_{xc}[n^1 + n_c]} + \overline{E_H[n^1]} \\ &+ \int_{\Omega_r} v_H[n_{n_{Zc}}] n^1(\mathbf{r}) d\mathbf{r}. \end{aligned} \tag{15}$$

and the final term is,

$$\begin{aligned} \tilde{E}^1 &= \sum_{ij} \rho_{ij} \langle \tilde{\phi}_i | -\frac{1}{2}\nabla^2 | \tilde{\phi}_j \rangle + \overline{E_{xc}[\tilde{n}^1 + \hat{n} + \tilde{n}_c]} + \overline{E_H[\tilde{n}^1 + \hat{n}]} \\ &+ \int_{\Omega_r} v_H[\tilde{n}_{Zc}][\tilde{n}^1(\mathbf{r}) + \hat{n}(\mathbf{r})] d\mathbf{r}, \end{aligned} \tag{16}$$

In all these three energy terms, the electrostatic potential $v_H[n]$ and electrostatic energy $E_H[n]$ of charge density n is given by:

$$v_H[n](\mathbf{r}) = \int \frac{n(\mathbf{r}')}{|\mathbf{r}-\mathbf{r}'|} d\mathbf{r}', \tag{17}$$

$$E_H[n] = \frac{1}{2}\int d\mathbf{r} \int d\mathbf{r}' \frac{n(\mathbf{r})n(\mathbf{r}')}{|\mathbf{r}-\mathbf{r}'|}. \tag{18}$$

In the total energy functional the smooth part $\tilde{E}$ is evaluated on a regular grids in Fourier or real space, and the two one-center contributions $E^1$and $\tilde{E}^1$ are evaluated on radial grids for each sphere individually.

2.3. Compensation Charge Density

The compensation charge density $\hat{n}$ is defined such that $\tilde{n}^1 + \hat{n}$ has exactly the same moments as the AE charge density n^1 has, within the augmentation region, which then requires,

$$\int_{\Omega_r} (n^1 - \tilde{n}^1 - \hat{n})|\mathbf{r}-\mathbf{R}|^l Y_L^*(\widehat{\mathbf{r}-\mathbf{R}}) d\mathbf{r} = 0. \tag{19}$$

The charge difference Q_{ij} between the AE and PS partial wave for each channel (i,j) within the augmentation region is defined by,

$$Q_{ij}(\mathbf{r}) = \phi_i^*(\mathbf{r})\phi_j(\mathbf{r}) - \tilde{\phi}_i^*(\mathbf{r})\tilde{\phi}_j(\mathbf{r}), \tag{20}$$

and their moments q_{ij}^L are,

$$q_{ij}^L = \int_{\Omega_r} Q_{ij}(\mathbf{r})|\mathbf{r}-\mathbf{R}|^l Y_L^*(\widehat{\mathbf{r}-\mathbf{R}}) d\mathbf{r}, \tag{21}$$

which has non zero values only for certain combinations of L, i and j. Then the compensation charge density can be rewritten as,

$$\hat{n} = \sum_{i,j,L} \rho_{ij} \hat{Q}_{ij}^L(\mathbf{r}), \tag{22}$$

where,

$$\hat{Q}_{ij}^L = q_{ij}^L g_l(|\mathbf{r}-\mathbf{R}|) Y_L(\widehat{\mathbf{r}-\mathbf{R}}), \tag{23}$$

where $g_l(r)$ are the functions for which the moment is equal to 1.

2.4. Operators

Let us consider some operator $\mathcal{O}$, so its expectation value $\langle\mathcal{O}\rangle = \sum_n f_n \langle\Psi_n|\mathcal{O}|\Psi_n\rangle$, where n is the band index and f_n is the occupation of the state. As in the PAW method we work with the PS wave function, we need to obtain observables as the expectation values of PS wave function. Applying the form of the transformation τ and using $\sum_i |\tilde{\phi}_i\rangle\langle\tilde{p}_i| = 1$ within the augmentation region Ω_R and $|\tilde{\phi}_i\rangle = |\phi_i\rangle$ outside the augmentation region, for quasilocal operators, the PS operator $\tilde{\mathcal{O}}$ has the following form[1]:

$$\begin{aligned} \tilde{\mathcal{O}} &= \tau^\dagger \mathcal{O} \tau \\ &= \mathcal{O} + \sum_{ij} |\tilde{p}_i\rangle(\langle\phi_i|\mathcal{O}|\phi_j\rangle - \langle\tilde{\phi}_i|\mathcal{O}|\tilde{\phi}_j\rangle)\langle\tilde{p}_j|. \end{aligned} \tag{24}$$

Overlap operator: The PS wave function obey orthogonality condition,

$$\langle\tilde{\Psi}_n|\mathcal{S}|\tilde{\Psi}_m\rangle = \delta_{nm}, \tag{25}$$

where $\mathcal{S}$ is the overlap operator in the PAW approach. The overlap matrix in the AE representation is given by the matrix elements of unitary operator. Therefore, $\mathcal{S}$ has the form given by,

$$\mathcal{S}[\{\mathbf{R}\}] = 1 + \sum_i |\tilde{p}_i\rangle[\langle\phi_i|\phi_j\rangle - \langle\tilde{\phi}_i|\tilde{\phi}_j\rangle]\langle\tilde{p}_j|. \tag{26}$$

Hamiltonian operator: The Hamiltonian operator is defined as the first derivative of the total energy functional with respect to the density operator, $\tilde{\rho} = \sum_n f_n|\tilde{\Psi}_n\rangle\langle\tilde{\Psi}_n|$,

$$H = \frac{dE}{d\tilde{\rho}}, \tag{27}$$

where the PS density operator $\tilde{\rho}$ depends on the PS charge density $\tilde{n}$ and on the occupancies of each augmentation channel ρ_{ij}. So the variation of the total energy functional can be rewritten as,

$$\frac{dE}{d\tilde{\rho}} = \frac{\partial E}{\partial\tilde{\rho}} + \int \frac{\delta E}{\delta\tilde{n}(\mathbf{r})}\frac{\partial\tilde{n}(\mathbf{r})}{\partial\tilde{\rho}}d\mathbf{r} + \sum_{i,j}\frac{\partial E}{\partial\rho_{ij}}\frac{\partial\rho_{ij}}{\partial\tilde{\rho}}. \tag{28}$$

The partial derivative of $\tilde{E}$ with respect to the PS density operator is the kinetic energy operator and the variation with respect to $\tilde{n}(\mathbf{r})$ leads to the usual one-electron potential $\tilde{v}_{eff}$,

$$\frac{\partial\tilde{E}}{\partial\tilde{\rho}} = -\frac{1}{2}\nabla^2, \tag{29}$$

and,

$$\frac{\delta\tilde{E}}{\delta\tilde{n}(\mathbf{r})} = \tilde{v}_{eff} = v_H[\tilde{n} + \hat{n} + \tilde{n}_{Zc}] + v_{xc}[\tilde{n} + \hat{n} + \tilde{n}_c]. \tag{30}$$

As in the smooth part of the total energy functional, $\tilde{E}$, the occupancies ρ_{ij} enter only through the compensation charge density $\hat{n}$, the variation of $\tilde{E}$ with respect to ρ_{ij} is given by,

$$\begin{aligned} \hat{D}_{ij} = \frac{\partial\tilde{E}}{\partial\rho_{ij}} &= \int \frac{\delta\tilde{E}}{\delta\hat{n}(\mathbf{r})}\frac{\partial\hat{n}(\mathbf{r})}{\partial\rho_{ij}}d\mathbf{r} \\ &= \sum_L \int \tilde{v}_{eff}(\mathbf{r})\hat{Q}_{ij}^L(\mathbf{r})d\mathbf{r}. \end{aligned} \tag{31}$$

[1]The kinetic energy operator $-1/2\nabla^2$ and the real space projector operators $|r\rangle\langle r|$ are quasi local operators. For truely nonlocal operators an extra term must be added to the $\tilde{\mathcal{O}}$ expression.

In the remaining two terms E^1 and $\tilde{E}^1$ in the energy functional, ρ_{ij} enters directly via kinetic energy or through n^1, $\tilde{n}^1$ and $\hat{n}$. Now the variation of E^1 with the occupancies ρ_{ij} is given by,

$$D^1_{ij} = \frac{\partial E^1}{\partial \rho_{ij}} = \langle \phi_i| - \frac{1}{2}\nabla^2 + v^1_{eff}|\phi_j\rangle, \tag{32}$$

where,

$$v^1_{eff}[n^1] = v_H[n^1 + n_{Zc}] + v_{xc}[n^1 + n_c],$$

and the variation of $\tilde{E}^1$ is given by,

$$\tilde{D}^1_{ij} = \frac{\partial \tilde{E}^1}{\partial \rho_{ij}} = \langle \tilde{\phi}_i| - \frac{1}{2}\nabla^2 + \tilde{v}^1_{eff}|\tilde{\phi}_j\rangle + \sum_L \int_{\Omega_r} d\mathbf{r}\tilde{v}^1_{eff}(\mathbf{r})\hat{Q}^L_{ij}(\mathbf{r}), \tag{33}$$

where,

$$\tilde{v}^1_{eff}[\tilde{n}^1] = v_H[\tilde{n}^1 + \hat{n} + \tilde{n}_{Zc}] + v_{xc}[\tilde{n}^1 + \hat{n} + \tilde{n}_c].$$

The onsite terms D^1_{ij} and $\tilde{D}^1_{ij}$ are evaluated on radial grid within each augmentation region and they restore the correct shape of the AE wave function within the sphere. The final form of the Hamiltonian operator:

$$H[\rho, \{\mathbf{R}\}] = -\frac{1}{2}\nabla^2 + \tilde{v}_{eff} + \sum_{i,j} |\tilde{p}_i\rangle(\hat{D}_{ij} + D^1_{ij} - \tilde{D}^1_{ij})\langle\tilde{p}_j|. \tag{34}$$

2.5. Forces in the PAW Method

Forces are usually defined as the total derivative of the energy with respect to the ionic position $\mathbf{R}$,

$$\mathbf{F} = -\frac{dE}{d\mathbf{R}}. \tag{35}$$

The total derivative consists of two terms, first one is the forces on the ionic core and the second term is the correction due to the change of AE wave functions for fixed PS wave functions when ions are moved. This correction term comes because augmentation depends on the ionic positions and are known as Pulay force [20]. When calculating this Pulay correction one must consider the overlap between the wave functions due to the change in ionic position. According to Goedecker and Maschke[21], the total derivative can be written as,

$$\frac{dE}{d\mathbf{R}} = \frac{\partial U(\mathbf{R}, Z_{ion})}{\partial \mathbf{R}} + \sum_n f_n \left\langle \tilde{\Psi}_n \left| \frac{\partial (H[\rho, \{\mathbf{r}\}] - \epsilon_n S[\{\mathbf{R}\}])}{\partial \mathbf{R}} \right| \tilde{\Psi}_n \right\rangle, \tag{36}$$

where the first term is the forces between the ionic cores and ϵ_n are the Kohn-Sham eigenvalues, and the PS wave functions assumed to satisfy the orthogonality condition $\langle\tilde{\Psi}_n|S|\tilde{\Psi}_m\rangle = \delta_{nm}$ and the corresponding Kohn-Sham equation, reads $H|\tilde{\Psi}_n\rangle = \epsilon_n S|\tilde{\Psi}_n\rangle$.

The first contribution, $\mathbf{F}^1$ to the second term of the total derivative comes from the change in the effective local potential $\tilde{v}_{eff}$ when the ions are moved and $\tilde{v}_{eff}$ depends explicitly on the ionic positions via $\hat{n}_{Zc}$,

$$\mathbf{F}^1 = -\int \left[\frac{\delta Tr[H\tilde{\rho}]}{\delta v_H[\tilde{n}_{Zc}](\mathbf{R})} \frac{\partial v_H[\tilde{n}_{Zc}](\mathbf{R})}{\partial \mathbf{R}} \right] d\mathbf{R}. \tag{37}$$

This equation can be further simplified to,

$$\begin{aligned}\mathbf{F}^1 &= -\int \left(\tilde{n}(\mathbf{r}) + \sum_{i,j,L} \hat{Q}_{ij}^L(\mathbf{r})\rho_{ij}\right) \frac{\partial v_H[\tilde{n}_{Zc}](\mathbf{r})}{\partial \mathbf{R}} d\mathbf{r} \\ &= -\int [\tilde{n}(\mathbf{r}) + \hat{n}(\mathbf{r})] \frac{\partial v_H[\tilde{n}_{Zc}](\mathbf{r})}{\partial \mathbf{R}} d\mathbf{r}. \end{aligned} \tag{38}$$

The second contribution arise from $\hat{D}_{ij}$ due to the changes in the compensation charge density $\hat{n}$, when ions are moved,

$$\mathbf{F}^2 = -\sum_{i,j,L} \int \tilde{v}_{eff}(\mathbf{r})\rho_{ij} q_{ij}^L \frac{\partial g_l(|\mathbf{r}-\mathbf{R}|)Y_L(\widehat{\mathbf{r}-\mathbf{R}})}{\partial \mathbf{R}} d\mathbf{r}. \tag{39}$$

These two terms, $\mathbf{F}^1$ and $\mathbf{F}^2$, together describe the electrostatic contributions to the force. The third term comes due to the change in the projector function $\tilde{p}_i$ as ions are moved,

$$\mathbf{F}^3 = -\sum_{n,i,j} (\hat{D}_{ij} + D_{ij}^1 - \tilde{D}_{ij}^1 - \epsilon_n q_{ij}) f_n \left\langle \tilde{\Psi}_n \left| \frac{\partial |\tilde{p}_i\rangle\langle\tilde{p}_j|}{\partial \mathbf{R}} \right| \tilde{\Psi}_n \right\rangle, \tag{40}$$

where $q_{ij} = \langle\phi_i|\phi_j\rangle - \langle\tilde{\phi}_i|\tilde{\phi}_j\rangle$.

As the exchange-correlation potential depends on the nonlinear core corrections $\tilde{n}_c$ gives an additional contribution, which can be obtained from the total energy functional,

$$\mathbf{F}^{\text{nlcc}} = -\int v_{xc}[\tilde{n} + \hat{n} + \tilde{n}_c] \frac{\partial \tilde{n}_c(\mathbf{r})}{\partial \mathbf{R}} d\mathbf{r}. \tag{41}$$

All the differences between the PAW method and ultra-soft pseudopotential are automatically absorbed in the in the definition of $(\hat{D}_{ij} + D_{ij}^1 - \tilde{D}_{ij}^1)$. $D_{ij}^1 - \tilde{D}_{ij}^1$ are constant for US-PP, where they are calculated once and forever whereas in PAW method they vary during the calculation of the electronic ground state.

3. Computational Method

The electronic and magnetic calculations of the pure and doped- manganese clusters were carried out using density functional theory (DFT), within the pseudopotential plane wave method [16, 17], using projected augmented wave (PAW) formalism [18, 19] and Perdew-Burke-Ernzerhof (PBE) exchange-correlation functional [22] for the spin-polarized generalized gradient approximation (GGA), as it is implanted in VASP code [23]. The wave functions are expanded in a plane wave basis set with the kinetic energy cutoff equal to 337.3 eV and reciprocal space integrations were carried out at the Γ point. The $3d$, $4s$ for Mn-atom and $4s$, $4p$ orbitals for As-atom were treated as valence states. Symmetry unrestricted geometry optimizations were performed using quasi Newtonian and conjugate gradient methods until all the force components are less than 0.005 eV/ Å. Simple cubic supercells are used with neighbouring clusters separated by at least 12 Å vacuum regions. Several initial structures were studied to ensure that the globally optimized geometry does not correspond to the local minima, as well as, for all clusters, we have explicitly considered *all possible* spin multiplicities to determine the magnetic ground state.

4. Results

4.1. Magnetic Transition: Pure Mn_n ($n \leq 20$) Clusters

We start with the magnetism in the pure manganese clusters. Theory and experimental reports are in contradiction for the Mn_2 dimer. More than 30 years ago, Nesbet [24] calculated binding energy, bond length and magnetic moment of the dimer at the restricted Hartree-Fock (RHF) level and predicted an antiferromagnetic (AFM) ground state with bond length 2.88 Å. Later on, unrestricted Hartree-Fock calculation was done by Shillady *et al.* [25] and found a ferromagnetic (FM) ground state with total spin 10 μ_B and bond length 3.50 Å. The experiments based on resonant Raman spectroscopy [26] and Electron Spin Resonance (ESR)[27] predicted an AFM ground state with a bond length 3.4 Å. However, all DFT calculations [28, 29, 8, 30] within different levels of approximation and using different exchange-correlation functionals, predict a FM ground state with much smaller bond length, $\sim$ 2.60 Å than the experimental bond length. A comparison among different levels of theory is given in the Table1.

Table 1. Summary of binding energy (E_b), bond length (R_e) and magnetic moment (μ) of Mn_2 reported by various authors.

Authors	Method	E_b (eV)	R_e (Å)	μ (μ_B)
Nesbet [24]	RHF + Heisenberg exchange	0.79	2.88	0
Wolf and Schmidtke [31]	RHF		1.52	0
Shillady *et al.* [25]	UHF	0.08	3.50	10
Fujima and Yamaguchi [32]	LSDA	0.70	3.40	0
Harris and Jones [33]	LSDA	1.25	2.70	10
Salahub and Baykara [34]	LSDA	0.86	2.52	0
Nayak *et al.* [28]	All-electron + GGA (BPW91)	0.91	2.60	10
Parvanova *et al.*[29]	All-electron + GGA (PBE)	0.98	2.60	10
Kabir *et al.* [8]	Pseudopotential + GGA (PBE)	1.06	2.58	10
Pederson *et al.* [30]	All-electron + GGA	0.99	2.61	10
Experiment	ESR in rare-gas matrix [27]	0.1 ± 0.1	3.4	0

Due to the filled $4s$ and half-filled $3d$ electronic configuration, as well as high (2.14 eV) $4s^2\,3d^5 \rightarrow 4s\,3d^6$ promotion energy Mn atoms do not bind strongly as two Mn atoms come closer to form Mn_2 dimer, and as a result Mn_2 is very weakly bond van der Walls dimer, which is also evident from the low experimental value, 0.01 $\pm$ 0.1 — 0.56 $\pm$ 0.26 eV/atom [6]. However, no experimental results available in the gas-phase and all the experiments are done in the gas matrix, and therefore, it is quite possible that the Mn atoms interact with the matrix, which could stretch R_e and could lead to the AFM ground state.

From the Table2, we see that R_e decreases monotonically as the net moment decreases. It is simply because the reduction of the atomic spacing leads to comparatively stronger overlap of the atomic orbitals which reduces the magnetic moment. The manganese trimer and tetramer is found to have FM ground state with 5 μ_B/atom magnetic moment. Resonance Raman spectroscopy study by Bier *et al.*[26] suggested a distorted D_{3h} structure with

Table 2. The binding energy and equilibrium bond length of Mn_2 dimer for all possible spins.

spin	Binding energy (eV)	Bond length
0 (AFM)	0.51	2.57
2	0.44	1.53
4	unbound	1.73
6	unbound	1.94
8	0.47	2.24
10	1.06	2.58

odd-integer magnetic moment for trimer. Manganese tetramer in solid silicon was studied by Ludwig *et al.* [35] and observed a 21-line hyperfine pattern which confirmed that the all four atoms are equivalent with a total magnetic moment 20 μ_B.

Calculated binding energy per atom, relative energies of the isomers, magnetic moments per atom with the experimental comparison are given in the Table3 (n = 2-10) and Table4 (n = 2-20) and the calculated magnetic moment per atom corresponding to the ground state are plotted in the Figure1. We see a FM $\rightarrow$ ferrimagnetic transition takes place at n=5. and remains the same for larger size clusters [8]. Figure1 shows very good agreement with the recent Stern-Gerlach experiments[4, 5]. However, we find, for all sizes there exists several

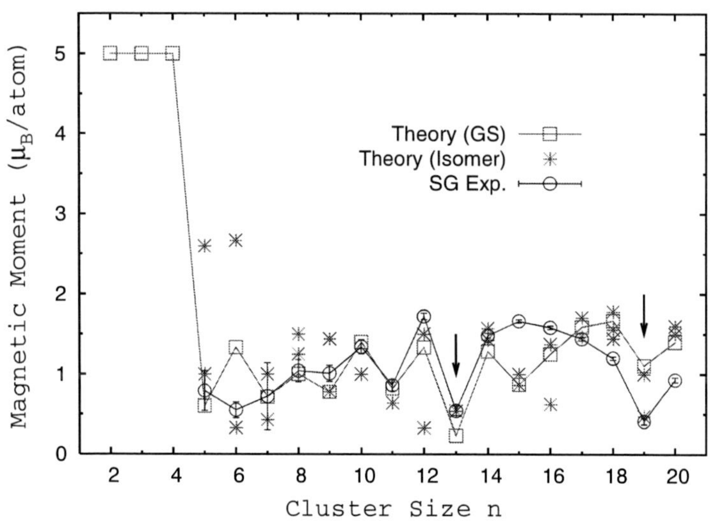

Figure 1. Variation of magnetic moment with cluster size for the size range $n = 2 - 20$. A comparison is made with the experimentally measured value [5]. For all sizes, the magnetic moments of closely lying isomers are also plotted.

Table 3. Binding energy per atom, relative energy to the ground state ($\triangle E = E - E_{GS}$) and per atom magnetic moment for Mn_n ($n \leq 10$) clusters. Predicted magnetic moments are compared with the Stern-Gerlach experiment.[4, 5]

Cluster	BE (eV/atom)	$\triangle E$ (eV)	Magnetic Moment (μ_B/atom) Theory[8]	Experiment[4, 5]
Mn_2	0.530	0.000	5.000	—
	0.255	0.531	0.000	
Mn_3	0.823	0.000	5.000	—
	0.808	0.046	1.667	
Mn_4	1.179	0.000	5.000	—
	1.160	0.078	2.500	
	1.130	0.196	0.000	
	1.130	0.195	2.000	
Mn_5	1.413	0.000	0.600	0.79 ± 0.25
	1.401	0.059	2.600	
	1.399	0.069	1.000	
	1.374	0.193	4.600	
Mn_6	1.567	0.000	1.333	0.55 ± 0.10
	1.564	0.017	0.333	
	1.559	0.045	2.667	
	1.543	0.142	4.333	
Mn_7	1.726	0.000	0.714	0.72 ± 0.42
	1.713	0.091	1.000	
	1.699	0.193	0.429	
Mn_8	1.770	0.000	1.000	1.04 ± 0.14
	1.770	0.000	1.500	
	1.765	0.039	1.250	
Mn_9	1.867	0.000	0.778	1.01 ± 0.10
	1.856	0.099	1.444	
	1.850	0.151	0.778	
	1.844	0.211	1.000	
Mn_{10}	1.936	0.000	1.400	1.34 ± 0.09
	1.934	0.015	1.000	
	1.935	0.007	0.400	
	1.935	0.009	0.400	

isomers with different magnetic structure[8].

In the SG experiment [5] it has been seen that the experimental uncertainty in measurement of magnetic moment generally decreases with the cluster size as the cluster production efficiency increases with the size. However, this uncertainty for Mn_7 is quite high, 0.72 ± 0.42 μ_B/atom[5], which is 58% of the measured value. Now we discuss what might be the possible reason for this. We predict [8] the Mn_7 to be a pentagonal bi-pyramidal (PBP) structure in its ground state with magnetic moment 5 μ_B, which is exactly the experimental value. In addition, we found another two PBP isomers, with magnetic moment 7 μ_B and 3 μ_B, to be close in energy to the ground state, which are shown in the Figure2. They lie 0.09

Table 4. Same as Table3 for Mn_n (n = 11— 20) clusters.

Cluster	BE (eV/atom)	$\triangle E$ (eV)	Magnetic Moment (μ_B/atom) Theory[8]	Magnetic Moment (μ_B/atom) Experiment[4, 5]
Mn_{11}	1.993	0.000	0.818	0.86 ± 0.07
	1.984	0.107	0.455	
	1.980	0.153	0.636	
Mn_{12}	2.081	0.000	1.333	1.72 ± 0.04
	2.077	0.052	0.333	
	2.071	0.114	1.500	
Mn_{13}	2.171	0.000	0.231	0.54 ± 0.06
	2.165	0.076	0.538	
Mn_{14}	2.171	0.000	1.286	1.48 ± 0.03
	2.170	0.021	1.429	
	2.167	0.055	1.571	
Mn_{15}	2.231	0.000	0.867	1.66 ± 0.02
	2.229	0.033	0.333	
	2.228	0.057	0.467	
	2.228	0.049	0.867	
	2.227	0.064	1.000	
	2.213	0.281	0.467	
Mn_{16}	2.274	0.000	1.250	1.58 ± 0.02
	2.273	0.017	1.375	
	2.271	0.055	0.625	
	2.268	0.097	0.500	
Mn_{17}	2.327	0.000	1.588	1.44 ± 0.02
	2.323	0.069	1.471	
	2.322	0.087	1.706	
Mn_{18}	2.351	0.000	1.667	1.20 ± 0.02
	2.350	0.018	1.556	
	2.350	0.020	1.444	
	2.348	0.064	1.778	
Mn_{19}	2.373	0.000	1.105	0.41 ± 0.04
	2.373	0.009	1.000	
	2.369	0.076	0.474	
Mn_{20}	2.370	0.000	1.400	0.93 ± 0.03
	2.370	0.003	1.500	
	2.368	0.055	1.600	
	2.367	0.067	0.800	

eV and 0.20 eV higher in energy. Therefore, the presence of these two isomers along with the ground state in the SG molecular beam could be a possible explanation to the observed large uncertainty. But it should be pointed out here that this might be a possible argument but never the certain reason, because for all sizes there exist more than one isomer with different magnetic structure, but the corresponding experimental uncertainty is not large.

Now we move to the sudden deep observed in the measured magnetic moment at n=13 and 19 compared to their neighbours[4, 5]. Through *ab initio* calculation, we reproduced

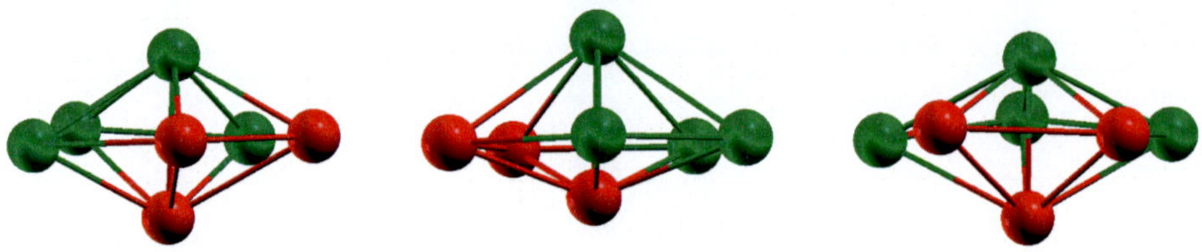

$5\ \mu_B$ $7\ \mu_B$ $3\ \mu_B$

Figure 2. Ground state and isomeric configuration for Mn_7. Note that for all configurations number of spin up atoms and spin down atoms are the same ($N_\uparrow = 4$, $N_\downarrow = 3$). Green(Red) ball represents up(down) atom and we follow the same convention throughout.

this behaviour[8]. For Mn_{13}, we studied icosahedral, cubooctahedral and hexagonal structures for all possible multiplicities and we found the icosahedral structure with magnetic moment 3 μ_B to be the ground state. The optimal cuboctahedral and hexagonal structure have 9 and 11 μ_B magnetic moment, which is much higher than the SG experimental value,0.54 $\pm$ 0.06 μ_B[4, 5], and they lie much higher in energy than the icosahedral ground state. They are shown in the Figure3. Similarly, the ground state of Mn_{19} is found to be double icosahedral structure with magnetic moment 21 μ_B (Figure4). The optimal FCC structure (Figure4) lie 1.53 eV higher in energy. So in our recent paper we concluded that the "closed" icosahedral geometric structure is responsible for the observed sudden deep.

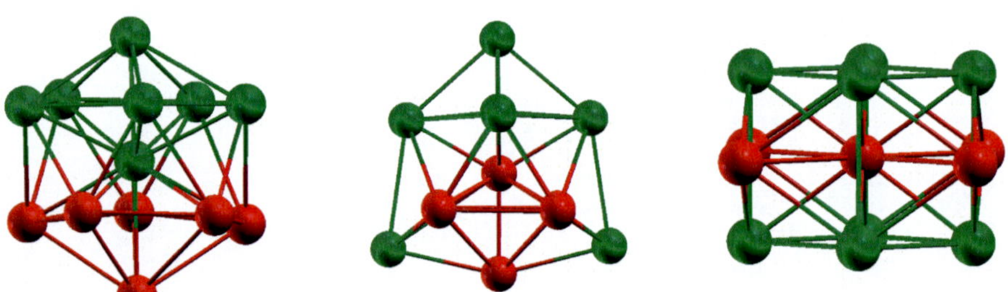

Icosahedral	Hexagonal	Cubooctahedral
$\triangle E = 0.00$ eV	$\triangle E = 0.89$ eV	$\triangle E = 1.12$ eV
$3\ \mu_B$	$9\ \mu_B$	$11\ \mu_B$

Figure 3. Spin configuration for the icosahedral ground state of Mn_{13}. Lowest energy spin configurations for hexagonal and cubooctahedral structure have also been shown.

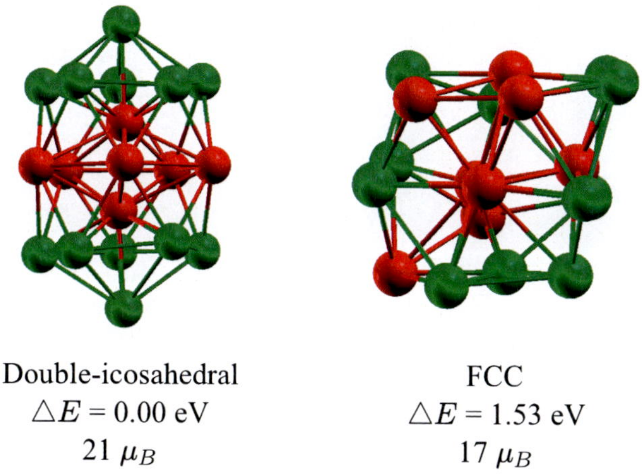

Figure 4. Ground state configuration for Mn_{19} is a double icosahedron. Optimal FCC structure lies much higher in energy.

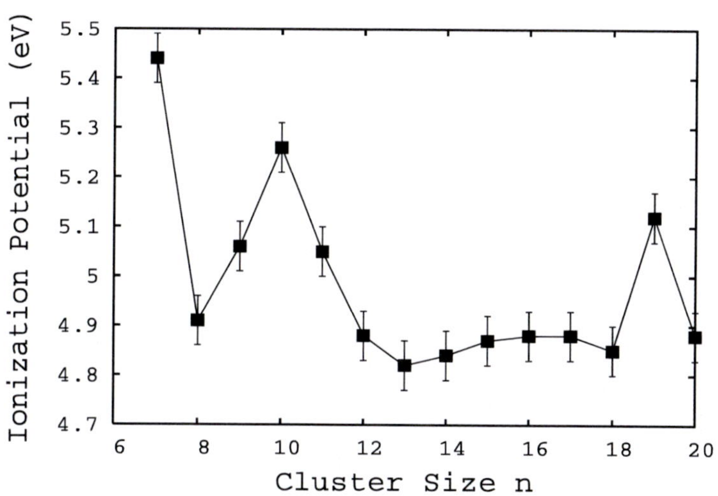

Figure 5. Experimentally measured ionization potential for Mn_n clusters in the size range n = 2-20.(From Ref.[37].)

4.2. Non-metal to Metal Transition?

Parks *et al.* produced Mn_n clusters containing up to 70 atoms by cooling the inert gas condensation source to -160^{0}C to study their reaction rate with the molecular hydrogen[9]. They found that the clusters with $n \leq 15$ are nonreactive toward H_2, whereas they form stable hydride above $n = 15$ and the reaction rate varies considerably with cluster size. It was thought that this might arise due to the possible abrupt change in the bonding character at $n = 16$: For small clusters $n \leq 15$, the bonding is weak and of van der Walls kind, which perhaps become metallic at $n = 16$ and remains the same for larger clusters. If it is so then once the cluster is metallic, it is energetically possible to transfer charge from metallic

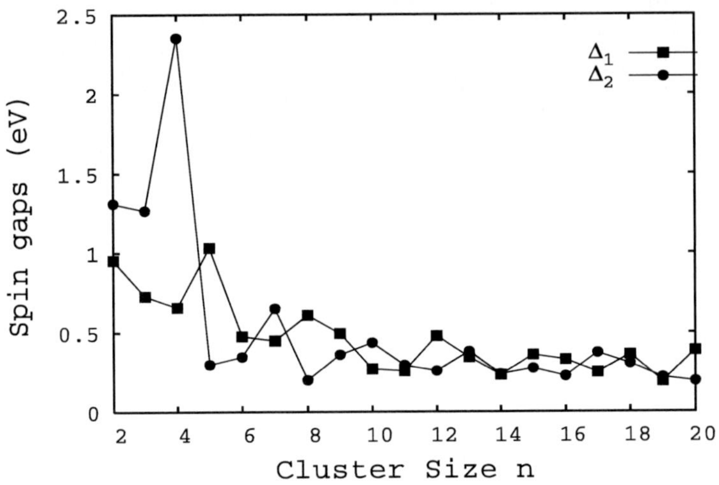

Figure 6. Variation of spin gaps with cluster size in the size range $n = 2$—20. (From Ref.[8].)

cluster to the antibonding state of the H_2 molecule and H-H bond would consequently break and H atom would attach to the metal cluster. If this is indeed the case, it is likely that a significant downward change in the ionization potential would be observed at Mn_{16}. The similar effect has observed in the free mercury clusters[36], where a steady decrease in the ionization potentials attribute to the nonmetal-metal transition. We also expect the closing up of the two spin gaps for this kind of transition. Koretsky *et al.* measured ionization potentials of Mn_n clusters in the size range $n = 7 — 64$[37]. However, no sudden decrease is observed at $n = 16$ (Figure5).

Spin gaps[2] are defined as follows:

$$\triangle_1 = - \left[\epsilon_{\text{HOMO}}^{\text{majority}} - \epsilon_{\text{LUMO}}^{\text{minority}} \right]$$
$$\triangle_2 = - \left[\epsilon_{\text{HOMO}}^{\text{minority}} - \epsilon_{\text{LUMO}}^{\text{majority}} \right], \quad (42)$$

and we plot them in the Figure6. We did not find any closing up of these spin gaps. Therefore, there is no evidence of nonmetal-metal transition and the observed sudden change in the reactivity is not due to any kind of structural transition at $n = 16$ either, as all the medium sized clusters adopt icosahedral growth pattern. So the reason for the abrupt change in the reaction rate of Mn_n clusters with H_2 at $n = 16$ is yet unknown.

4.3. Localization of $d-$electrons

In finite size clusters coordination number may very drastically for different sites and, therefore, they are very good system to study the fact that how the $d-$electrons get localized with

[2]Any spin arrangement for these Mn_n clusters is magnetically stable only if both the spin gaps are positive. This means that the lowest unoccupied molecular orbital (LUMO) of the minority(majority) spin lies above the highest occupied molecular orbital(HOMO) of the majority(minority) spin.

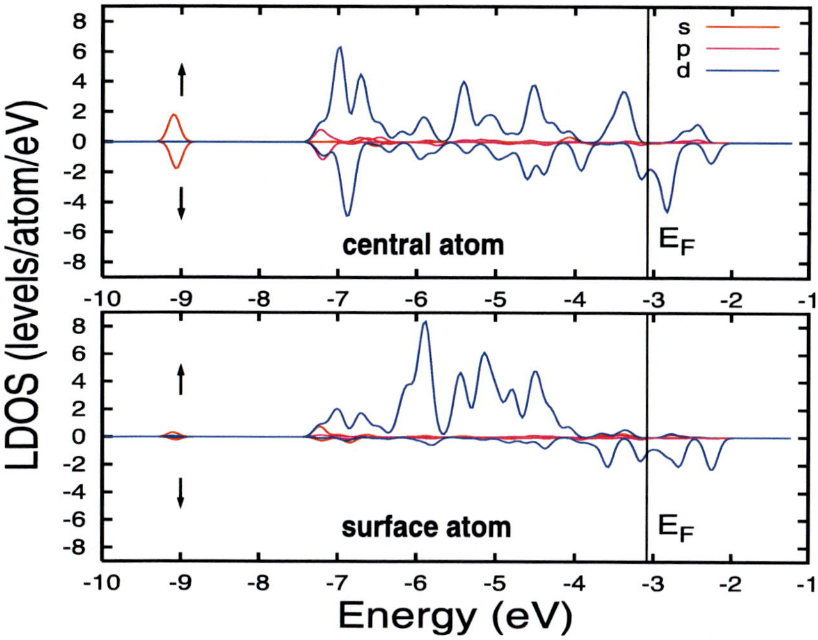

Figure 7. The local density of states (LDOS) for the central and a surface atom for Mn_{13} cluster.

the coordination. Take an example of icosahedral Mn_{13} cluster, where the bonding character of the 12-coordinated central atom is very much different from the the 6-coordinated surface atoms. Here to study the effect of localization, mainly of $d-$ electrons, on the coordination, we calculate the angular momentum projected local density of states, which is plotted in the Figure7 for Mn_{13} cluster. We see that the $d-$ projected local density of states (LDOS) of the central atom are broad for both the majority and minority spin states, whereas the same for the surface atoms are rather localized and the majority spin states are nearly fully occupied. This is also reflected from the local magnetic moments of the central and surface atoms, which is defined as follows:

$$\mathcal{M}_i = \int_0^R \left[\rho^{\uparrow}(\mathbf{r}) - \rho^{\downarrow}(\mathbf{r})\right] d\mathbf{r}, \tag{43}$$

where $\rho^{\uparrow}(\mathbf{r})$ and $\rho^{\downarrow}(\mathbf{r})$ are spin up and spin down densities, respectively and R is the radius of the integrating sphere centering each atom. The local magnetic moment of the central atom is small, 1.42 μ_B, compared to the surface atoms $\sim$ 3.60 μ_B due to the above mentioned reason.

5. Single As-doping in Mn_n Clusters: Mn_nAs ($n \leq 10$)

We already have discussed the pure Mn_n clusters and have seen that very small clusters containing up to 4 atoms are ferromagnetic and they undergo a magnetic (ferromagnetic to ferrimagnetic) transition at n=5. Now it will be interesting to see that what happens to the magnetic structure of those if we dope a single arsenic into it to form Mn_nAs species. In the previous section, we have also noticed that the binding energy of pure Mn_n clusters are small compared to other $3d-$transition metal clusters due to its $3d^5\,4s^2$ electronic structure. Here we will also discuss that what happens to the bonding due to single As-doping. Particularly magnetism in these Mn_nAs clusters are more interesting, as manganese-doped semiconductors, such as (GaMn)N, (GaMn)As and (InMn)As, have attracted considerable attention because of their carrier induced ferromagnetism[38, 39, 40, 41, 42]. The Mn dopants in these III-V dilute magnetic semiconductors serve the dual roles of provision of magnetic moments and acceptor production. A wide range of T_c, 10-940 K, has been reported for $Ga_{1-x}Mn_xN$ by varying Mn concentration (x ~7-14%) as well as by varying the growth conditions[43, 44, 45, 46]. But still it is unclear at present, whether all these reports of ferromagnetism are indeed intrinsic or arise due to the some kind of 'defects' originated during the growth, as magnetic atoms are not thermodynamically stable in the semiconductor host and tend to form 'defects'. Recent investigation[51] found that ferromagnetism persists up to $\sim$ 980 K for metastable Mn-doped ZnO and further heating transforms the metastable phase and kills the ferromagnetism. Earlier experimental results on $Ga_{1-x}Mn_xAs$ indicate a nonmonotonic behaviour of $T_c(x)$, first increases with the Mn concentration x, reaching a maximum of 110 K for x ~5% and then decreases with the further increase of x[47, 48, 49, 50]. However, recent experimental studies [52, 48, 53], under carefully controlled growth and annealing conditions, suggest that the 'metastable' nature and high 'defect' content (such as clustering of Mn) of low temperature of MBE grown $Ga_{1-x}Mn_xAs$ may be playing an important role in the magnetic properties and could enhance T_c with Mn content. Chiba *et al.* [52] have reported T_c as high as 160 K for 7.4 % Mn concentration, in a layered structure. These studies rule out the earlier suggestion that the T_c in $Ga_{1-x}Mn_xAs$ is fundamentally limited to 110 K and certainly opened up a hope for further possible increase in T_c for such metastable phase under different suitable growth and annealing conditions. Towards the end of the chapter, we will discuss this recently debated clustering phenomena, which at least puts some light on the issue.

5.1. Geometry and Bonding

We start with the MnAs dimer, which has much higher binding energy of 1.12 eV/atom and much shorter bond length, 2.21 Å compared with the pure Mn_2 cluster, which has the values 0.53 eV/atom and 2.58 Å, respectively[10]. The ground state of the Mn_2As is an isosceles triangle and the binding energy gets substantially increased to 1.63 eV/atom. The Mn-Mn bond, 2.59 Å, in this Mn_2As cluster is nearly equal to that of the Mn_2 dimer. As more Mn-atoms are added to the Mn_2As cluster, the resultant clusters adopt three dimensional shape and, moreover, the determination of the ground state is far more difficult as it is known that the number of local minima in the potential energy surface is an exponential function of the number of atoms present in that cluster.

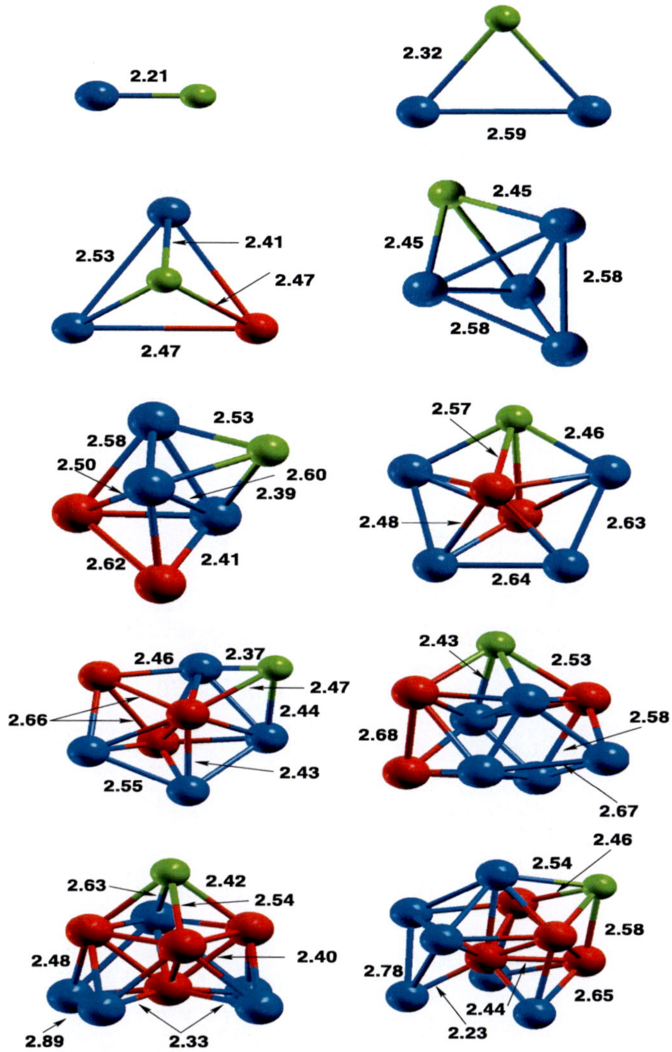

Figure 8. Ground state geometries and spin ordering of Mn_xAs clusters, x=1-10. Blue and red balls represent $Mn_\uparrow$ and $Mn_\downarrow$ atoms, respectively. Green ball represents the As atom. The bond lengths are given in Å. Magnetic polarization of As, is negative for MnAs - Mn_8As and, whereas, positive for Mn_9As and $Mn_{10}As$. Note, Mn atoms are coupled ferromagnetically only in Mn_2As and Mn_4As clusters.

Calculated binding energy, total magnetic moment, shortest Mn-Mn and Mn-As bond lengths and two spin gaps are given in the Table5 for the ground state along with the isomers for Mn_nAs cluster. The presence of an As-atom makes Mn_3As cluster tetrahedral, however, Mn_4As is a Mn_4 tetrahedra with As at a face cap. Generally, the ground state structures of Mn_nAs clusters can be seen as that of the corresponding pure Mn_n cluster with As capped at some face, i.e. As-doping does not give rise to any considerable structural change, but a moderate perturbation to the corresponding pure cluster structure. However, the Mn_6As

cluster is the only cluster whose geometry differs significantly from the pure Mn_6 cluster. Mn_6 is octahedral, whereas Mn_6As is a pentagonal bipyramid, where the As atom is trapped in the pentagonal ring. As n in the Mn_nAs increases, binding energy increases monotonically due to the increase in the coordination number. However, the increase rate is very slow after $n = 2$: Increases slowly from 1.73 eV/atom for Mn_3As to 2.22 eV/atom for $Mn_{10}As$. The shortest Mn-As bond length increases from 2.21 Å for MnAs dimer to 2.46 Å for $Mn_{10}As$ cluster, whereas the shortest Mn-Mn bond length in these clusters decreases with cluster size (2.59 Å for Mn_2As to 2.23 Å for $Mn_{10}As$). Generally, we find, the bonds between Mn atoms of opposite spin to be somewhat shorter (2.20-2.60 Å) than the bonds between Mn atoms of like spin (2.50-2.90 Å), whereas Mn-As distance varies between 2.20-2.60 Å. All the Mn-As-Mn bond angles in these clusters vary in between $\sim$ 60-70^0. All the clusters in the Figure8 and their respective isomers are magnetically stable i.e. both the spin gaps are positive. These two spin gaps, $\triangle_1$ and $\triangle_2$, for Mn_2As (0.83 and 1.34 eV) and Mn_4As (0.89 and 1.14 eV) are the highest among all clusters. As x increases, $\triangle_1$ and $\triangle_2$ decrease to a value 0.47 and 0.35 eV, respectively, for $Mn_{10}As$ (see Table5).

Now we discuss the important issue whether the Mn-clustering around the single As-atom are at all fafourable or not. First we look into the binding energy of pure Mn_n and Mn_nAs clusters, which is plotted in the Figure9. We see that due to single As-doping the binding energy of the Mn_nAs clusters are substantially enhanced from their respective pure counterpart. As it is discussed earlier that due to the half-filled 3 d and filled $4s$ shell and high $3d^5\,4s^2 \rightarrow 3d^6\,4s^1$ promotion energy (2.12 eV), Mn atoms do not bind strongly when they are brought together to form clusters. However, as a single As-atom is attached to the pure Mn_n clusters, the $4s^2$ electrons of the Mn-atom interact with the $4p^3$ electrons of As and results enhancement in the binding energy.

To understand this better, one can calculate the energy gains which are of two kind[10]:

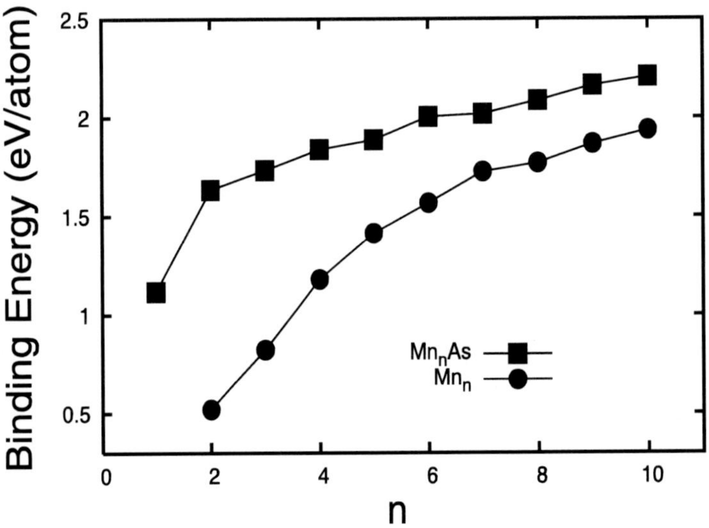

Figure 9. Plot of binding energy per atom for pure and doped manganese clusters, Mn_n and Mn_nAs, in the size range $n \leq 10$.

- The energy gain due to adding an As-atom to a pure Mn_n cluster to form Mn_nAs cluster,

$$\triangle^1 = -\left[E(\mathrm{Mn_nAs}) - \mathrm{E(Mn_n)} - \mathrm{E(As)}\right], \tag{44}$$

and,

- The energy gain due to adding a Mn-atom to a existing $Mn_{n-1}As$ cluster to from Mn_nAs cluster,

$$\triangle^2 = -\left[E(\mathrm{Mn_nAs}) - \mathrm{E(Mn_{n-1}As)} - \mathrm{E(Mn)}\right]. \tag{45}$$

These two energy gains are plotted as a function of Mn concentration in the Figure10. The $\triangle^1$ increases with n: The increment from $n = 1$ to $n = 3$ is monotonous and much stiff but afterwards it is not monotonic and tends to saturate. The other energy gain, $\triangle^2$, gives the number that how many Mn-atoms can be bonded to a single As-atom, which is significant, 2.65 eV, even for $Mn_{10}As$ cluster. Both the energy gains are positive and the energy gain in adding an As-atom is much greater than that of adding a Mn-atom ($\triangle^1 >> \triangle^2$) and, therefore, we can conclude that the clustering of Mn-atom around a single As is favourable[10]. So, the clustering of Mn is possible around As during the growth in Mn-doped GaAs/InAs. However, the other factors like lattice distortions and available space will play a role in determining the size of Mn clustering in these semiconductors.

5.2. Magnetic Order

Now we ask the next important question that what is the Mn-Mn magnetic coupling in these Mn_nAs clusters? The total magnetic moments of Mn_nAs clusters corresponding to

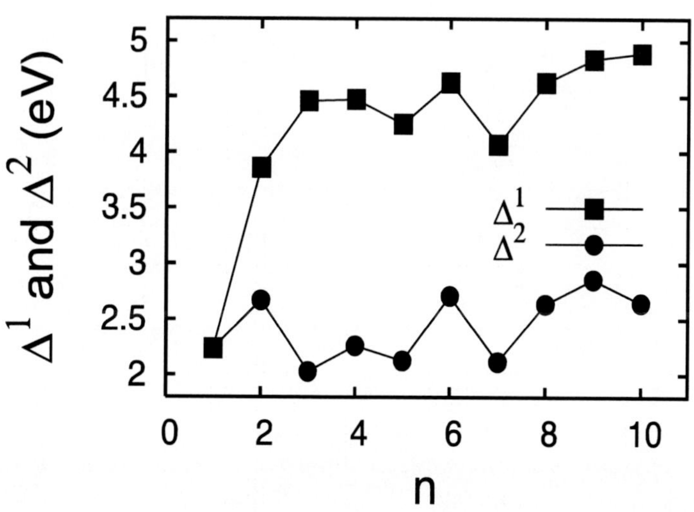

Figure 10. Plot of two different energy gains, $\triangle_1$ and $\triangle_2$ as a function of size.

Table 5. Binding energy, total moment, lowest Mn-Mn ($d_{\mathrm{Mn-Mn}}$) and Mn-As ($d_{\mathrm{Mn-As}}$) bond length and two spin gaps ($\triangle_1$ and $\triangle_2$) for Mn_nAs (n = 1-10) clusters for their respective ground and isomeric state.

Cluster	B.E. (eV/atom)	Total moment μ_B	$d_{\mathrm{Mn-Mn}}$ Å	$d_{\mathrm{Mn-As}}$ Å	$\triangle_1$ eV	$\triangle_2$ eV
MnAs	1.116	4	—	2.208	0.709	1.016
Mn_2As	1.634	9	2.591	2.316	0.826	1.342
	1.597	1	2.428	2.246	0.582	1.345
Mn_3As	1.733	4	2.473	2.413	0.651	0.504
	1.702	12	2.481	2.372	1.295	0.523
Mn_4As	1.838	17	2.573	2.453	0.891	1.144
Mn_5As	1.886	2	2.417	2.389	0.856	0.381
	1.871	12	2.401	2.464	0.620	0.665
Mn_6As	2.004	9	2.469	2.461	0.540	0.669
	1.974	1	2.362	2.435	0.014	0.623
Mn_7As	2.018	6	2.430	2.375	0.409	0.593
	2.014	14	2.405	2.436	0.470	0.611
	2.012	6	2.323	2.472	0.456	0.402
Mn_8As	2.086	7	2.307	2.424	0.405	0.347
	2.075	3	2.305	2.478	0.072	0.552
	2.074	5	2.227	2.468	0.468	0.378
Mn_9As	2.163	10	2.306	2.416	0.418	0.250
	2.158	12	2.308	2.449	0.159	0.826
$Mn_{10}As$	2.207	13	2.225	2.455	0.467	0.347
	2.204	3	2.200	2.501	0.484	0.372

Figure 11. Constant spin density surfaces for Mn_2As and Mn_4As corresponding to 0.04 e/Å^3. Red and blue surfaces represent positive and negative spin densities, respectively. Green ball is the As atom, which has negative polarization in both the structures. Note ferromagnetic coupling among Mn atoms in the both Mn_2As and Mn_4As clusters.

the ground state geometries are 4, 9, 4, 17, 2, 9, 6, 7, 10 and 13 μ_B, respectively, for n=1-10 (Table5). These large magnetic moments generally arise form the ferrimagnetic coupling between the moments at Mn sites with exceptions for Mn_2As and Mn_4As, where the magnetic coupling is found to be ferromagnetic. We note, generally, no change in the magnetic coupling between the Mn sites in Mn_nAs clusters from their respective Mn_n, however, for Mn_3As, the Mn-Mn coupling behaviour changes form ferromagnetic to ferrimagnetic, due to As doping in Mn_3. Another tetrahedral isomer with magnetic moment 12 μ_B is 0.12 eV higher, where all the Mn atoms are ferromagnetically ordered. Here we should point out, as mentioned in the Table3, a frustrated antiferromagnetic structure with total magnetic moment 5 μ_B is only 0.05 eV higher in energy than the ferromagnetic Mn_3 cluster. The ground state magnetic moments of Mn_nAs, for n=1, 2, 3, 5, 7, and 10 can be represented as $(\mu_n - 1)$ μ_B, whereas for n=4, 6, 8, 9, it can be expressed as $(\mu_n - 3)$ μ_B, where μ_n is the total magnetic moment of the Mn_n cluster corresponding to the ground state or the 'first' isomer. It should be pointed out here that, for n=6 and 7, the ground state magnetic moments of Mn_nAs clusters are higher than that of Mn_n, as due to As doping, Mn_nAs can favour the isomeric magnetic structure of the corresponding Mn_n cluster without changing the overall magnetic behaviour between the Mn sites. For MnAs dimer, the magnetic moment at Mn site, $\mathcal{M}_{\text{Mn}}$, is 3.72 μ_B and $\mathcal{M}_{\text{As}}$ is -0.26 μ_B for As. This large negative polarization of the anion, As, is due to the strong $p-d$ interaction. Polarized neutron diffraction study found a local magnetic moment -0.23 ± 0.05 μ_B at the As sites for NiAs-type MnAs [54], which is very close to the present value for MnAs dimer. In the Mn_2As cluster, Mn atoms are ferromagnetically coupled with $\mathcal{M}_{\text{Mn}}$=3.79 μ_B each, whereas the magnetic polarization is negative for As atom, -0.14 μ_B. Mn atoms show ferrimagnetic ordering with $\mathcal{M}_{\text{Mn}}$ 3.1, 3.1 and -3.9 μ_B for Mn_3As, whereas, As has $\mathcal{M}_{\text{As}}$=-0.21 μ_B, which gives a total magnetic moment 4 μ_B. For Mn_4As, Mn atoms are ferromagnetically arranged with average M_{Mn}=3.66 μ_B and are coupled antiferromagnetically with As atom, $\mathcal{M}_{\text{As}}$=-0.22 μ_B. Nature of magnetic coupling can be visualize through their respective spin iso-density plots, which is shown in the Figure11 for Mn_2As and Mn_4As clusters. For Mn_5As, $\mathcal{M}_{\text{Mn}}$ varies between 3.04-3.72 μ_B with a local magnetic moment -0.23 μ_B at As site. We observe, the negative polarization of As, $\mathcal{M}_{\text{As}}$, decreases sharply to -0.08 μ_B for Mn_6As and further decreases monotonically to $-$ 0.02 μ_B for Mn_8As, however it become positive, 0.04 and 0.02, for Mn_9As and $Mn_{10}As$, respectively. For all Mn_xAs clusters, x =6-10, Mn atoms are coupled ferrimagnetically, where $\mathcal{M}_{\text{Mn}}$ varies between 0.8 - 3.7 μ_B. These are unlike the study of Rao and Jena [55], where Mn atoms were found to be coupled ferromagnetically for all sizes, n=1-5, in nitrogen doped Mn_n clusters.

6. Exchange Coupling

To determine the exchange interactions J_{ij}'s between the atoms i and j, we map the magnetic energy onto a classical Heisenberg Hamiltonian:

$$\mathcal{H} = -\sum_{ij} J_{ij}(r)\, \mathbf{S}_i \cdot \mathbf{S}_j, \tag{46}$$

where the sum is over all distinct magnetically coupled atoms, $\mathbf{S}_i(\mathbf{S}_j)$ denotes the localized magnetic moments at $i(j)$-th Mn-site. $r = |\mathbf{R}_i - \mathbf{R}_j|$ is the Mn-Mn spatial separation. Here, rather than the conventional $|\mathbf{S}_i| = 1$ consideration, it is taken as $|\mathcal{M}_i|/2$, where

$$\mathcal{M}_i = \int_0^{R'} [\rho_i^{\uparrow}(\mathbf{r}') - \rho_i^{\downarrow}(\mathbf{r}')] \, d\mathbf{r}'. \tag{47}$$

The reason for this kind of consideration is that, for Mn_nAs clusters Mn atoms have different environment and so as the bonding, and therefore, it is very much likely that $|\mathbf{S}_i|$ have (slightly or much) different value at different site i. In the Eqn 1.47, R' is the radius of the integrating sphere centering each atom and taken as the half of the lowest Mn-Mn separation in the cluster to make sure that no two spheres overlap. $\rho^{\uparrow}(\mathbf{r}')$ and $\rho^{\downarrow}(\mathbf{r}')$ are the up and down spin densities, respectively, at a point $\mathbf{r}'$. It should be noted that in the construction of the Hamiltonian $\mathcal{H}$ in the Eqn.1.46, the signs (but not the magnitudes) are already absorbed in the definition of J_{ij}: positive for ferromagnetic and negative for antiferromagnetic coupling. We can calculate J_{ij}'s from Eqn.1.46, by computing the total energy for a judicious choice of spin configurations with inequivalent combinations of pair correlation functions $\mathbf{S}_i \cdot \mathbf{S}_j$, which results in a set of linear equations for the J_{ij}'s.

We can then compare the calculated J_{ij}'s with the Ruderman-Kittel-Kasuya-Yosida (RKKY)-like theory. This kind of theory describes the magnetic interaction between the localized Mn moments induced by the free carrier spin polarization. This indirect Mn-Mn exchange interaction arise from the local Zener coupling (or the $p-d$ hybridization) between the holes and the Mn d-levels, which then leads to the effective Mn-Mn RKKY interaction:

$$J_{ij}^{\text{RKKY}}(r) \propto r^{-4} \left[\sin(2k_F r) - 2k_F r \cos(2k_F r)\right], \tag{48}$$

or, in the simplest approximation it reads,

$$J_{ij}^{\text{RKKY}}(r) \propto r^{-3} \cos(2k_F r), \tag{49}$$

k_F is the Fermi wave vector. In dilute magnetic semiconductors, RKKY-like models predict that J_{ij} increases with concentration as $n^{1/3}$ at 0K and for fixed n, are independent of environment. We will check the case for As-doped Manganese (Mn_nAs) clusters.

For the smallest possible cluster Mn_2As, the Mn-Mn exchange coupling J can be calculated from the energy difference between the two different spin orientations: parallel ($\uparrow\uparrow$) and antiparallel ($\uparrow\downarrow$) as,

$$E_{\uparrow\downarrow}(r) - E_{\uparrow\uparrow}(r) = J(r)\left[|S_1^{\uparrow}|\,|S_2^{\downarrow}| + |S_1^{\uparrow}|\,|S_2^{\uparrow}|\right], \tag{50}$$

which is the energy measure to flip either one of the moments. Spatial ($r_{\text{Mn}-\text{Mn}}$) dependence of $E_{\uparrow\downarrow}$, $E_{\uparrow\uparrow}$ and exchange coupling J is plotted in the Figure12. Calculated J is compared with the simplest RKKY analytic form $J^{\text{RKKY}}(r) \propto r^{-3}\cos(2k_F r)$. Exchange coupling J oscillates between positive and negative with r favouring FM and AFM solutions, respectively and dies down as $1/r^3$ — a typical RKKY type behaviour.

The case of Mn_3As is very interesting because of the possible magnetic frustration. The Mn moment orientation in the ground state of Mn_3As nanostructure is $\uparrow\uparrow\downarrow$ with a total moment 4 μ_B (Figure13). Energies of different Mn-spin orientations are given in the Table6.

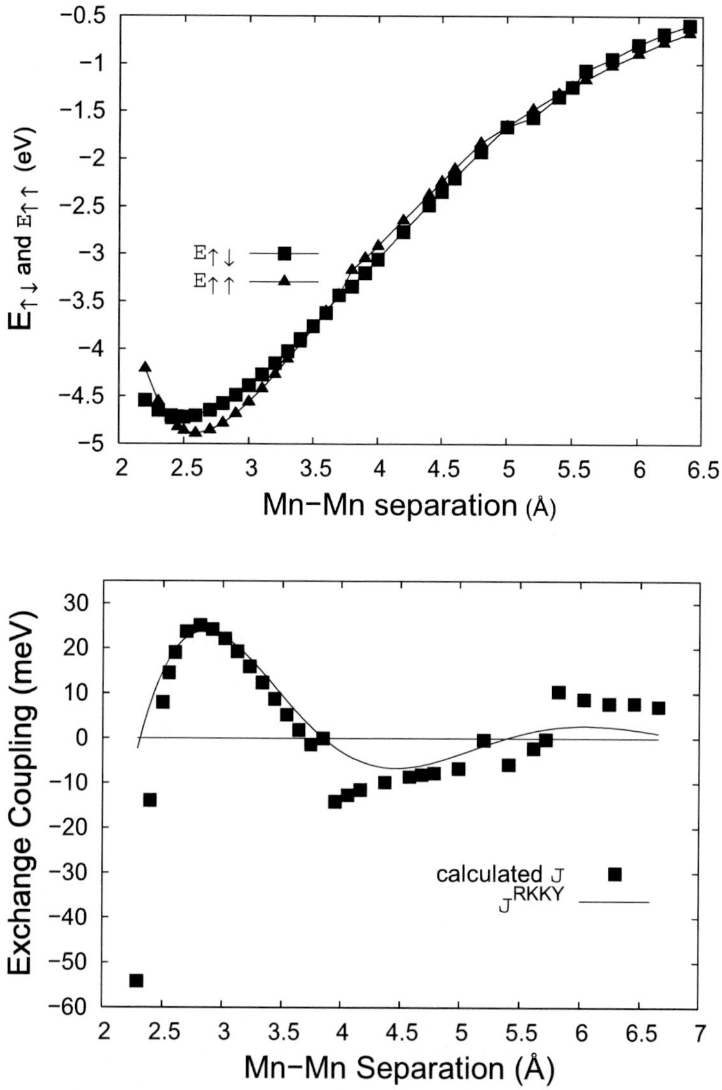

Figure 12. (upper) Spatial dependence of total energy for ↑↑ and ↑↓ configuration of Mn moment in Mn_2As cluster. (lower) Plot of Mn-Mn exchange coupling J with $r_{\mathrm{Mn-Mn}}$. and compared with the simplistic RKKY-like form fitted with $k_F = 1.02\,\text{Å}^{-1}$.

However, instead of two negative and one positive values, all the computed exchange couplings J_{ij}'s turn out to be negative ($J_{12} = -11$ meV and $J_{23}, J_{13} = -14$ meV) indicating the magnetic frustration of Mn-spins.

Now, it is interesting to see, how exchange couplings behave with the increasing cluster size n in Mn_nAs clusters. We plot averaged exchange coupling $\bar{J}_{ij}$ in the Figure 14, which behaves quite differently form RKKY-like theory: $\bar{J}_{ij}$ decreases as n increases with an exceptional increase for ferromagnetic Mn_4As cluster and it has a strong environment dependency.

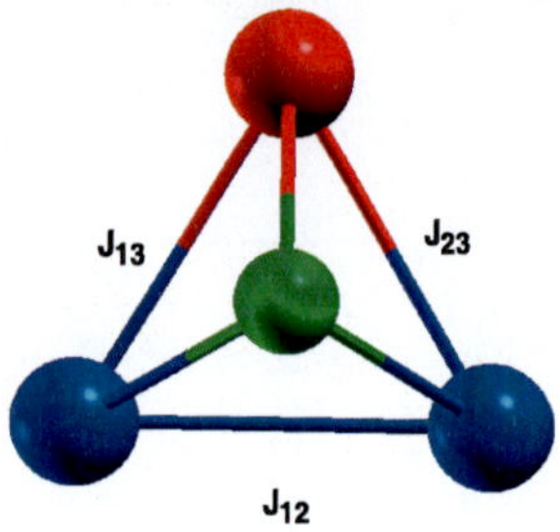

Figure 13. Ground state spin configuration for Mn_3As. Red(Blue) ball(s) represent Mn(↓) and Mn(↑) spin. The structure has a total magnetic moment $4\mu_B$.

7. Summary

In this chapter we discussed the electronic and magnetic properties of pure and doped manganese , Mn_n (n = 2 - 20) and Mn_nAs (n = 1 - 10), clusters from the density functional theory within the pseudopotential scheme. Here we use projector augmented wave method [19] which has been discussed briefly in the Section2.. The spin polarized generalized gradient approximation has been used to treat the exchange-correlation energy. To determine the ground state structure (both geometrical and magnetic), many different initial geometrical structures have been considered and all the possible spin multiplicities have also been investigated for each geometrical conformation. For both pure and doped clusters, we found many isomers with different magnetic structures are possible for a particular size.

Very small pure clusters, Mn_n, up to 4 atoms show Mn-Mn ferromagnetic coupling with 5 μ_B/atom magnetic moment. An addition of a single Mn atom then makes the magnetic coupling ferrimagnetic for Mn_5 and remains the same for the entire size range n = 5–20. These results show very good agreement with the recent experiments[4, 5]. The extraordinarily large experimental uncertainty for Mn_7 cluster[5] is explained due to the possible presence of the isomers in the experimental SG beam. However, the present density functional theory calculation predicts that for almost all the sizes there exist more than one isomer with different magnetic structure, but we do not observe any sufficiently large uncertainty corresponding to the size. The sudden deep in the experimentally measured

Table 6. Total energies of the Mn_3As clusters for different spin orientations.

Mn-spin configuration	Total energy (eV)
↑↑↑	-6.734
↑↑↓	-6.926
↑↓↑	-6.904
↓↑↑	-6.904

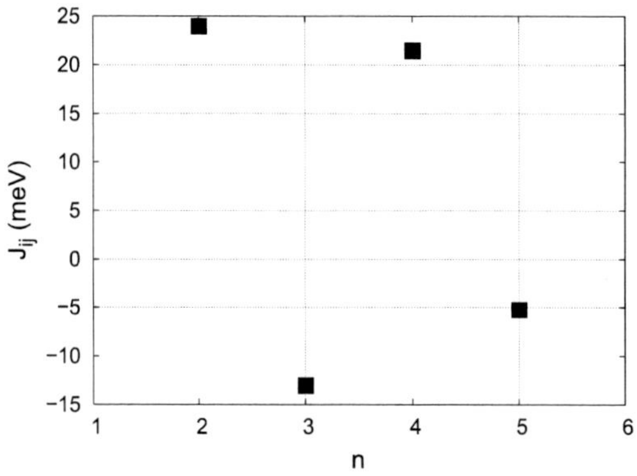

Figure 14. Plot of average exchange coupling $\bar{J}_{ij}$ with the number of Mn-atoms in the cluster n in Mn_nAs.

magnetic moment at Mn_{13} and Mn_{19} [4, 5] is well reproduced within the density functional calculation. This is attributed to their closed icosahedral structure[8].

Previously it was argued that the the sudden increase in the reaction rate of the Mn_n clusters with molecular hydrogen [9] is attributed to the nonmetal-to-metal transition at $n = 16$. If this is indeed the case then there must be a large decrease in the ionization potential and a closing-up of spin gaps at that particular size $n = 16$. However, such kind of downward discontinuity in the ionization potential [37] or any closing-up of spin gaps [8] are not observed at $n = 16$ and the reason for this abrupt change in the reaction rate remains unknown.

Finite systems like cluster are very good system to study the effect of coordination on the electron localization. For an example we discussed Mn_{13} cluster: The $d-$ electrons of the 12-coordinated central atom are more delocalized than those of the 6-coordinated surface atoms, i.e. the $d-$ electron localization decreases with the increase in coordination[8].

Next we discussed the issue how electronic and magnetic structure of these Mn_n clusters do change due to single As-doping. The geometrical structures do not change substantially due to As-doping, but only a moderate perturbation to the corresponding pure Mn_n cluster, i.e. a Mn_nAs cluster, more or less, could be viewed as pure Mn_n cluster with a As-atom caped somewhere. However, the magnetic structure of the Mn_nAs cluster change substantially from their pure counterpart. Only in Mn_2As and Mn_4As clusters, the Mn-Mn coupling is ferromagnetic, and for all other size range this coupling is ferrimagnetic. It is seen that in Mn_nAs clusters the exchange couplings are anomalous and behave quite differently from the RKKY-type predictions[10]. It is known that due to $4s^2$, $3d^5$ electronic structure as well as for high enough $4s^2$, $3d^5 \rightarrow 4s$, $3d^6$ promotion energy of the isolated Mn-atom , they do not bind strongly and as a result Mn_n clusters have lowest binding energies among all other $3d-$transition metal clusters. This scenario has been changed when a single As-atom is attached to them to form a resultant Mn_nAs cluster. The binding energy of the resultant Mn_nAs clusters are enhanced by the hybridization of the $3d-$electrons of the Mn-atom with the $4p-$electrons of the As-atom. The corresponding energy gain in adding an As-atom to a existing Mn_n cluster ($\triangle^1$) is much larger than that of adding an Mn-atom to an existing $Mn_{1-n}As$ cluster ($\triangle^2$), which finally points out that the clustering of the Mn-atoms around an As-atom is energetically favourable.

This point is very important and should be discussed in the context of the observed ferromagnetism in the Mn-doped GaAs/InAs semiconductors. From the free cluster calculations, one come to the conclusion that As-atom act as a nucleation centre for Mn-atoms and it is very much likely that Mn-clusters might be present in those low temperature molecular beam epitaxy grown (Ga, Mn)As/(In, Mn)As samples. Now how the presence of these clusters influence the observed ferromagnetism and Curie temperature is a debatable question for a long time. The ferromagnetic ordering of Mn-atoms are intrinsic for Mn_2As and Mn_4As clusters and are the source of effective internal magnetic field, which influences the energy structure and transport properties. They provide (high temperature) ferromagnetic contribution to the total magnetization of the GaMnAs samples. Although the Mn_2As and/or Mn_4As phases can be mostly responsible for the observed ferromagnetic behaviour of GaMnAs, it is not excluded that predicted carrier-induced ferromagnetic or other mechanism leading to ferromagnetic behaviour can be effective. The presence of ferromagnetic Mn_2As and/or Mn_4As phases in the GaMnAs samples, which have large magnetic moment and very high exchange coupling would consequently enhance the Curie temperature. On the other hand, the presence of the other sized clusters would eventually lead to the low Curie temperature due to their low magnetic moment and low exchange coupling value.

Finally, it should be pointed out that in this chapter we talked about collinear spin arrangements. However, it will be interesting so see whether these clusters show noncollinear magnetic structure or not. Indeed, for larger clusters both the pure and As-doped manganese clusters with more that 5 manganese atoms into it might have noncollinear magnetic structure with substantially different total magnetic moment [8, 10].

References

[1] A. J. Cox, J. G. Louderback and L. A. Bloomfield, *Phys. Rev. Lett.* **71**, 923 (1993); A. J. Cox, J. G. Louderback, S. E. Apsel and L. A. Bloomfield, *Phys. Rev. B* **49**, 12295 (1994).

[2] D. M. Cox, D. J. Trevor, R. L. Wheetten, E. A. Rohlfing and A. Kaldor, *Phys. Rev. B* **32**, 7290 (1985).

[3] L. A. Bloomfield, J. Deng, H. Zhang and J. W. Emmert, in *Proceedings of the International Symposium on Cluster and Nanostructure Interfaces* , edited by P. Jena, S. N. Khanna and B. K. Rao (World Publishers, Singapore, 2000), p.131

[4] M. B. Knickelbein, *Phys. Rev. Lett.* **86**, 5255 (2001).

[5] M. B. Knickelbein, *Phys. Rev. B* **70**, 14424 (2004).

[6] M. D. Morse, *Chem. Rev.* **86**, 1049 (1986).

[7] D. A. Young, *Phase Diagrams of the Elements* (Berkeley, Los Angeles, CA: University of California Press).

[8] M. Kabir, A. Mookerjee and D. G. Kanhere, (To be published.)

[9] E. K. Parks, G. C. Nieman and S. J. Riley, *J. Chem. Phys.* **104**, 3531 (1996).

[10] M. Kabir, D. G. Kanhere and A. Mookerjee, physics/0503009, (2005).

[11] O.K. Andersen, *Phys. Rev. B* **12**, 3060 (1975).

[12] J.C. Slater, *Phys. Rev.* **51**, 846 (1937). .

[13] P.M. Marcus, *Int. J. Quantum. Chem.* **1S**, 567 (1967).

[14] J. Korringa, *Physica* **13**, 392 (1947).

[15] W. Kohn and N. Rostocker, *Phys. Rev.* **94**, 111 (1954).

[16] D.R. Hamann, M. Schlüter and C. Chiang, *Phys. Rev. Lett.* **43**, 1494 (1979).

[17] D. Vanderbilt, *Phys. Rev. B* **41**, 7892 (1985) .

[18] P. E. Blöchl, *Phys. Rev. B* **50**, 17953 (1994).

[19] G. Kresse and D. Joubert, *Phys. Rev. B* **59**, 1758 (1999).

[20] P. Pulay, in *Modern Theoretical Chemistry*, edited by H. F. Schaefer (Plenum, New York, 1977); Mol Phys. **17**, 197 (1969).

[21] S. Goedecker and K. Maschke, *Phys. Rev. B* **45**, 1597 (1992).

[22] J. P. Perdew, K. Burke and M. Ernzerhof, *Phys. Rev. Lett.* **77**, 3865 (1996).

[23] G. Kresse and J. Furthmuller, *Phys. Rev. B* **54**, 11169 (1996).

[24] R. K. Nesbet, *Phys. Rev.* **135**, A460 (1964).

[25] D. D. Shillady, P. Jena, B. K. Rao, and M. R. Press, *Int. J. Quantum Chem.* **22**, 231 (1998).

[26] K. D. Bier, T. L. Haslett, A. D. Krikwood and M. Moskovits, *J. Chem. Phys.* **89**, 6 (1988).

[27] R. J. Van Zee and W. Weltner,Jr., *J. Chem. Phys.* **89**, 4444 (1988); C. A. Bauman, R. J. Van Zee, S. Bhat, and W. Weltner,Jr., *J. Chem. Phys.* **78**, 190 (1983); R. J. Van Zee, C. A. Baumann and W. Weltner, *J. Chem. Phys.* **74**, 6977 (1981).

[28] S. K. Nayak, B. K. Rao and P. Jena, *J. Phys.: Condens Matter* **10**, 10863 (1998).

[29] P. Bobadova-Parvanova, K. A. Jackson, S. Srinivas and M. Horoi, Phys. Rev. A **67**, 61202 (2003); *J. Chem. Phys* **122**, 14310 (2005).

[30] M. R. Pederson, F. Ruse and S. N. Khanna, *Phys. Rev. B* **58**, 5632 (1998).

[31] A. Wolf and H-H. Schmidtke, *Int. J. Quantum Chem.* **18**, 1187 (1980).

[32] N. Fujima and T. Yamaguchi, *J. Phys. Sols Japan* **64**, 1251 (1995).

[33] J. Harris and R. O. Jones, *J. Chem. Phys.* **70**, 830 (1979).

[34] D. R. Salahub and N. A. Baykara, *Surf. Sci.* **156**, 605 (1985).

[35] G. W. Ludwig, H. H. Woodbury and R. O. Carlson, *J. Phys. Chem. Solids* **8**, 490 (1959).

[36] K. Rademann, B. Kaiser, U. Even, and F. Hensel, *Phys. Rev. Lett.* **59**, 2319 (1987).

[37] G. M. Koretsky and M. B. Knickelbein, *J. Chem. Phys.* **106**, 9810 (1997).

[38] H. Ohno, *J. Magn. Magn. Mater.* **200**, 110 (1999).

[39] H. Ohno and F. Matsukura, *Solid State Commun.* **117**, 179 (2001).

[40] T. Dietl, H. Ohno, F. Matsukura, J. Cibert, and D. Ferrand, *Science* **287**, 1019 (2000).

[41] H. Ohno, H. Munekata, T. Penney, S. Von Molnar, and L. L. Chang, *Phys. Rev. Lett.* **68**, 2664 (1992).

[42] H. Ohno, A. Shen, F. Matsukura, A. Oiwa, A. Endo, S. Katsumoto, and Y. Iye, *Appl. Phys. Lett.* **69**, 363 (1996).

[43] M. E. Overberg, C. R. Abernathy, and S. J. Pearton, *Appl. Phys. Lett.* **79**, 1312 (2001).

[44] M. L. Reed, N. A. El-Masry, H. H. Stadelmaier, M. K. Ritums, M. J. Reed, C. A. Parker, J. C. Roberts, and S. M. Bedair, *Appl. Phys. Lett.* **79**, 3473 (2001).

[45] S. Sonoda, S. Shimizu, T. Sasaki, Y. Yamamoto, and H. Hori, cond-mat/0108159, (2001).

[46] A. F. Guillermet and G. Grimvall, *Phys. Rev. B* **40**, 10582 (1989).

[47] H. Shimizu, T. Hayashi, T. Nishinaga, and M. Tanaka, *Appl. Phys. Lett.* **74**, 398 (1999).

[48] S. J. Potashnik, K. C. Ku, R. Mahendiran, S. H. Chun, R. F. Wang, N. Samarth, and P. Schier, *Phys. Rev. B* **66**, 012408 (2002).

[49] S. J. Potashnik, K. C. Ku, S. H. Chun, J. J. Berry, N. Samarth, and P. Schier, *Appl. Phys. Lett.* **79**, 1495 (2001).

[50] T. Hayashi, Y. Hashimoto, S. Katsumoto, and Y. Iye, *Appl. Phys. Lett.* **78**, 1691 (2001).

[51] Darshan C. Kundaliya, S. B. Ogale, S. E. Lofland, S. Dhar, C. J. Metting, S. R. Shinde, Z. Ma, B. Varughese, K.V. Ramanujachary, L. Salamanca-Riba, T. Venkatesan, *Nature Materials* **3**, 709 (2004).

[52] D. Chiba, K. Tankamura, F. Matsukura and H. Ohno, *Appl. Phys. Lett.* **82**, 3020 (2003).

[53] K. W. Edmonds, K. Y. Wang, R. P. Campion, A. C. Neumann, N. R. S. Farley, B. L. Gallagher and C. T. Foxon, *Appl. Phys. Lett.* **81**, 4991 (2002); K. C. Ku, S. J. Potashnik, R. F. Wang, S. H. Chun, P. Schiffer, N. Samarth, M. J. Seong, A. Mascarenhas, E. Johnston-Halperin, R. C. Myers, A. C. Gossard, and D. D. Awschalom, *ibid* **82**, 2302 (2003).

[54] Y. Yamaguchi and H. Watanabe, *J. Magn. Magn. Mater.* **31-34**, 619 (1983).

[55] B. K. Rao and P. Jena, *Phys. Rev. Lett.* **89**, 185504 (2002).

In: Astrophysics and Condensed Matter
Editor: Thomas G. Hardwell, pp. 137-158
ISBN: 978-1-60456-332-0

Chapter 4

TOPOLOGICAL DEFECTS IN FERROMAGNETIC, ANTIFERROMAGNETIC AND CYCLIC SPINOR CONDENSATES – A HOMOTOPY THEORY

***Yunbo Zhang*[1,2], *Harri Mäkelä*[1] and *Kalle-Antti Suominen*[1]**
[1]Department of Physics, University of Turku, FIN-20014 Turun yliopisto, Finland
[2]Department of Physics and Institute of Theoretical Physics,
Shanxi University, Taiyuan 03006, P. R. China

Abstract

We apply the homotopy group theory in classifying the topological defects in atomic spin-1 and spin-2 Bose-Einstein condensates. The nature of the defects depends crucially on the spin-spin interaction between the atoms. We find the topologically stable defects both for spin-1 ferromagnetic and anti-ferromagnetic states, and for spin-2 ferromagnetic and cyclic states. With this rigorous approach we clarify the previously controversial identification of symmetry groups and order parameter spaces for the spin-1 anti-ferromagnetic state, and show that the spin-2 cyclic case provides a rare example of a physical system with non-Abelian line defects, like those observed in biaxial nematics. We also show the possibility to produce vortices with fractional winding numbers of $1/2$, $1/3$ and their multiples in spinor condensates.

1. Introduction

The all-optical trapping of Bose-Einstein condensates (BEC) [1, 2] has opened up a new direction in the study of dilute atomic gases, i.e., the spinor condensates with degenerate internal degrees of freedom of the hyperfine spin F. For alkali atoms with $F = 1$, both experiments and theories have shown two possible kinds of spin correlations in the atom species, namely ferromagnetic (e.g. ^{87}Rb [3, 4, 5]) or antiferromagnetic (e.g. ^{23}Na [3, 4, 6]). With the experimental success of condensing alkali bosons with $F > 1$ such as ^{85}Rb [7] and ^{133}Cs [8], and the unusual stability of the $F = 2$ state (against spin-exchange) in ^{87}Rb [9], one expects that defects with much richer structure can be created in the future. A remarkable feature here is that both the gauge symmetry $\mathrm{U}(1)$ and the spin symmetry $\mathrm{SO}(3)$ are involved, a situation similar to superfluid ^{3}He where three different continuous symmetries (orbital, spin and gauge) are broken either independently or in a connected fashion [10, 11].

Topological defects and excitations in the spinor BECs have been studied theoretically by several groups [3, 4, 12, 13, 14, 15, 16, 17, 18, 19, 20]. Stoof and Khawaja [12] showed that ferromagnetic condensates have long-lived Skyrmion excitations, which are nonsingular but topologically nontrivial pointlike spin textures. Moreover, they also found that spin-1 Bose-Einstein antiferromagnets have singular pointlike topological spin textures [13], which are analogous to the 't Hooft-Polyakov magnetic monopoles in particle physics. Coreless vortices were demonstrated to be thermodynamically stable in ferromagnetic $F = 1$ spinor condensates under rotation [14, 16] and were phase imprinted in a $F = 1$ sodium condensate experimentally [21]. Yip [17] has performed a systematic study on vortex structures and presented several axisymmetric and non-axisymmetric vortices for $F = 1$ antiferromagnetic BEC. Martikainen et al. [15] proposed and demonstrated numerically a method to create monopoles in three dimensional two-component condensates. Linear defects were studied by Leonhardt and Volovik [18], who pointed out the existence of Alice strings in the condensate of ^{23}Na.

Most of the work on this subject is based on the original identification of the order parameter spaces by Ho [3]. After the original studies it was also claimed by Zhou that a discrete symmetry of Z_2 type was missed in the case of antiferromagnetic spin-1 condensate [19, 20] and therefore the topological defects would manifest totally different structures. In this article we present a rigorous topological study that both solves this spin-1 controversy, and reveals interesting aspects of spin-2 systems. The phases of spin-2 spinor condensates are characterized by a pair of parameters $|\langle \mathbf{F} \rangle|$ and $|\Theta|$ describing the ferromagnetic order and the formation of singlet pairs, respectively [22, 23, 24, 25]. It turns out that for the so called cyclic phase the fundamental group that determines the nature of possible stable topological defects is *non-Abelian*. The only known physical example of such a system so far has been the biaxial nematic liquid crystal.

The organization of this paper is as follows: In the following section, we shall review the basic physics of the spinor condensate and discuss the possible ground states for hyperfine spin $F = 1$ and $F = 2$. In Section 3, we give a brief introduction of the homotopy theory of the defect classification, taking the nematic liquid crystal and superfluid ^{3}He as examples. We present our calculation of the homotopy groups for spinor condensates in Sections 4-7. The non-Abelian fundamental group for the cyclic phase and its indications are discussed in detail and the order parameter spaces are easily identified in a correct way following our procedure of symmetry breaking. We summarize our results in Section 8.

2. Spinor Condensate

Neutral atomic gases can be confined in conventional magnetic traps with the availability of hyperfine states being restricted by the requirement that the trapped atoms remain in weak-field seeking states. Alkali atoms with a nuclear spin of $I = 3/2$, such as ^{87}Rb and ^{23}Na, have three weak-field seeking states at small field. A far-off-resonant optical trap, however, confines atoms regardless of their hyperfine state. Thus, the atomic spin is liberated from the requirements of magnetic trapping and becomes a new degree of freedom. In particular, all atoms in the lower hyperfine manifold, for example the $F = 1$ hyperfine manifold of sodium, can be stably trapped simultaneously. Such multi-component optically trapped condensates are represented by an order parameter which is a vector in hyperfine spin space, and are thus called spinor Bose-Einstein condensates. The spin relaxation collisions in spinor condensates allow for population exchange among hyperfine states without trap loss. Theoretical studies started with the determination of the ground state structure in mean field theory for both spin-1 [3, 4] and spin-2 [22, 23, 24, 25, 26, 27] cases. Law et al. [28, 29] investigated the spin correlation beyond mean-field limit and the spin-mixing dynamics due to the nonlinear interaction in the spinor condensate. The dynamics is sensitive to the relative phase and particle number distribution among the individual components of the condensate. Ho and Yip [30] later found that the ground state of a spin-1 Bose gas with an antiferromagnetic interaction was a fragmented condensate in uniform magnetic fields. Zhou [19, 20] showed that the low energy spin dynamics in the system can be mapped into an $o(n)$ nonlinear sigma model. The formation of ground state spin domains, metastable states and quantum tunneling were observed in a series experiments at MIT [6, 31, 32, 33]. The discussions in this paper, however, mainly concern the possible ground states in mean-field theory.

2.1. Spin-1 Case

The ground states of the spinor condensate are determined through the minimization of the energy functional with the constraint of the conservation of the atom number and magnetization [33]. An $F = 1$ spinor Bose-Einstein condensate is described by a three-component order parameter $\psi(\mathbf{r}) = \left\langle \hat{\Psi}(\mathbf{r}) \right\rangle = (\psi_{+1}, \psi_0, \psi_{-1})^T$. In second quantized notation, the Hamiltonian describing a weakly-interacting Bose gas can be obtained from the Gross-Pitaevskii theory [3]

$$\begin{aligned} H &= \int d^3\mathbf{r} \Big\{ \hat{\Psi}_i^\dagger(\mathbf{r}) \left(-\frac{\hbar^2\nabla^2}{2m} + U(\mathbf{r}) \right) \hat{\Psi}_j(\mathbf{r}) \delta_{ij} \\ &\quad + \frac{1}{2} g_0 \hat{\Psi}_i^\dagger(\mathbf{r}) \hat{\Psi}_j^\dagger(\mathbf{r}) \hat{\Psi}_i(\mathbf{r}) \hat{\Psi}_j(\mathbf{r}) \\ &\quad + \frac{1}{2} g_2 \hat{\Psi}_i^\dagger(\mathbf{r}) \hat{\Psi}_j^\dagger(\mathbf{r}) \left(F_a\right)_{ik} \left(F_a\right)_{jl} \hat{\Psi}_k(\mathbf{r}) \hat{\Psi}_l(\mathbf{r}) \Big\} \end{aligned} \tag{1}$$

where $\hat{\Psi}_i(\mathbf{r})$ is the field annihilation operator for an atom with mass m in hyperfine state $|1, i\rangle$ at position r with $i = +1, 0, -1$ and $U(\mathbf{r})$ is the trapping potential. Here the repeated indices are summed. The scattering lengths a_0 and a_2 characterize collisions between atoms through the total spin 0 and 2 channels, respectively, $g_0 = \frac{4\pi\hbar^2}{m}\frac{a_0+2a_2}{3}$ is interaction strength through the "density" channel, and $g_2 = \frac{4\pi\hbar^2}{m}\frac{a_2-a_0}{3}$ is that through the "spin" channel.

It is convenient to express the order parameter as $\psi(\mathbf{r}) = \sqrt{n(\mathbf{r})}\zeta(\mathbf{r})$ where $n(\mathbf{r})$ is the atomic density and $\zeta(\mathbf{r})$ is a three-component spinor $\zeta(\mathbf{r}) = (\zeta_{+1}, \zeta_0, \zeta_{-1})^T = \left(x_+ e^{i\theta_+}, x_0 e^{i\theta_0}, x_- e^{i\theta_-}\right)^T$ of normalization $|\zeta|^2 = 1$. Here x and θ are the amplitudes and phases of the components. The spinor determines the average local spin by means of

$\langle \mathbf{F} \rangle = \zeta^\dagger(\mathbf{r})\mathbf{F}\zeta(\mathbf{r})$, and $\mathbf{F}$ are the usual spin-1 matrices with

$$F_x = \frac{1}{\sqrt{2}}\begin{pmatrix} 0 & 1 & 0 \\ 1 & 0 & 1 \\ 0 & 1 & 0 \end{pmatrix}, F_y = \frac{i}{\sqrt{2}}\begin{pmatrix} 0 & -1 & 0 \\ 1 & 0 & -1 \\ 0 & 1 & 0 \end{pmatrix}, F_z = \begin{pmatrix} 1 & 0 & 0 \\ 0 & 0 & 0 \\ 0 & 0 & -1 \end{pmatrix}$$

which obey the commutation relations $[F_a, F_b] = i\epsilon_{abc}F_c$. We thus obtain the energy functional

$$\begin{aligned} K &= \int d^3\mathbf{r} \left\{ \psi^\dagger \left(-\frac{\hbar^2\nabla^2}{2m} \right) \psi + (U(\mathbf{r}) - \mu)\, n + \frac{n^2}{2}\left(g_0 + g_2 \langle \mathbf{F} \rangle^2 \right) \right\} \\ &= \int d^3\mathbf{r} \left(K_0 + n^2 g_2 \langle \mathbf{F} \rangle^2 /2 \right) \end{aligned} \tag{2}$$

where K_0 is the density-dependent part and the chemical potential μ determines the number of atoms in the condensate. It is obvious that all spinors related to each other by gauge transformation $e^{i\theta}$ and spin rotations $\mathcal{U} = e^{-iF_z\alpha}e^{-iF_y\beta}e^{-iF_z\gamma}$ are energetically degenerate in zero external magnetic field, where (α, β, γ) are the Euler angles. The ground-state spinor is determined by minimizing the spin-dependent mean-field interaction energy, $n^2 g_2 \langle \mathbf{F} \rangle^2 /2$. There are two distinct states depending on the sign of the interaction parameter g_2:

- $g_2 > 0$ (i.e. $a_2 > a_0$, e.g. ^{23}Na): anti-ferromagnetic or polar state as the condensate lowers its energy by minimizing its average spin, i.e. by making $\langle \mathbf{F} \rangle = 0$. The ground state spinor is then one of a degenerate set of spinors, the “polar” states, corresponding to all possible rotations of the hyperfine state $m_F = 0$, i.e.

$$\zeta(\mathbf{r}) = e^{i\theta}\mathcal{U}\begin{pmatrix} 0 \\ 1 \\ 0 \end{pmatrix} = e^{i\theta}\begin{pmatrix} -\frac{1}{\sqrt{2}}e^{-i\alpha}\sin\beta \\ \cos\beta \\ \frac{1}{\sqrt{2}}e^{i\alpha}\sin\beta \end{pmatrix} \tag{3}$$

- $g_2 < 0$ (i.e. $a_2 < a_0$, e.g. ^{87}Rb): ferromagnetic as the condensate lowers its energy by maximizing its average spin, i.e. by making $\langle \mathbf{F} \rangle = 1$. In this case the ground state spinors correspond to all rotations of the hyperfine state $m_F = 1$, i.e.

$$\zeta(\mathbf{r}) = e^{i\theta}\mathcal{U}\begin{pmatrix} 1 \\ 0 \\ 0 \end{pmatrix} = e^{i(\theta-\gamma)}\begin{pmatrix} e^{-i\alpha}\cos^2\frac{\beta}{2} \\ \sqrt{2}\cos\frac{\beta}{2}\sin\frac{\beta}{2} \\ e^{i\alpha}\sin^2\frac{\beta}{2} \end{pmatrix} \tag{4}$$

2.2. Magnetic Field

One can tailor the ground state structure with an external magnetic field and the effects of field inhomogeneities and quadratic Zeeman shifts modify the spin-dependent interaction energy into [6]

$$K_{spin} = \left(c\langle \mathbf{F} \rangle^2 - p\langle F_z \rangle + q\left\langle F_z^2 \right\rangle \right) n \tag{5}$$

where $c = g_2 n/2$. The linear Zeeman shift $p = g\mu_B Bz + p_0$, where g is the Landé g-factor and μ_B is the Bohr magneton, comes from the field gradient B along the long axis z of the condensate, while the last term gives the quadratic Zeeman shift from homogeneous field which is always positive for spin-1 condensate in a weak field. Assuming conservation of total spin, we have included a Lagrange multiplier p_0 into p. For a system with zero total spin, p_0 cancels the linear Zeeman shift due to a homogeneous bias B_0, yielding $p = 0$. Positive (negative) values of p are achieved for condensates with a positive (negative) overall spin. The parameters p and q can be related to the individual level shifts by (energies in units of the hyperfine splitting E_{HFS})

$$\begin{aligned} 2p &= E_- - E_+ \\ 2q &= E_- + E_+ - 2E_0 \end{aligned} \tag{6}$$

where the Zeeman energies E_+, E_0 and E_- of the $m_F = +1, 0, -1$ can be expressed according to the Breit-Rabi formula [34] as

$$\begin{aligned} E_+ &= -\frac{1}{8} - \frac{1}{2}\sqrt{1 + x + x^2} \\ E_0 &= -\frac{1}{8} - \frac{1}{2}\sqrt{1 + x^2} \\ E_- &= -\frac{1}{8} - \frac{1}{2}\sqrt{1 - x + x^2} \end{aligned} \tag{7}$$

with $x = g\mu_B B/E_{HFS}$.

Including the non-diagonal terms of the mean field interaction, we may minimize the energy functional

$$\begin{aligned} K_{spin}/n &= c\left(x_+^2 - x_-^2\right)^2 + 2cx_0^2\left(x_+^2 + x_-^2 + 2x_+x_-\cos\phi\right) \\ &\quad -p\left(x_+^2 - x_-^2\right) + q\left(x_+^2 + x_-^2\right) \end{aligned} \tag{8}$$

by means of the Lagrange multiplier method subjected to the constraint of normalization

$$g = x_+^2 + x_0^2 + x_-^2 - 1 = 0 \tag{9}$$

where $\phi = \theta_+ + \theta_- - 2\theta_0$. The solutions to the first derivatives of the Lagrange multiplier function $X = K_{spin}/n - \lambda g$ can be classified into the following table of spinors with their corresponding energies

	spinors	energies
1	$e^{i\theta_{+1}}(1,0,0)$	$c - p + q$
2	$e^{i\theta_{-1}}(0,0,1)$	$c + p + q$
3	$e^{i\theta_0}(0,1,0)$	0
4	$\left(e^{i\theta_{+1}}\sqrt{\frac{2c+p}{4c}}, 0, e^{i\theta_{-1}}\sqrt{\frac{2c-p}{4c}}\right)$	$q - \frac{p^2}{4c}$
5	$\left(e^{i\theta_{+1}}\sqrt{\frac{2c-p+q}{2c}}, e^{i\theta_0}\sqrt{\frac{p-q}{2c}}, 0\right)$	$\frac{(2c-p+q)^2}{4c}$
6	$\left(0, e^{i\theta_0}\sqrt{\frac{-p-q}{2c}}, e^{i\theta_{-1}}\sqrt{\frac{2c+p+q}{2c}}\right)$	$\frac{(2c+p+q)^2}{4c}$
7	$\left(e^{i\theta_{+1}}\sqrt{\frac{q^2+4cq-p^2}{16cq^3}}(q+p)^2, e^{i\theta_0}\sqrt{\frac{-q^2+4cq-p^2}{8cq^3}}(q^2-p^2), e^{i\theta_{-1}}\sqrt{\frac{q^2+4cq-p^2}{16cq^3}}(q-p)^2\right)$	$\frac{\left(q^2+4cq-p^2\right)^2}{16cq^2}$

We notice that spinors 4-7 are only well-defined in some specific regions in the p–q plane, i.e., the quantities under the square root must be non-negative. For example, spinor 7 may only exist for $q^2 + 4cq - p^2 < 0$ and $q^2 > p^2$, and in addition we must have $\phi = 0$ or π. The ground state spinors obtained by minimizing the energy functional can be indicated in the so-called spin-domain phase diagrams (Figure 1 in ref. [6]). For $c = 0$, the Zeeman energy causes the cloud to separate into three pure domains with $m_F = +1, 0, -1$ and with boundaries at $|p| = q$. For $c > 0$, a spin domain with mixed $m_F = \pm 1$ components, i.e., spinor 4, appears in the anti-ferromagnetic phase diagram. For $c < 0$, all three components are generally miscible and have no sharp boundaries, which corresponds to spinor 7.

2.3. Conservation of Magnetization

Although conservation of the magnetization was included in the above section, it was not separately discussed. Consequently the results do not easily apply to systems with fixed values of the magnetization $\mathcal{M}$. The ground state structures as given in [6] correspond to the actual ground state as realized through an $\mathcal{M}$ non-conserving evaporation process (e.g. in the presence of a non-zero B-field) that serves as a reservoir for condensate magnetization. On the other hand, the phase diagram for fixed values of $\mathcal{M}$ was also explicitly discussed [35], which could physically correspond to experimental ground states (with/without a B-field) due to an $\mathcal{M}$ conserving evaporation process. This requires the introduction of two Lagrange multipliers during the minimization subjected to conservation constraints for both the atomic number N and magnetization $\mathcal{M}$, which in the mean-field approximation are given by

$$\begin{aligned} N &= \int d^3\mathbf{r} n(\mathbf{r}) \left(x_+^2(\mathbf{r}) + x_0^2(\mathbf{r}) + x_-^2(\mathbf{r})\right), \\ \mathcal{M} &= \int d^3\mathbf{r} n(\mathbf{r}) \left(x_+^2(\mathbf{r}) - x_-^2(\mathbf{r})\right). \end{aligned} \tag{10}$$

We restrict the discussion here to the situation that equation (9) and

$$h = x_+^2 - x_-^2 - m = 0 \tag{11}$$

are satisfied where $m = \mathcal{M}/N$. With the definition $x = x_+^2 + x_-^2$, we can assort the possible spinors minimizing the Lagrange multiplier function $X = K_{spin}/n - \lambda g - \delta h$ into the following classes (where the energy zero point has been moved to pm):

	spinors	energies
4	$\left(e^{i\theta_{+1}}\sqrt{\frac{1+m}{2}}, 0, e^{i\theta_{-1}}\sqrt{\frac{1-m}{2}}\right)$	$cm^2 + q$
5	$\left(e^{i\theta_{+1}}\sqrt{m}, e^{i\theta_0}\sqrt{1-m}, 0\right)$	$-cm^2 + (2c+q)\,m$
6	$\left(0, e^{i\theta_0}\sqrt{1+m}, e^{i\theta_{-1}}\sqrt{-m}\right)$	$-cm^2 - (2c+q)\,m$
7	$\left(e^{i\theta_{+1}}\sqrt{\frac{x_m+m}{2}}, e^{i\theta_0}\sqrt{1-x_m}, e^{i\theta_{-1}}\sqrt{\frac{x_m-m}{2}}\right)$	$cm^2 + g_\pm(x_m) + qx_m$

We still have the spinors 1-3 which are the same as in above section, however, they only exist for special values $m = +1, -1, 0$, respectively. While spinor 5(6) is confined to the positive (negative) values of m, 4 and 7 may exist for the whole region $-1 \leq m \leq 1$. In spinor 7 with three nonzero components, the phase convention remains $\phi = 0$ or π and the minimum is reached when $x = x_m$ where x_m is determined by

$$g'_{\pm}(x_m) + q = 0 \tag{12}$$

with $g_{\pm}(x) = 2c(1-x)\left(x \pm \sqrt{x^2 - m^2}\right)$ for ferromagnetic(+) or anti-ferromagnetic(−) interaction, respectively. The ground state spinor phase diagram for a homogeneous condensate may be determined in the m–q plane, as indicated for positive m case in Figure 4 of ref. [35]. For $c = 0$, spinor 5 will always dominate except that on the boundary $q = 0$ we have spinor 7. For $c < 0$, spinor 7 will dominate. For $c > 0$, a curve $q = 2c(1 - \sqrt{1 - m^2})$ divides spinors 4 and 7.

2.4. Spin-2 Case

For ^{23}Na and ^{87}Rb with regular hyperfine multiplets, the lower hyperfine state $F = 1$ has lower energy than the upper state $F = 2$. Experimentally only atoms in the lower hyperfine states can be confined in the optical trap. Those in the upper hyperfine states will leave the trap by spin-flip scattering. Since spin-flip scattering is strong in ^{23}Na, only the high-field seeking stretched state $|2, -2\rangle$ exhibits reasonable stability, experiments with more complex spinor condensate do not seem to be possible [36]. On the other hand, optically trapped ^{87}Rb has proved to be a candidate for spin-2 Bose gas [37] with rich spin dynamics and magnetization conservation was also observed during the mixing [5]. In the case of ^{85}Rb, the lowest multiplet has spin $F = 2$ and a negative s-wave scattering length in zero field. With the success to Bose condense ^{85}Rb in magnetic traps [7], it is conceivable that an $F = 2$ spinor condensate might be trapped optically in lower hyperfine states, provided that the three particle losses when the field is reduced through the Feshbach resonance are not too large.

Bose systems require that the total angular momentum of two colliding spin-2 particles is restricted to 0, 2, and 4. The effective low-energy Hamiltonian including the interaction energy describing binary collisions via the s-wave scattering can be generally expressed as [22, 23, 24]

$$\begin{aligned} H &= \int d^3\mathbf{r}\Big\{ \hat{\Psi}_i^\dagger(\mathbf{r})\left(-\frac{\hbar^2\nabla^2}{2m} + U(\mathbf{r})\right)\hat{\Psi}_j(\mathbf{r})\delta_{ij} \\ &\quad + \frac{1}{2}c_0\hat{\Psi}_i^\dagger(\mathbf{r})\hat{\Psi}_j^\dagger(\mathbf{r})\hat{\Psi}_i(\mathbf{r})\hat{\Psi}_j(\mathbf{r}) \\ &\quad + \frac{1}{2}c_1\hat{\Psi}_i^\dagger(\mathbf{r})\hat{\Psi}_j^\dagger(\mathbf{r})\left(F_a\right)_{ik}\left(F_a\right)_{jl}\hat{\Psi}_k(\mathbf{r})\hat{\Psi}_l(\mathbf{r}) \\ &\quad + \frac{1}{2}5c_2\hat{\Psi}_i^\dagger(\mathbf{r})\hat{\Psi}_j^\dagger(\mathbf{r})\left\langle 2i;2j|00\right\rangle\left\langle 00|2k;2l\right\rangle\hat{\Psi}_k(\mathbf{r})\hat{\Psi}_l(\mathbf{r})\Big\} \end{aligned} \tag{13}$$

where $\langle 00|2k; 2l\rangle$ is the Clebsch-Gordan coefficient for combining two spin-2 particles with

$m_F = k$ and l into a spin singlet $|0,0\rangle$. The parameters

$$\begin{aligned} c_0 &= \frac{4\pi\hbar^2}{m}\frac{4a_2+3a_4}{7}, \\ c_1 &= \frac{4\pi\hbar^2}{m}\frac{a_4-a_2}{7}, \\ 5c_2 &= \frac{4\pi\hbar^2}{m}\frac{3a_4-10a_2+7a_0}{7} \end{aligned} \tag{14}$$

describe the density-density interaction, spin-spin interaction, and formation of the singlet pair, respectively. The spinor $\zeta(\mathbf{r})$ with five components $\zeta(\mathbf{r}) = (\zeta_{+2}, \zeta_{+1}, \zeta_0, \zeta_{-1}, \zeta_{-2})$ normalized to unity, determines the average local spin as $\langle \mathbf{F} \rangle = \zeta^\dagger(\mathbf{r})\mathbf{F}\zeta(\mathbf{r})$, and $\mathbf{F}$ are the 5×5 spin-2 matrices which obey the same commutation relations $[F_a, F_b] = i\epsilon_{abc}F_c$. In the mean-field approach the properties of a spinor condensate are determined by the spin-dependent energy functional

$$K_{spin} = \left(c_1 \langle \mathbf{F} \rangle^2 + c_2 |\Theta|^2 - p\langle F_z \rangle + q\langle F_z^2 \rangle\right) n \tag{15}$$

where $\Theta = 2\zeta_{+2}\zeta_{-2} - \zeta_{+1}\zeta_{-1} + \zeta_0^2$ represents a singlet pair of identical spin-2 particles and is invariant under any rotation. The parameters p and q are related to the individual level shifts by

$$\begin{aligned} p &= \frac{1}{12}(E_{+2} - E_{-2}) + \frac{2}{3}(E_{-1} - E_{+1}) \\ q &= -\frac{1}{24}(E_{+2} + E_{-2}) + \frac{2}{3}(E_{-1} + E_{+1}) - \frac{5}{4}E_0 \end{aligned} \tag{16}$$

The Breit-Rabi formula [34] in the case of ^{23}Na or ^{87}Rb ($F = 2$ is the upper hyperfine state with higher energy) gives

$$\begin{aligned} E_{+2} &= -\frac{1}{8} + \frac{1}{2}(1+x) \\ E_{+1} &= -\frac{1}{8} + \frac{1}{2}\sqrt{1+x+x^2} \\ E_0 &= -\frac{1}{8} + \frac{1}{2}\sqrt{1+x^2} \\ E_{-1} &= -\frac{1}{8} + \frac{1}{2}\sqrt{1-x+x^2} \\ E_{-2} &= -\frac{1}{8} + \frac{1}{2}(1-x) \end{aligned}$$

In weak field, the quadratic Zeeman splitting is always negative, i.e., $q = -\frac{1}{16}x^2 + O(x^4)$. In the case of ^{85}Rb ($I = 5/2$) the lowest multiplet has spin $F = 2$. From the individual level shift

$$
\begin{aligned}
E_{+2} &= -\frac{1}{12} - \frac{1}{2}\sqrt{1 + \frac{4}{3}x + x^2} \\
E_{+1} &= -\frac{1}{12} - \frac{1}{2}\sqrt{1 + \frac{2}{3}x + x^2} \\
E_0 &= -\frac{1}{12} - \frac{1}{2}\sqrt{1 + x^2} \\
E_{-1} &= -\frac{1}{12} - \frac{1}{2}\sqrt{1 - \frac{2}{3}x + x^2} \\
E_{-2} &= -\frac{1}{12} - \frac{1}{2}\sqrt{1 - \frac{4}{3}x + x^2}
\end{aligned}
$$

we easily see q is always positive for ^{85}Rb at small field, $q = \frac{1}{36}x^2 + O\left(x^4\right)$. Unlike in the case of spin-1, the whole p–q plane is accessible experimentally for a spin-2 condensate.

The ground state magnetization must be aligned with the external field, i.e. along z-axis, implying $\langle \mathbf{F} \rangle^2 = \langle F_z \rangle^2$ in eq. (15). Minimization of the spin dependent energy functional using the similar Lagrange multiplier method leads to three possible phases, one more compared to the spin-1 case. These phases are characterized by a pair of parameters $|\langle \mathbf{F} \rangle|$ and $|\Theta|$ describing the ferromagnetic order and the formation of singlet pairs, respectively. For convenience, we only consider the linear Zeeman shift p due to the magnetic field:

- Polar/Anti-ferromagnetic phases

$$
\begin{aligned}
P &: \sqrt{\frac{1}{2}}\left(e^{i\theta_{+2}}\sqrt{1 + \frac{p}{4c_1 - c_2}}, 0, 0, 0, e^{i\theta_{-2}}\sqrt{1 - \frac{p}{4c_1 - c_2}}\right) \\
P_1 &: \sqrt{\frac{1}{2}}\left(0, e^{i\theta_{+1}}\sqrt{1 + \frac{p}{2(c_1 - c_2)}}, 0, e^{i\theta_{-1}}\sqrt{1 - \frac{p}{2(c_1 - c_2)}}, 0\right) \\
P_0 &: e^{i\theta_0}\left(0, 0, 1, 0, 0\right)
\end{aligned}
$$

  with energies $c_2 - p^2/\left(4c_1 - c_2\right), c_2 - p^2/4\left(c_1 - c_2\right)$ and c_2 respectively. Here θ_i are arbitrary phases for the corresponding components. These states are energetically degenerate in the absence of the external field with energy c_2 and parameters $\langle \mathbf{F} \rangle = 0$ and $|\Theta| = 1$.

- Ferromagnetic phases

$$
\begin{aligned}
F &: e^{i\theta_{+2}}\left(1, 0, 0, 0, 0\right) \\
F' &: e^{i\theta_{+1}}\left(0, 1, 0, 0, 0\right)
\end{aligned}
$$

  with energies $4c_1 - 2p$ and $c_1 - p$ respectively. This phase has a non-vanishing parameter $|\langle \mathbf{F} \rangle| = 1$ indicating the ferromagnetic order and $|\Theta| = 0$.

- Cyclic phase

$$
C : \frac{1}{2}\left(e^{i\phi}\left(1 + \frac{p}{4c_1}\right), 0, \sqrt{2 - \frac{p^2}{8c_1^2}}, 0, e^{-i\phi}\left(-1 + \frac{p}{4c_1}\right)\right)
$$

with energy $-p^2/4c_1$ and ϕ an arbitrary phase. This is a nonmagnetic phase which has no spin-1 analog and was referred to as the cyclic state because of its close analog to the d-wave BCS superfluids. Both parameters are zero, $\langle \mathbf{F} \rangle = 0$ and $|\Theta| = 0$.

Recent experiments observed clear evidence of polar behaviour for $F = 2$ spinor condensate of ^{87}Rb, and the slow dynamics of prepared cyclic ground states showed the $F = 2$ state to be close to the cyclic phase [26, 37]. However, the nature of the spinor condensate which depends on the s-wave scattering lengths for the total spins 0, 2, and 4, may be changed into other phases by an offset magnetic field.

3. Outline of the Homotopy Theory of Defects

We sketch out the procedure which has been widely used in the study of topological defects in ordered media such as liquid crystals, superfluid ^{3}He and heavy-fermion superconductors. The explicit use of homotopy for topological classification of defects was made by some French [38, 39, 40, 41] and Russian authors [42, 43]. The results were well summarized in two review articles [39, 44]. The central feature of the classification scheme of the defects emerges from examining the mappings of closed curves in physical space into the order-parameter space (OPS).

The order parameter of a system has associated with it a group of transformations G. The set of all transformations in G that leave the reference order parameter f (chosen arbitrarily but thereafter fixed) unchanged is known as the isotropy group $H = \{g \in G | gf = f\}$. The OPS can then be taken to be the space of cosets of H in G: $M = G/H$. In terms of broken symmetry, the fact that the ordering breaks the underlying symmetry is expressed in the fact that H is only a subgroup of G. The description that follows will be valid for any group G that *acts transitively* on M, i.e., if f_1 and f_2 are possible values of the order parameter, then there is a transformation g in G which takes f_1 into f_2: $f_2 = gf_1$.

Homotopy groups of the order-parameter space describe physical defects [44]. The n-th homotopy group $\pi_n(M)$ of the space M consists of the equivalence classes of continuous maps from n-dimensional sphere S_n to the space M. Two maps are equivalent if they are homotopic to one another. In three dimensional space, the first homotopy group, also called the fundamental group, $\pi_1(M)$ describes singular line defects and domain walls, which are non-singular defects. The second homotopy group $\pi_2(M)$ describes singular point defects and non-singular line defects. These can be calculated with the help of the fundamental theorem: Let G be a *connected, simply connected* continuous group and H_0 be the set of points in H that can be connected to the identity by a continuous path lying entirely in H. Then we have the isomorphisms

$$\pi_1(M) = H/H_0, \qquad \pi_2(M) = \pi_1(H_0). \tag{17}$$

For the theorem to hold, it is necessary that $\pi_0(G) = \pi_1(G) = \pi_2(G) = 0$, meaning that G has only one connected piece, any loop in G can be shrunk continuously to a point, and G has a vanishing second homotopy group. While the second homotopy groups are always Abelian, the fundamental groups can either be Abelian (each element constitutes a conjugacy class), or non-Abelian (the line defects are characterized by the conjugacy classes

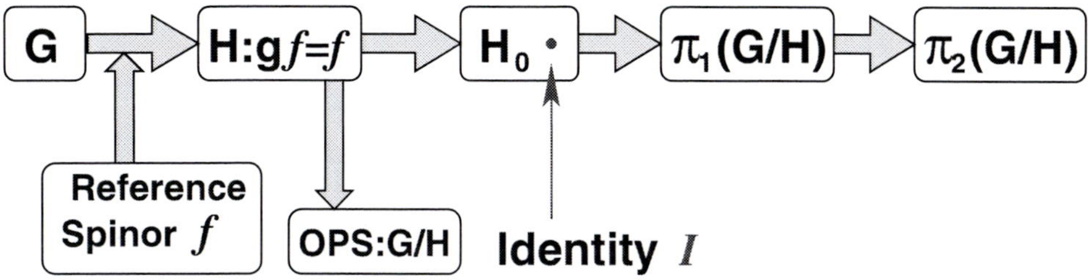

Figure 1. A schematic description of the procedure for calculation of homotopy groups.

instead of the elements). In Figure 1 we give a schematic description of the procedure for calculation of homotopy groups.

A ready example for illustrating the above procedure is the biaxial nematics, whose symmetry is that of a rectangular box (proper point group D_2). If G is taken to be SO(3) then the isotropy subgroup H is the four-element group consisting of the identity and $180°$ rotations about three mutually perpendicular axes (D_2). Order parameter space is thus identified as $M = \mathrm{SO}(3)/D_2$. If, however, we take G to be $\mathrm{SU}(2)$, the universal covering group of $\mathrm{SO}(3)$, then H is expanded to the non-Abelian quaternion group Q (known as the *lift* or *double group*) with eight elements

$$Q = \{\pm 1, \pm i\sigma_x, \pm i\sigma_y, \pm i\sigma_z\}. \tag{18}$$

The natural representation for the order parameter space of a biaxial nematic turns out to be $M = \mathrm{SU}(2)/Q$. Since it is a discrete subgroup of SU(2), $H/H_0 = H$. Thus $\pi_1(M) = Q$, and $\pi_2(M) = 0$. There are no stable point defects in biaxial nematics and the line defects are characterized by five conjugacy classes of group Q

$$\begin{aligned} C_0 &= \{1\}, \overline{C_0} = \{-1\}, \\ C_x &= \{\pm i\sigma_x\}, C_y = \{\pm i\sigma_y\}, C_z = \{\pm i\sigma_z\}. \end{aligned} \tag{19}$$

The class C_0 contains removable trivial defects; $\overline{C_0}$ contains defects in which the object rotates about $360°$ as the defect line is encircled; the other three classes contain defects in which the rotation is through $180°$ about each of the three distinct symmetry axes. The defects here are non-commutative, providing an example with non-Abelian fundamental group.

Another illustrative example is the dipole-free A-phase of ^{3}He, which affords an unusual example of a case where G must be bigger than $\mathrm{SO}(3)$. The order parameter is the product of an arbitrary unit 3-vector $\hat{n}$ and a complex 3-vector of the form $\hat{u} + i\hat{v}$, where $\hat{u}$ and $\hat{v}$ are an orthonormal pair. The orientations of $\hat{n}$ and $\hat{u} + i\hat{v}$ are uncoupled. Take the reference order parameter to be $A_{ij} = z_i(x_j + iy_j)$, the group G can be taken to be the direct product of SO(3) with itself: $G = \mathrm{SO}(3) \times \mathrm{SO}(3)$, elements of G consisting of pairs (R, R') of distinct rotations. The isotropy group H consists of elements of the form $(R(\hat{z}, \theta), 1)$ and $(R(\hat{u}, \pi), R(\hat{z}, \pi))$ for any axis $\hat{u}$ in the x–y plane. To construct a simply connected G, we must replace each $\mathrm{SO}(3)$ by $\mathrm{SU}(2)$. Determining the *lift* of H from $\mathrm{SO}(3) \times \mathrm{SO}(3)$ to its covering group $\mathrm{SU}(2) \times \mathrm{SU}(2)$, we find the isotropy group consists of 4 pieces $\{H_0, gH_{0,}g^2H_0, g^3H_0\}$ with the connected component of the identity

$H_0 = \{(u(\hat{z},\theta),1)\}$ and $g = (u(\hat{x},\pi), u(\hat{z},\pi))$. In this article, the notations R and u represent the rotations in SO(3) and SU(2), respectively. The fundamental group is thus isomorphic to the cyclic group of order 4, $\pi_1(M) = Z_4$ and $\pi_2(M) = Z$.

For the spinor condensate it seems natural to identify the underlying symmetry group as $\mathrm{U}(1) \times \mathrm{SO}(3)$, the groups in the direct product representing the gauge and spin degrees of freedom respectively. This group is *not* simply connected, i.e., $\pi_1(\mathrm{U}(1) \times \mathrm{SO}(3)) \neq 0$. To apply the theorem, however, it is again essential that one chooses the group G to be simply connected. We proceed by specifying the symmetry group as its universal covering group $R \times \mathrm{SU}(2)$, with the group of real numbers R representing any translation $\theta \in (-\infty, +\infty)$ in the phase of the condensate. For $F = 1$, we use the 3D representation of the group SU(2) in order to obtain the isotropy group, e.g., a rotation $u(z,\alpha)$ around axis z by angle α takes the form of a diagonal matrix $Diag\left(e^{-i\alpha}, 1, e^{i\alpha}\right)$, a rotation $u(y,\beta)$ around axis y by angle β takes the form of

$$\begin{pmatrix} \frac{1}{2}(1+\cos\beta) & \frac{-\sin\beta}{\sqrt{2}} & \frac{1}{2}(1-\cos\beta) \\ \frac{\sin\beta}{\sqrt{2}} & \cos\beta & \frac{-\sin\beta}{\sqrt{2}} \\ \frac{1}{2}(1-\cos\beta) & \frac{\sin\beta}{\sqrt{2}} & \frac{1}{2}(1+\cos\beta) \end{pmatrix}.$$

The two elements $\pm u(z,\alpha)$ are represented by the same matrix $Diag\left(e^{-i\alpha}, 1, e^{i\alpha}\right)$ in this even representation of SU(2), though we know (and should always bear in mind) that $u(z,\alpha+2\pi) = -u(z,\alpha)$ while $u(z,\alpha+4\pi) = u(z,\alpha)$.

4. Calculation of the Homotopy Groups

There are two possible ground states in $F = 1$ case. For the ferromagnetic state, the isotropy group H is constructed by the set of transformations which leave the reference order parameter $(1,0,0)^T$ invariant. From the degenerate family of the ground state spinor eq.(4) we know immediately that the angles should satisfy

$$\beta = 0, \theta - \alpha - \gamma = 2n\pi \tag{20}$$

with n an integer. The elements in group H are the combination of a translational part and a rotational part $H = \{(\theta, u(z,\theta)), (\theta, u(z,\theta+2\pi))\} = \{(\theta, \pm u(z,\theta))\}$. Evidently this group includes two disconnected components—the connected component of the identity $H_0 = \{(\theta, u(z,\theta))\}$ is isomorphic to R. The group H/H_0 is isomorphic to the integers modulo 2, i.e., Z_2. The second homotopy group π_2 is trivial and we arrive at the same result as that in Ref. [12]

$$\pi_1(M) = Z_2, \qquad \pi_2(M) = 0, \text{ (spin-1 FM state).} \tag{21}$$

A ferromagnetic spin-1 condensate may have therefore only singular vortices with winding number one while the point-like defects are topologically unstable. Alternatively we may take the symmetry group as SU(2) because we can produce all possible gauge transformations by absorbing θ into the Euler angle γ. The isotropy group is discrete and isomorphic to Z_2, which gives exactly the same result.

The polar state emerges if the atoms in the condensate interact anti-ferromagnetically. In the ground state eq. (3), the reference parameter $(0,1,0)^T$ is left invariant for just those elements with

$$\beta = 0, \theta = 2n\pi \quad \text{or} \quad \beta = \pi, \theta = (2n+1)\,\pi. \tag{22}$$

Thus the isotropy group H includes now the transformations in which both the rotation and the translation leave the spinor unchanged, and those in which the rotation takes the reference spinor $(0,1,0)^T$ to $(0,-1,0)^T$ and the translation takes it back, i.e., a π rotation about arbitrary axis perpendicular to $\hat{z}$ combined with a π translation in θ (or any odd multiples of π). The latter invariance is identical to the Ising gauge symmetry emphasized in eq. (14) of Ref. [20]. The full isotropy group is the union of these two sets, $H = \{(2n\pi, u(z,\alpha)), ((2n+1)\pi, gu(z,\alpha))\}$ where $g = u(y,\pi)$. There are infinitely many discrete components in H, while the connected component of the identity $H_0 = \{(0, u(z,\alpha))\}$ is isomorphic to $\mathrm{U}(1)$. The elements with an even translational parity are of the form $(2n\pi, I)H_0$, and those with an odd parity are of the form $((2n+1)\pi, g)\,H_0$. The group H/H_0 is therefore isomorphic to the group of integers Z through the isomorphism $\left((2n+j)\pi, g^j\right) H_0 \mapsto 2n+j$ for $j = 0, 1$. We recover the conclusion that line and point defects in spin-1 polar state can be classified by integer winding numbers,

$$\pi_1(M) = Z, \qquad \pi_2(M) = Z, \text{(spin-1 Polar state)}. \tag{23}$$

Thus the Z_2 term does not appear in the homotopy group. We argue that the identification of the OPS in Ref. [19, 20] is also incorrect (see below). Physically there are indeed infinite number of line defects corresponding to integer and half-integer vortices (eq. (27) in Ref. [20]). On the other hand, it is the Ising symmetry that leads to half-vortices ($j = 1$), which have been shown to be the unique linear defects in polar condensate in addition to the usual integer vortices ($j = 0$) [18]. If we move around a closed path in the condensate we note that when we return to the starting point the angle θ has changed by some amount. If we define the change in this angle divided by 2π to be the winding number, we see from the elements of H/H_0 that the winding number can be either an integer n or a half-integer $n + 1/2$.

5. Spin-2 Bose Condensate

We next apply the same approach to the BEC of spin-2 bosons. The defects which may be created in spin-2 condensate exhibit even more elaborate structures due to quantum correlations among bosons. For $F = 2$ we have to use the 5D representation of $SU(2)$, e.g., the rotation $u(z,\alpha)$ is represented by matrix $Diag\left(e^{-2i\alpha}, e^{-i\alpha}, 1, e^{i\alpha}, e^{2i\alpha}\right)$ and $u(y,\beta)$ takes the form of

$$\begin{pmatrix} \cos^4\frac{\beta}{2} & -\sin\beta\cos^2\frac{\beta}{2} & \frac{\sqrt{6}}{4}\sin^2\beta & -\sin\beta\sin^2\frac{\beta}{2} & \sin^4\frac{\beta}{2} \\ \sin\beta\cos^2\frac{\beta}{2} & \frac{\cos\beta+\cos2\beta}{2} & -\frac{\sqrt{6}}{4}\sin2\beta & \frac{\cos\beta-\cos2\beta}{2} & -\sin\beta\sin^2\frac{\beta}{2} \\ \frac{\sqrt{6}}{4}\sin^2\beta & \frac{\sqrt{6}}{4}\sin2\beta & \frac{1+3\cos2\beta}{4} & -\frac{\sqrt{6}}{4}\sin2\beta & \frac{\sqrt{6}}{4}\sin^2\beta \\ \sin\beta\sin^2\frac{\beta}{2} & \frac{\cos\beta-\cos2\beta}{2} & \frac{\sqrt{6}}{4}\sin2\beta & \frac{\cos\beta+\cos2\beta}{2} & -\sin\beta\cos^2\frac{\beta}{2} \\ \sin^4\frac{\beta}{2} & \sin\beta\sin^2\frac{\beta}{2} & \frac{\sqrt{6}}{4}\sin^2\beta & \sin\beta\cos^2\frac{\beta}{2} & \cos^4\frac{\beta}{2} \end{pmatrix}$$

The calculations of the degenerate family of the ground state spinors and the corresponding homotopy groups are straightforward and some results have been reported in [47]. Here we pick up some interesting features in our results, focusing on the symmetry properties of the defects in comparison with those in other ordered media. We first consider the defects in the absence of an external field and the effect of magnetic field will be discussed later.

We start with the case of the ferromagnetic state F. Equating the general expression for the ground state spinor

$$\zeta = e^{i(\theta-2\gamma)} \begin{pmatrix} e^{-2i\alpha}\cos^4\frac{\beta}{2} \\ e^{-i\alpha}\sin\beta\cos^2\frac{\beta}{2} \\ \frac{\sqrt{6}}{4}\sin^2\beta \\ e^{i\alpha}\sin\beta\sin^2\frac{\beta}{2} \\ e^{2i\alpha}\sin^4\frac{\beta}{2} \end{pmatrix} \tag{24}$$

with the reference spinor $(1,0,0,0,0)^T$ leads to the requirement for the isotropy group H

$$\beta = 0, \theta - 2\alpha - 2\gamma = 2n\pi. \tag{25}$$

We see that taking $n = 0,1,2,3$ is enough for all possible transformations, with the translational part being arbitrary and the rotational part containing the rotations around axis z by $\theta/2, \theta/2+\pi, \theta/2+2\pi, \theta/2+3\pi$ respectively. Hence the group H is composed of four pieces $H = \{(\theta, u(z, n\pi+\theta/2))\}$. Here it is important to show that the four components are not connected: there does not exist a continuous path in H which connects one component to another, though the rotational parts themselves are connected. The connected component of the identity $H_0 = \{(\theta, u(z,\theta/2))\}$ is again isomorphic to R. If we define an element g of the group $R\times \mathrm{SU}(2)$ by $(0, u(z,\pi))$, we see that the quotient group H/H_0 has the same structure as the cyclic group of order 4, i.e., $\{e, g, g^2, g^3\}$ and we conclude that

$$\pi_1(M) = Z_4, \qquad \pi_2(M) = 0, \text{ (spin-2 } F \text{ state)}. \tag{26}$$

It is interesting to check how the group Z_4 characterizes vortices for state F. In spin-1 case there is only one topologically stable line defect, that is, a vortex with winding number one. Equation (26) shows that there are three stable vortices for spin-2 condensates. We can set $\theta - 2\gamma = 2m\varphi$, $-\alpha = m\varphi$, $\beta = \pi t$ in the ground state for F state, Eq. (24), which leads to a family of spinor states parametrized by a parameter t between 0 and 1. Here $m > 0$ is an integer, φ is the azimuthal angle. When t evolves from 0 to 1, the $4m\varphi$ vortex state $\zeta(t=0) = \left(e^{i4m\varphi},0,0,0,0\right)^T$ evolves continuously to the vortex free state $\zeta(t=1) = (0,0,0,0,1)^T$. This shows that vortices with winding number $4m$ are topologically unstable. Similarly, by multiplying factors $e^{ik\varphi}(k=1,2,3)$ one obtains the following correspondences

$$e^{i(4m+k)\varphi}(1,0,0,0,0)^T \to e^{ik\varphi}(0,0,0,0,1)^T \tag{27}$$

i.e., the vortices with winding numbers $4m+k$ may evolve into vortices with winding numbers k, respectively. There are thus three classes of topologically stable line defects.

Together with the uniform state, they form the fundamental group Z_4. Non-trivial vortices are those in which the reference spinor rotates through 180 °, 360° or 540° about the z-axis when the defect line is circulated. Straightforwardly for ferromagnetic condensates with spin F, the fundamental group $\pi_1(M) = Z_{2F}$ characterizes $(2F-1)$ classes of stable line defects.

Spin variations in the ferromagnetic states in general lead to superflows [3, 25]. To illustrate the coreless (or Skyrmion) vortices in spin-2 case, we set $\theta - 2\tau = 2\varphi, \alpha = \varphi$ in the spinor degenerate family (24) and consider the condensate

$$\zeta(\mathbf{r}) = \begin{pmatrix} \cos^4 \frac{\beta}{2} \\ e^{i\varphi} \sin\beta \cos^2 \frac{\beta}{2} \\ e^{i2\varphi} \frac{\sqrt{6}}{4} \sin^2 \beta \\ e^{i3\varphi} \sin\beta \sin^2 \frac{\beta}{2} \\ e^{i4\varphi} \sin^4 \frac{\beta}{2} \end{pmatrix} \tag{28}$$

where $\beta = \beta(r)$ is an increasing function of r starting from $\beta = 0$ at $r = 0$.The superfluid velocity does not depend on z and it is cylindrically symmetric

$$\mathbf{v}_s = \frac{\hbar}{M}\left[2\nabla\varphi - 2\cos\beta\nabla\varphi\right] = \frac{2\hbar}{Mr}\left(1-\cos\beta\right)\hat{\varphi} \tag{29}$$

i.e., the coreless vortex may exist in the spin-2 case, with only the velocity doubling its value compared to the spin-1 case [3]. The velocity vanishes instead of diverging at $r = 0$ because $\beta(0) = 0$. This is called a coreless vortex. For a Mermin-Ho vortex [45], the bending angle β must be $\pi/2$ at the boundary of the condensate, while for an Anderson-Toulouse [46] vortex β must be π, i.e.

$$\begin{aligned} \beta(R) &= \pi/2, \text{ for Mermin-Ho} \\ \beta(R) &= \pi, \text{ for Anderson-Toulouse.} \end{aligned} \tag{30}$$

6. Non-Abelian Homotopy Groups

Media with non-Abelian fundamental groups are especially interesting from the topological point of view. The only illustrative example in ordered media so far have been biaxial nematic liquid crystals [48]. Their multiplication table has been verified experimentally [49].

We have found that the cyclic state C provides another physically realistic example in which the fundamental group is non-commutative. A rotation and a gauge transformation of the reference spinor $\frac{1}{2}\left(e^{i\phi}, 0, \sqrt{2}, 0, -e^{-i\phi}\right)^T$ in zero field produce the following degenerate family

$$\zeta = \frac{1}{2}e^{i\theta}\begin{pmatrix} e^{-2i\alpha}\left(\cos^4\frac{\beta}{2}e^{i\phi-2i\tau} + \frac{\sqrt{3}}{2}\sin^2\beta - \sin^4\frac{\beta}{2}e^{-i\phi+2i\tau}\right) \\ e^{-i\alpha}\sin\beta\left(\cos^2\frac{\beta}{2}e^{i\phi-2i\tau} - \sqrt{3}\cos\beta + \sin^2\frac{\beta}{2}e^{-i\phi+2i\tau}\right) \\ \frac{\sqrt{6}}{4}\sin^2\beta e^{i\phi-2i\tau} + \frac{\sqrt{2}}{4}(1+3\cos 2\beta) - \frac{\sqrt{6}}{4}\sin^2\beta e^{-i\phi+2i\tau} \\ e^{i\alpha}\sin\beta\left(\sin^2\frac{\beta}{2}e^{i\phi-2i\tau} + \sqrt{3}\cos\beta + \cos^2\frac{\beta}{2}e^{-i\phi+2i\tau}\right) \\ e^{2i\alpha}\left(\sin^4\frac{\beta}{2}e^{i\phi-2i\tau} + \frac{\sqrt{3}}{2}\sin^2\beta - \cos^4\frac{\beta}{2}e^{-i\phi+2i\tau}\right) \end{pmatrix}$$

The reference spinor is left invariant by the elements of three sets characterized by the translations in the phase of the condensate θ :

- For $\theta = 2n\pi$, one must have $\beta = 0, \alpha + \tau = m\pi$, or $\beta = \pi, \alpha - \tau = -\phi + \frac{\pi}{2} + m\pi$;
- For $\theta = \frac{2\pi}{3} + 2n\pi$, one must have $\beta = \frac{\pi}{2}, \alpha + \tau = -\frac{\pi}{2} + m\pi, \alpha - \tau = -\phi + m'\pi$;
- For $\theta = \frac{4\pi}{3} + 2n\pi$, one must have $\beta = \frac{\pi}{2}, \alpha + \tau = \frac{\pi}{2} + m\pi, \alpha - \tau = -\phi + m'\pi$.

Here m and m' are integers satisfying $m + m' = odd$. For all possible transformations we need take the values $m, m' = 0, 1, 2, 3$ so that there are eight possibilities

$$\begin{aligned} m &= 0, m' = 1, 3 \\ m &= 1, m' = 0, 2 \\ m &= 2, m' = 1, 3 \\ m &= 3, m' = 0, 2 \end{aligned}$$

This gives the isotropy group

$$\begin{aligned} H = \{&\pm I, \pm a, \pm b, \pm c, \\ &\pm d, \pm e, \pm f, \pm g, \\ &\pm d^2, \pm e^2, \pm f^2, \pm g^2\}. \end{aligned} \tag{31}$$

The spin rotations $a = u(z, \pi), b = u(y, \pi)u(z, \phi + \pi/2)$ and $c = ba$ satisfy $a^2 = b^2 = c^2 = -I$, while $d = u(z, \pi/4 + \phi/2)u(y, \pi/2)u(z, \pi/4 - \phi/2)$, $e = -da$, $f = -ad$ and $g = -ada$ satisfy $d^3 = e^3 = f^3 = g^3 = -I$. Each element in the first, second, third row is associated with an additional phase change $2n\pi, 2\pi/3 + 2n\pi, 4\pi/3 + 2n\pi$ respectively. It is a discrete group, and H_0 consists of the identity $(0, I)$ only. The fundamental theorems identify that

$$\pi_1(M) = H, \qquad \pi_2(M) = 0, \text{ (spin-2 } C \text{ state).} \tag{32}$$

The elements in the fundamental group are non-commutative, for example $ab = -c \neq ba$.

The criterion for the topological equivalence of defects applies in the most general case in terms of conjugacy classes of the fundamental group. Two line defects are topologically equivalent if and only if they are characterized by the same conjugacy class. Defects can still be labelled by the elements of the first homotopy group, but if these elements belong to the same conjugacy class, corresponding defects can be continuously transformed to one another. However, if they belong to different conjugacy classes this is not possible. It is thus necessary to classify the group into the following conjugacy classes:

$$\begin{aligned} C_0(n) &= \{I\}_n, \quad \overline{C_0}(n) = \{-I\}_n, \quad C_2(n) = \{\pm a, \pm b, \pm c\}_n, \\ C_3(n+1/3) &= \{d, e, f, g\}_{n+1/3}, \quad \overline{C_3}(n+1/3) = \{-d, -e, -f, -g\}_{n+1/3}, \\ C_3^2(n+2/3) &= \{d^2, e^2, f^2, g^2\}_{n+2/3}, \quad \overline{C_3^2}(n+2/3) = \{-d^2, -e^2, -f^2, -g^2\}_{n+2/3} \end{aligned}$$

with the subscripts standing for the winding numbers of the defects. Physically this indicates the feasibility of creating not only vortices with any integer winding number but also

Table 1. The multiplication table of the conjugacy classes of the cyclic phase.

	$\overline{C_0}$	C_2	C_3	$\overline{C_3}$	C_3^2	$\overline{C_3^2}$
$\overline{C_0}$	C_0					
C_2	C_2	$6C_0 + 6\overline{C_0} + 4C_2$				
C_3	$\overline{C_3}$	$3\left(C_3 + \overline{C_3}\right)$	$3\overline{C_3^2} + C_3^2$			
$\overline{C_3}$	C_3	$3\left(C_3 + \overline{C_3}\right)$	$\overline{C_3^2} + 3C_3^2$	$3\overline{C_3^2} + C_3^2$		
C_3^2	$\overline{C_3^2}$	$3\left(C_3^2 + \overline{C_3^2}\right)$	$4\overline{C_0} + 2C_2$	$4C_0 + 2C_2$	$3C_3 + \overline{C_3}$	
$\overline{C_3^2}$	C_3^2	$3\left(C_3^2 + \overline{C_3^2}\right)$	$4C_0 + 2C_2$	$4\overline{C_0} + 2C_2$	$C_3 + 3\overline{C_3}$	$3C_3 + \overline{C_3}$

fractional quantum vortices. The class $C_0(n)$ describes defects in which the phase of the spinor is changed by $2\pi n$ as the defect line is encircled. Note that only $C_0(0)$ corresponds to trivial defects. In the case of $\overline{C_0}(n)$ phase change of $2\pi n$ is accompanied by a $360°$ rotation about z-axis. The element a with winding number n in the class $C_2(n)$ depicts a defect in which the spinor rotates through $180°$ about the z-axis and changes phase by $2\pi n$ as the line is encircled. The multiplication table of conjugacy classes is shown in table 1. Only half of table is shown because the class multiplication is commutative. Winding numbers have been omitted for clarity. When two classes are multiplied the winding number of the resulting class is the sum of the individual winding numbers. It shows that, for example, when we combine defect $C_2(n)$ with $C_2(-n)$ they can either annihilate each other ($C_0(0)$) or form defect $\overline{C_0}(0)$ or $C_2(0)$, the result depending on how they are brought together. Interesting features of this non-Abelian fundamental group include the topological instability of the defects and their interaction, i.e., entanglement when two of them are brought to cross with each other [44].

The defects can be further grouped into classes, which form an Abelian group isomorphic to the first homology group of the order parameter space [50]. This coarser classification is more general than the homotopic one because two defects are considered equivalent also, if they can be transformed into each other via a *catalyzation* process consisting of splitting a line singularity into two and recombining them beyond a third one. All elements labelled by elements of the commutator subgroup D of $\pi_1(M)$ can be catalyzed away by this procedure. D is generated by the commutators $\delta\tau\delta^{-1}\tau^{-1}$ of all pairs of elements $\delta, \tau \in \pi_1(M)$. The elements of $\pi_1(M)/D$ are unions of conjugacy classes. In our case D is the union of the conjugacy classes with winding number zero, $D = C_0(0) \cup \overline{C_0}(0) \cup C_2(0)$ and the first homology group is

$$\pi_1(M)/D = \left\{C_0 \cup \overline{C_0} \cup C_2, C_3 \cup \overline{C_3}, C_3^2 \cup \overline{C_3^2}\right\} \tag{33}$$

The homology theory assembles the conjugacy classes further into three sets for each n, in which the defects are labeled by the winding numbers $n, n + 1/3, n + 2/3$ respectively [47]. Two defects in the same conjugacy classes can be continuously converted into one another by local surgery, while two defects in the same homology class can be deformed into one another by the *catalyzation* procedure.

Table 2. Comparison of the OPS and fundamental groups for spin-1 polar condensate

	OPS	$\pi_1(M)$
Ho, Stoof, etc.	$\mathrm{U}(1) \times S^2$	Z
Zhou	$\mathrm{U}(1) \times S^2/Z_2$	$Z \times Z_2$
This paper	$(\mathrm{U}(1) \times \mathrm{SO}(3))/\mathrm{O}(2)$	Z

7. Order Parameter Spaces

Like the quaternion group Q for biaxial nematics, the fundamental group (31) is the *lift* of a point group in $R \times \mathrm{SU}(2)$. To find the remaining discrete symmetry group for the cyclic state, and, in addition, to clarify the controversial identification of the OPS for spin-1 case, in the remaining of this paper we turn to describe the system in terms of rotations in $\mathrm{SO}(3)$, e.g., two elements $\pm u(z,\alpha)$ in $\mathrm{SU}(2)$ are mapped into one $R(z,\alpha)$ in $\mathrm{SO}(3)$ with $R(z,\alpha+2\pi) = R(z,\alpha)$.

The OPS for $F=1$ polar state was identified as $\mathrm{U}(1) \times S^2$ in Refs. [3, 12]. An extra Z_2 symmetry was claimed in Ref. [19] so the author concluded the OPS as $\mathrm{U}(1) \times S^2/Z_2$. Here we show that previous studies are incorrect. Taking the group G as $\mathrm{U}(1)_G \times \mathrm{SO}(3)_S$ where the subscripts stand for the gauge and spin symmetries respectively, we see that the isotropy group H consists of two separate parts, $\left\{\left(e^{i0}, R(z,\alpha)\right)\right\}$ and $\left\{\left(e^{i\pi}, R(y,\pi)R(z,\alpha)\right)\right\}$. The rotations in the first part constitute the group $\mathrm{SO}(2)$, while the elements in the second part are just those in the group $\mathrm{O}(2)$ but not in $\mathrm{SO}(2)$ with determinants -1. The combination of these two parts gives the full isotropy group as $\mathrm{O}(2)$ where both gauge and spin symmetries are involved. The OPS is the quotient $G/H = \left(\mathrm{U}(1)_G \times \mathrm{SO}(3)_S\right)/\mathrm{O}(2)_{G+S}$ and here it is not possible to apply the fundamental theorem for G is not any more simply connected. One may wonder if we can factorize the OPS further as

$$
\begin{aligned}
G/H &= \left(\mathrm{U}(1)_G \times \mathrm{SO}(3)_S\right)/\mathrm{O}(2)_{G+S} \\
&= \left(\mathrm{U}(1)_G \times \mathrm{SO}(3)_S\right)/(\mathrm{SO}(2) \times \mathrm{Z}_2)_{G+S} \\
&= \mathrm{U}(1) \times S^2/Z_2
\end{aligned}
$$

However it is incorrect because though in 3 dimensional space we have $\mathrm{O}(3) = \mathrm{SO}(3) \times \mathrm{Z}_2$ but it is not true in 2 dimensional case, i.e. $\mathrm{O}(2) \neq \mathrm{SO}(2) \times \mathrm{Z}_2$. The spin and gauge symmetries are broken in a connected fashion just as in the system of ^{3}He [10, 11]. Table 2 summarizes our result in comparison with the previous studies.

For the ferromagnetic state the group H may be obtained if one notices that the 2π difference in the rotational angle does not give another component as it did in the case of $\mathrm{SU}(2)$. We have $H = \left\{\left(e^{i\theta}, R(z,\theta)\right)\right\}$ which is isomorphic to $\mathrm{U}(1)_{G+S}$. This means that there is a remaining symmetry $\mathrm{U}(1)$ in the symmetry broken system. The OPS is thus factorized as $\left(\mathrm{U}(1)_G \times \mathrm{SO}(3)_S\right)/\mathrm{U}(1)_{G+S} = \mathrm{SO}(3)_{S+G}$.

The discrete symmetry group of defects in the spin-2 cyclic state C can be shown to be isomorphic to the tetrahedral group T. We continue to represent G as $\mathrm{U}(1)_G \times \mathrm{SO}(3)_S$. The isotropy group (31) is shrunk to a group of 12 elements if one understands the rotation in the sense of $\mathrm{SO}(3)$ (i.e., $a = R(z,\pi)$),

$$
H = \{I, a, b, c, \varepsilon d, \varepsilon e, \varepsilon f, \varepsilon g, \varepsilon^2 d^2, \varepsilon^2 e^2, \varepsilon^2 f^2, \varepsilon^2 g^2\}, \tag{34}
$$

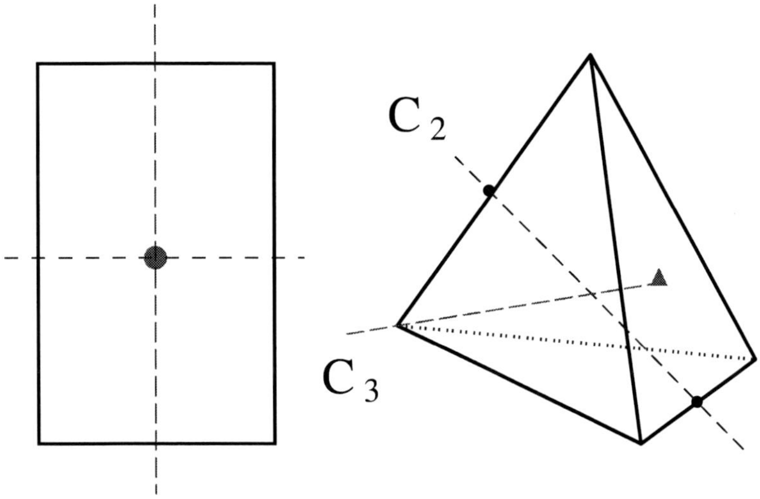

Figure 2. Symmetries of the defects in biaxial nematics (D_2) and cyclic state C in spin-2 condensate (T). The dot at the center of the rectangle stands for axis z. The dashed lines represent 2-fold axes, except that with a triangle for 3-fold axis.

where $\varepsilon = \exp(2\pi i/3)$ comes from the gauge transformation and εd, for instance, is an abbreviation for the element (ε, d). Three 2-fold rotational axes are z and 2 lines in xy plane perpendicular to each other (which lie on axes x and y if we choose the arbitrary phase $\phi = \pi/2$). The elements $\varepsilon d, \varepsilon e, \varepsilon f, \varepsilon g$ are four 3-fold axes. The symmetries remaining in the symmetry broken states for biaxial nematics and spin-2 cyclic state are shown in Figure 2. The OPS for state C can be identified as $\left(\mathrm{U}(1)_G \times \mathrm{SO}(3)_S\right)/T_{G+S}$.

It should be noted that an applied magnetic field B changes the defect structure severely by reducing the degenerate family of the spinor. We take again the cyclic state as an example. The symmetry group in this case is $\mathrm{U}(1) \times \mathrm{SO}(2)$ because the magnetic field chooses its direction automatically as the quantization axis. From the spinor

$$\frac{1}{2}\begin{pmatrix} (1+p)e^{i\phi_1} \\ 0 \\ \sqrt{2-2p^2}e^{i(\phi_1+\phi_2)/2} \\ 0 \\ (-1+p)e^{i\phi_2} \end{pmatrix}$$

where $\phi_{1,2}$ are two arbitrary phases and $p \sim B$, we easily see the possibility to create vortices in any of the three nonzero components with winding number for i−th component n_i confined by $n_1 + n_5 = 2n_3$.

8. Summary

Our main findings are summarized in Table 3. We have determined the nature of the topological defects in spin-1 and spin-2 condensates. The order parameter spaces are identified as the spaces of the coset of the isotropy group H in the transformation group G. Topologically stable vortices with winding numbers larger than unity may be created in the

Table 3. Main results on calculation of the OPS and homotopy groups

	OPS	π_1	π_2
Spin-1 FM	SO(3)	Z_2	0
Spin-1 AFM	$(\mathrm{U}(1) \times \mathrm{SO}(3))/\mathrm{O}(2)$	Z	Z
Spin-2 F	$\mathrm{SO}(3)/Z_2$	Z_4	0
Spin-2 F'	SO(3)	Z_2	0
Spin-2 C	$(\mathrm{U}(1) \times \mathrm{SO}(3))/T$	H eq.(31)	0

ferromagnetic state for condensates with $F > 1$, up to the value $(2F - 1)$. The line defects in the spin-2 cyclic state C exhibit non-commutative features, resulting e.g. in line defects with winding numbers of $1/3$ and its multiples. It also turns out that in the zero field $\mathrm{U}(1) \times \mathrm{SO}(3)$ does not act transitively on the order-parameter space of the polar phase and thus the defect structure remains unsolved.

Acknowledgements

The authors acknowledge the Academy of Finland (Project No 206108) for financial support. YZ is also supported by NSF of China (Grant Nos. 10175039 and 90203007). Discussions with O. Kärki and J. Hietarinta are appreciated.

References

[1] D. M. Stamper-Kurn, M. R. Andrews, A. P. Chikkatur, S. Inouye, H.-J. Miesner, J. Stenger, and W. Ketterle, *Phys. Rev. Lett.* **80**, 2027 (1998).

[2] M. D. Barrett, J. Sauer, and M. S. Chapman, *Phys. Rev. Lett.* **87**, 010404 (2001).

[3] T.-L. Ho, *Phys. Rev. Lett.* **81**, 742 (1998).

[4] T. Ohmi and K. Machida, *J. Phys. Soc. Jpn.* **67**, 1822 (1998).

[5] M. S. Chang, C. D. Hamley, M. D. Barrett, J. A. Sauer, K. M. Fortier, W. Zhang, L. You, and M. S. Chapman, cond-mat/0309164, to appear in *Phys. Rev. Lett.*

[6] J. Stenger, S. Inouye, D. M. Stamper-Kurn, H.-J. Miesner, A. P. Chikkatur, and W. Ketterle, *Nature* **396**, 345 (1998).

[7] S. L. Cornish, N. R. Claussen, J. L. Roberts, E. A. Cornell, and C. E. Wieman, *Phys. Rev. Lett.* **85**, 1795 (2000).

[8] T. Weber, J. Herbig, M. Mark, H.-C. Nägerl, and R. Grimm , *Science* **299**, 232 (2003).

[9] K.-A. Suominen, E. Tiesinga, and P. S. Julienne, *Phys. Rev. A* **58**, 3983 (1998).

[10] D. Vollhardt and P. Wölfle, *The superfluid phases of Helium 3* (Taylor & Francis, London, 1990).

[11] M. M. Salomaa and G. E. Volovik, *Rev. Mod. Phys.* **59**, 533 (1987).

[12] U. Al Khawaja and H. T. C. Stoof, Phys. Rev. A **64**, 043612 (2001); *Nature (London)* **411**, 918 (2001).

[13] H. T. C. Stoof, E. Vliegen, and U. Al Khawaja, *Phys. Rev. Lett.* **87**, 120407 (2001).

[14] T. Mizushima, K. Machida, and T. Kita, Phys. Rev. Lett. **89**, 030401 (2002); *Phys. Rev. A* **66**, 053610 (2002).

[15] J.-P. Martikainen, A. Collin, and K.-A. Suominen, *Phys. Rev. Lett.* **88**, 090404 (2002).

[16] J.-P. Martikainen, A. Collin, and K.-A. Suominen, *Phys. Rev. A* **66**, 053604 (2002).

[17] S.-K. Yip, *Phys. Rev. Lett.* **83**, 4677 (1999).

[18] V. Leonhardt and G. E. Volovik, *JETP Lett.* **72**, 46 (2000).

[19] F. Zhou, *Phys. Rev. Lett.* **87**, 080401 (2001).

[20] F. Zhou, *Int. J. Mod. Phys. B* **17**, 2643 (2003), cond-mat/0108473.

[21] A. E. Leanhardt, Y. Shin, D. Kielpinski, D. E. Pritchard, and W. Ketterle, *Phys. Rev. Lett.* **90**, 140403 (2003).

[22] C. V. Ciobanu, S.-K. Yip, and T.-L. Ho, *Phys. Rev. A* **61**, 033607 (2000).

[23] M. Koashi and M. Ueda, *Phys. Rev. Lett.* **84**, 1066 (2000).

[24] M. Ueda and M. Koashi, *Phys. Rev. A* **65**, 063602 (2002).

[25] J.-P. Martikainen and K.-A. Suominen, *J. Phys. B: At. Mol. Opt. Phys.* **34**, 4091 (2001).

[26] N. N. Klausen, J. L. Bohn, and C. H. Greene, *Phys. Rev. A* **64**, 053602 (2001).

[27] T.-L. Ho and L. Yin, *Phys. Rev. Lett.* **84**, 2302 (2000).

[28] C. K. Law, H. Pu, and N. P. Bigelow, *Phys. Rev. Lett.* **81**, 5257 (1998).

[29] H. Pu, C. K. Law, S. Raghavan, J. H. Eberly, and N. P. Bigelow, *Phys. Rev. A* **60**, 1463 (1999).

[30] T.-L. Ho and S. K. Yip, *Phys. Rev. Lett.* **84**, 4031 (2000).

[31] H.-J. Miesner, D.M. Stamper-Kurn, J. Stenger, S. Inouye, A.P. Chikkatur, and W. Ketterle, *Phys. Rev. Lett.* **82**, 2228 (1999).

[32] D.M. Stamper-Kurn, H.-J. Miesner, A.P. Chikkatur, S. Inouye, J. Stenger, and W. Ketterle, *Phys. Rev. Lett.* **83**, 661 (1999).

[33] D. M. Stamper-Kurn and W. Ketterle, Spinor Condensates and Light Scattering from Bose-Einstein Condensates, *Proc. Les Houches 1999 Summer School, Session LXXII*, (Springer, New York, 2001), cond-mat/0005001.

[34] G. K. Woodgate, *Elementary Atomic Sturcture* (Mcgraw-Hill, London/New York).

[35] W.-X. Zhang, S. Yi, and L. You, *New Journal of Physics* **5**, 77.1-77.12 (2003).

[36] A. Görlitz, T. L. Gustavson, A. E. Leanhardt, R. Löw, A. P. Chikkatur, S. Gupta, S. Inouye, D. E. Pritchard, and W. Ketterle, *Phys. Rev. Lett.* **90**, 090401 (2003).

[37] H. Schmaljohann, M. Erhard, J. Kronjäger, M. Kotte, S. van Staa, L. Cacciapuoti, J. J. Arlt, K. Bongs, and K. Sengstock, *Phys. Rev. Lett.* **92**, 040402 (2004).

[38] M. Kléman, L. Michel, and G. Toulouse, *J. Physique Lett.* **38,** L195 (1977).

[39] L. Michel, Rev. Mod. Phys. **52**, 617 (1980).

[40] G. Toulouse and M. Kléman, *J. Phys. (Paris) Lett.* **37,** L149 (1976).

[41] V. Poénaru, *Ill-condensed Matter (Les Houches, Session XXXI)*, eds. R. Balian, R. Maynard and G. Toulouse (North-Holland, Amsterdam, 1979), p. 263

[42] G. E. Volovik and V. P. Mineev, JETP Lett. **24** 561 (1976); JETP **72**, 2256 (1977).

[43] V. P. Mineev, *Topologically Stable Defects and Solitons in Ordered Media* (Harwood Academic, Amsterdam, 1988).

[44] N. D. Mermin, *Rev. Mod. Phys.* **51**, 591 (1979).

[45] N. D. Mermin and T.-L. Ho, *Phys. Rev. Lett.* **36**, 594 (1976).

[46] P. W. Anderson and G. Toulouse, *Phys. Rev. Lett.* **38**, 508 (1977).

[47] H. Mäkelä, Y. Zhang and K.-A. Suominen, *J. Phys. A: Math. Gen.* **36**, 8555 (2003).

[48] L. J. Yu and A. Saupe, *Phys. Rev. Lett.* **45**, 1000 (1980).

[49] T. De'Neve, M. Kléman and P. Navard, *J. Phys. II France* **2**, 187 (1992).

[50] H.-R. Trebin, Adv. Phys. **31**, 195 (1982); *Phys. Rev. B.* **30**, 4338 (1984).

In: Astrophysics and Condensed Matter
Editor: Thomas G. Hardwell, pp. 159-173

ISBN: 978-1-60456-332-0

Chapter 5

MUON COLLIDERS AND HIGGS BOSON PHYSICS

*M. S. Berger**
Physics Department, Indiana University
Bloomington, Indiana 47405

Abstract

The technology necessary for constructing muon colliders is currently under investigation around the world. These machines would employ muon beams rather than electron beams, and would require new cooling and acceleration methods to succeed. If the technical obstacles to construction of these novel colliders can be overcome, one can envision the possibility that muon colliders will vastly extend for decades to come the energy frontier that can be probed experimentally in particle physics.

Muon colliders can also play a unique role in determining the properties of Higgs bosons and fully understanding the Higgs sector. The much larger mass of the muon compared to the mass of the electron means the s-channel production rates are much larger at a muon collider. A muon collider with energy tuned to the mass of a Higgs boson would offer possibilities in measuring parameters of the Higgs boson that is highly complementary to Higgs studies at other planned colliders such as the Large Hadron Collider (LHC) or a future Linear Collider (LC). In this article, I detail the precision measurements of the Higgs boson mass and total width, and probe the ability of the muon collider to distinguish between the Standard Model Higgs boson and the lightest Higgs boson of supersymmetric extensions. The ability to determine the properties of a general extended Higgs sector, possibly with CP-violating couplings, is explored. The detection and the mass separation of heavy supersymmetric Higgs bosons is described. These supersymmetric Higgs bosons satisfy mass sum rules and I analyze the attractive possibility that very precise tests of radiative corrections in the Higgs sector can be performed. Finally the method of performing threshold scans of ZH production at a muon collider to determine the mass and spin of a Higgs boson is described.

1 Introduction

The discovery of a narrow state known as the Higgs boson, which couples to the Standard Model particles proportional to their mass, has become increasingly likely in the last decade.

*E-mail address: berger@indiana.edu

The electroweak precision measurements allow us to constrain the unknown physics by considering their action as virtual particles in loop diagrams in the Standard Model. These precision studies are consistent the existence of such a state just beyond experimental reach. If a Higgs boson is discovered, then it will be imperative to consider the best possible way to study its properties. Already the potential of various future collider technologies are being considered and studied in this context.

The next significant step in extending our experimental reach to higher center-of-mass energies will be provided by the CERN Large Hadron Collider (LHC) which is a proton-proton collider operating at 14 TeV. The constituent quarks of the proton will undergo effective interactions around 2 TeV. If a Standard Model-like Higgs boson exists, then the expectation is that it will be found after a year or two of running at the LHC. Following such a discovery, additional particle colliders will be necessary to study in detail the properties of the observed state. The most widely discussed future colliders are the electron-positron collider and the muon collider.

Electron-positron colliders have been a central tool of particle physicists for decades. The precision studies of the Z boson were conducted at CERN were performed with a electron-positron collider, and the majority of the precision measurements of the Standard Model have been performed at electron-positron colliders. The experiments at CERN concluded without finding convincing evidence for a Higgs boson. An obstacle to building electron-positron colliders of even higher energy is the radiative energy losses that are inversely proportional to the electron mass to the fourth power. To combat this effect the highest energy colliders are planned to be linear colliders. Clearly using a muon as a "heavy electron" will reduce the problem of radiative energy losses dramatically. Using muon beams one can accelerate to higher energies and use smaller storage rings[1].

Muons can be produced in large quantities from the decay of pions produced in proton-nucleus interactions. These muons occupy a large phase space volume which must be reduced by a process called cooling. The required amount of cooling would reduce the phase space volume by a factor 10^6 for a collider. Since muons decay very quickly one must employ a novel technique called ionization cooling. This technique is uniquely suited to muons because of their minimal interaction with matter.

Muon colliders offer some significant physics advantages. The small radiative losses permit very small beam-energy spreads to be achieved; for a low energy collider energy spreads as low as $R = \delta E/E = 0.003\%$ are possible. In addition the beam energy of a muon collider can be determined with extraordinary precision using the time-dependent decay asymmetry from the (naturally) polarized muons in the beam[2]. These two properties of a muon collider are ideal for measuring the masses and widths of any new resonances that can be produced in s-channel production. Furthermore a Higgs boson couples to particles proportional to their mass, so increased production rates in the s-channel are possible at a muon collider relative to an electron-positron collider. Finally the decaying muons are a source of large numbers of high-energy neutrinos[3, 4, 5]. In a straight section of a muon beam, the emitted neutrinos will effectively form a high luminosity neutrino beam whose flavor content is precisely known.

The physics potential of muon colliders is exciting. In this article the role of muon colliders in studying the Higgs sector is described.

2 Muon Colliders as Higgs Factories

Muon colliders may play a leading role in particle physics in the era following the Large Hadron Collider (LHC) and a future linear collider. In fact muon colliders may be the preferred route to extending the energy frontier. The muon mass is much larger than the electron mass, and the absence of significant bremsstrahlung means that a muon collider can be built with a circular acceleration scheme and with a circular storage ring. As mentioned in the introduction current designs for the next generation of electron-positron colliders require on the other hand that the electrons are accelerated in a linear fashion. There are therefore practical limitations on what energies can be obtained in future electron-positron colliders since increasing the energy necessarily requires increasing the size of the collider. Muon colliders may offer a path to ever higher energies and may be the necessary option to continue to probe shorter distance scales. The technology necessary for constructing a muon collider is an area of intense research[1].

In addition to this desire to extend our accelerators beyond the existing energy frontier, there are applications for muon colliders at energies that are attainable in hadron and electron-positron colliders. The large mass of the muon (it is about 200 times heavier than the electron) leads to the possibility of extremely narrow beam energy spreads. This may be narrow because the expected width of light Higgs bosons (h) might be quite small. This together with the fact that Higgs bosons (which couple to particles proportional to their mass) implies that a muon collider could be the ideal place to produce and study Higgs bosons in the s-channel production process. The rate for Higgs production at an electron-positron collider is too small to allow s-channel production as a feasible option, and current interest centers on using the Higgs-strahlung process $e^+e^- \to Zh$ or the WW fusion process as the production mechanisms. A Higgs boson with little or no coupling to the gauge bosons W and Z can still be produced at a muon collider provided it has Standard Model-like or enhanced couplings to muons[1]. In a muon collider the narrow energy spread and the enhanced coupling of the Higgs boson to muons allows the determination of the Higgs boson mass to very high precision[6, 7, 8, 9]. Since the production on a resonance has correspondingly large rates, a muon collider operating in the s-channel production mode is often called a "Higgs factory."

There is a trade-off between small beam energy spread and luminosity. Studies have shown that, even for small Higgs width, the s-channel production rate is maximized by choosing the energy spread $R = \delta E/E$ to be such that the spread δE around or even less than the small Higgs width. For a Standard Model Higgs boson, this means that $R \sim 0.003\%$ for a Higgs mass lighter than about 120 GeV.

3 Higgs Production

A linear collider can perform accurate measurements of Higgs parameters such as its mass, its couplings to gauge bosons, and its Yukawa couplings to heavy quarks. These tests form important tests of the symmetry breaking and mass generation mechanism in the Standard

[1]Important examples of Higgs bosons with these properties are the heavy Higgs bosons of the Minimal Supersymmetric Standard Model to be described in detail in a later section.

Model. The Standard Model should be viewed as the simplest parameterization of the electroweak symmetry breaking sector which consists of a single complex Higgs doublet. After spontaneous symmetry breaking, three of the four Higgs states are absorbed into the longitudinal polarizations of the heavy gauge bosons $W^{\pm}$ and Z, leaving a single physical state called the Higgs bosons. More complicated Higgs sectors will be characterized by multiple physical Higgs bosons, and these Higgs bosons will in general have properties that deviate from those expected for a Standard Model Higgs boson. If such deviations are found and measured sufficiently accurately, they can be used to constrain the parameter space of the extended Higgs sectors. It might then be conceivable that the masses of the heavier Higgs bosons could be estimated thereby allowing improved searches to be devised for them.

The production of a Higgs boson (generically denoted h) in the s-channel with interesting rates is a unique feature of a muon collider [6, 7]. The resonance cross section is

$$\sigma_h(\sqrt{s}) = \frac{4\pi\Gamma(h\to\mu\bar{\mu})\,\Gamma(h\to X)}{\left(s-m_h^2\right)^2 + m_h^2\left(\Gamma_{\rm tot}^h\right)^2}\,. \tag{1}$$

In practice, however, there is a Gaussian spread ($\sigma_{\sqrt{s}}$) to the center-of-mass energy and one must compute the effective s-channel Higgs cross section after convolution assuming some given central value of $\sqrt{s}$:

$$\begin{aligned} \bar{\sigma}_h(\sqrt{s}) &= \frac{1}{\sqrt{2\pi}\,\sigma_{\sqrt{s}}}\int \sigma_h(\sqrt{\hat{s}})\,\exp\left[\frac{-\left(\sqrt{\hat{s}}-\sqrt{s}\right)^2}{2\sigma_{\sqrt{s}}^2}\right] d\sqrt{\hat{s}} \\ &\overset{\sqrt{s}=m_h}{\simeq} \frac{4\pi}{m_h^2}\,\frac{\mathrm{BF}(h\to\mu\bar{\mu})\,\mathrm{BF}(h\to X)}{\left[1+\frac{8}{\pi}\left(\frac{\sigma_{\sqrt{s}}}{\Gamma_h^{\rm tot}}\right)^2\right]^{1/2}}\,. \end{aligned} \tag{2}$$

BF denotes the branching fraction for h decay; also, note that $\bar{\sigma}_h \propto 1/\sigma_{\sqrt{s}}$ for $\sigma_{\sqrt{s}} > \Gamma_{\rm tot}^h$. At $\sqrt{s}=m_h \approx 110$ GeV, the $b\bar{b}$ rates are

$$\text{signal} \approx 10^4 \text{ events} \times \mathrm{L}(\mathrm{fb}^{-1})\,, \tag{3}$$

$$\text{background} \approx 10^4 \text{ events} \times \mathrm{L}(\mathrm{fb}^{-1})\,, \tag{4}$$

assuming a b-tagging efficiency $\epsilon \sim 0.5$. It is convenient to express $\sigma_{\sqrt{s}}$ in terms of the root-mean-square (rms) Gaussian spread of the energy of an individual beam, R:

$$\sigma_{\sqrt{s}} = (2\ \mathrm{MeV})\left(\frac{R}{0.003\%}\right)\left(\frac{\sqrt{s}}{100\ \mathrm{GeV}}\right)\,. \tag{5}$$

It is clear from Eq. (1) that a resolution $\sigma_{\sqrt{s}} < \Gamma_h^{\rm tot}$ is needed to be sensitive to the Higgs width. Furthermore, Eq. (2) indicates that $\mathrm{BF}(h\to\mu\bar{\mu})$ must not be extremely suppressed for there to be large event rates for Higgs production. The width of a light Standard Model-like Higgs is very small (e.g. a few MeV for $m_{h_{SM}} \sim 110$ GeV), implying the need for R values as small as $\sim 0.003\%$ for studying a light Standard Model-like h. In addition to the very small beam energy spread, one must also be able to determine very accurately the beam energy to perform a scan over such a narrow resonance. This can be accomplished by utilizing the spin precession of the muon as noted above.

The significance of the signal from a scan over the Higgs resonance is impacted by two physical processes:

- For a Higgs mass near the Z-pole there is a significant background from $\mu^+\mu^- \to Z \to b\bar{b}$. However the most recent experimental results from LEP have pushed the Standard Model Higgs mass bound well above 91 GeV.

- For a Higgs mass $>$ 130 GeV, the Higgs width $\Gamma_h^{\rm tot}$ becomes much larger as the $WW^\star$ decay channel opens up.

The Higgs bosons in supersymmetric models are in general detectable at muon colliders. If the masses of the supersymmetric particles are large, the Higgs sector typically exhibits decoupling behavior in which the lightest supersymmetric Higgs boson will be very similar to the Standard Model Higgs boson when the other Higgs bosons are heavy, and the supersymmetric production rates will be very similar to the Standard Model rates. On the other hand, the heavier Higgs bosons in a typical supersymmetric model decouple from pairs of gauge bosons VV at large mass and remain reasonably narrow ($<$ 1 GeV unless the $t\bar{t}$ decay mode is open). As a result, their s-channel production rates remain large, and a muon collider can avoid the production channels that depend on a sizeable coupling to gauge bosons.

The LHC and a linear collider should be able to measure m_h and $\Gamma_h^{\rm tot}$ of a Standard Model-like h with mass sufficiently above M_Z and below the $WW^\star$ decay threshold. The mass should be determined to at least 100 MeV at the LHC and perhaps a factor 3 better at a linear collider (assuming an integrated luminosity of 500 fb^{-1}). The Higgs total width can be determined at a linear collider to something like 10-20%. The scenario envisioned for study of the Higgs boson at a muon collider is as follows:

- The Higgs boson will be discovered at the LHC and its mass determined to about $\Delta m_h \sim$ 100 MeV at a linear collider.

- A scan is performed over the threshold region with energies separated by 1-2 MeV with about $L \sim 1$ pb^{-1} of integrated luminosity per scan point. A muon collider ring design would be optimized to run at energy $\sqrt{s} = m_h$. For an initial Higgs-mass uncertainty of $\Delta m_h \sim$ 100 MeV, the maximum number of scan points required to locate the s-channel resonance peak at the muon collider is

$$n = \frac{2\Delta m_h}{\sigma_{\sqrt{s}}} \approx 100 \tag{6}$$

  for a $R = 0.003\%$ resolution of $\sigma_{\sqrt{s}} \approx 2$ MeV.

- A dedicated scan corresponding to a few years of operation involving three energies[6] near the resonance peak using $L \sim 0.2$ fb^{-1} of integrated luminosity.

For a Higgs boson sufficiently below the $WW^{\star}$ threshold, the total width $\Gamma_h^{\rm tot}$ is only a few MeV, and the only process that can directly measure it is s-channel production at a muon collider. At a linear collider the total width must be determined in an indirect way by measuring a partial width for some final state and the corresponding branching fraction, and then computing a total width from the ratio[10].

A linear collider becomes competitive if the Higgs boson mass is greater than 120 GeV when the $WW^{\star}$ decay rates dominate the decay width and can be measured. The usefulness of a muon collider in directly measuring the total decay width of a light Higgs boson therefore hinges on what the Higgs mass turns out to be. The current bound on the Standard Model Higgs is now only slightly below this range, while the supersymmetric Higgs boson is somewhat less. This means the range of allowed Higgs boson parameters that result in a very narrow (a few MeV) Higgs boson is becoming excluded. Intriguingly the last results from LEP[11, 12, 13, 14, 15] indicated some evidence for a Higgs boson around 115 GeV. Searches involved the decay channels $hZ \to b\bar{b}q\bar{q}$ (four jet), $hZ \to b\bar{b}\nu\bar{\nu}$ (missing energy), $hZ \to b\bar{b}\tau^+\tau^-$, $q\bar{q}\tau^+\tau^-$ (tau), and $hZ \to b\bar{b}\mu^+\mu^-$, $b\bar{b}e^+e^-$ (leptonic). While by no means definitive, if this experimental indication turns out to having been the first evidence for a Higgs boson with mass of 115 GeV, then the direct measurement of the total width of this particle may require a muon collider. A Higgs boson of this mass sits well above the Z-pole where a large background to the decay $h \to b\bar{b}$ exists. The decay width is still quite small because the $WW^{\star}$ decay does not open up until the Higgs boson becomes heavier than 120 GeV or so. A precise determination of this width would be an important test of the Standard Model, and any deviation from the Standard Model expectation would indicate a nonminimal Higgs sector or other new physics.

In summary, if a Higgs boson is discovered at the LHC as appears increasingly likely, there will be considerable interest in measuring the Higgs boson properties as accurately as possible at a linear collider and/or a muon collider. The goal will be to detect the presence of deviations from Standard Model expectations that require an extended Higgs sector or other new physics. A Higgs factory of modest luminosity can provide powerful constraints on the parameter space. An accurate determination of the rate for $\mu^+\mu^- \to b\bar{b}$ as well as the Higgs total width will be possible. Since these quantities are calculable in the Standard Model, any deviation from the theoretical value could indicate a Higgs sector more complicated than in the Standard Model. If the Higgs sector is extended and many more Higgs bosons are present perhaps with somewhat larger masses and smaller couplings, then a muon collider will be an excellent tool for finding and studying these new states. Recent discussions of Higgs factories that include some more details on the present status of the accelerator technologies can be found in Refs. [16, 17, 18].

4 The Usefulness of a Muon Collider/Higgs Factory

For many physics processes a linear collider and a muon collider with similar luminosities give similar rates. Apart from detector issues that might arise from using beams of decaying particles as is the case with muons, the cross sections for most processes do not depend on the mass of the initial state particles (the most important exception is the s-channel Higgs production process). But present expectations are that a linear collider will have perhaps a few hundred inverse femtobarns of integrated luminosity, while the more difficult technology needed for a muon collider might provide two orders of magnitude less. Nevertheless, a muon collider of limited luminosity can remain competitive with a linear collider for discriminating a Standard Model Higgs boson from some more exotic Higgs that is very Standard Model-like. The precise measurement of the process $\mu^+\mu^- \rightarrow h \rightarrow b\bar{b}$ at a muon collider might provide the best single discriminator between the two cases[19]. A muon collider can therefore provide additional information even if a linear collider has spent considerable time studying a light Higgs boson.

Finally the role of a Higgs factory as a stepping stone to higher energy muon colliders should not be ignored. The initial goal of extending the center-of-mass energy of particle colliders will be achieved by first constructing a smaller working collider.

5 Higgs Decays to Tau

An important decay channel of the Higgs boson at a muon collider is the decay into tau leptons, $\mu^+\mu^- \rightarrow h \rightarrow \tau^+\tau^-$. The production of tau leptons can also proceed via exchange of a photon or Z boson. The signal process involves the exchange of the spin-0 Higgs boson, while the background process involves the exchange of the spin-1 photon or Z boson. The angular distribution of the tau leptons would establish the spin of the Higgs boson [2].

We summarize here the conclusions of a study of this decay mode at a muon collider[20]. A narrow Higgs boson (as in the Standard Model) would require an excellent beam energy resolution for a clear signal. A moderate beam polarization would not help for the signal identification. The final state spin correlation could confirm the nature of the Higgs boson spin. Finally the relative decay rates of the Higgs boson into b and τ would determine the ratio of the couplings of the Higgs boson into these states. The role of the $\tau^+\tau^-$ final state in the study of CP violation arising in the Higgs sector is described in the next section.

6 CP Violation in the Higgs Sector

CP violation has been experimentally measured in the K and B systems, and the results are consistent with Standard Model expectations where the CP violation arises from a phase in the Cabibbo-Kobayashi-Maskawa (CKM) matrix. While there is no need at the moment to find new sources of CP violation to account for any experimental data, there are at least a couple of reasons for doing so. First, as mentioned before, the Standard Model is a simple parameterization of the electroweak symmetry breaking sector, and an extended Higgs sector may exist in which additional sources of CP violation naturally occur. Second, the observed baryon asymmetry of the Universe (predominance of matter over antimatter) requires a source of CP violation beyond that in the CKM matrix. The calculated asymmetry resulting from just the Standard Model CKM phase is insufficient to account for the asymmetry.

Using the s-channel production of Higgs bosons offers the further possibility for studying the CP properties of Higgs bosons. Such studies might be necessary to fully understand the Higgs sector. Measurements of correlation in the $\tau^+\tau^-$ final state (and possibly in the $t\bar{t}$

[2]See also the section on Higgs threshold measurements below for another way to measure the spin.

final state) can be performed. The heavy Higgs bosons H and A can be almost degenerate in the Minimal Supersymmetric Standard Model (MSSM)[3], and if there are CP-violating phases in the neutral Higgs potential, the CP-even H and the CP-odd A can mix. This can result in a sizeable mass-difference between the mass eigenstates, and the two states can be separated in the s-channel production process at a muon collider. This mass-difference can be combined with the mass-squared sum rule to be a powerful probe of new physics entering radiatively in the Higgs sector.

The best method for determining the CP-properties of the Higgs boson is by using transversely polarized muon beams. The muon coupling to the Higgs boson can be described by $\overline{\mu}(a + ib\gamma_5)\mu h$, and the cross section takes the form[21]

$$\begin{aligned}
\overline{\sigma}_h(\zeta) &= \overline{\sigma}_h^0 \left(1 + P_L^+ P_L^- + P_T^+ P_T^- \left[\frac{a^2 - b^2}{a^2 + b^2}\cos\zeta - \frac{2ab}{a^2 + b^2}\sin\zeta\right]\right) \\
&= \overline{\sigma}_h^0 \left[1 + P_L^+ P_L^- + P_T^+ P_T^- \cos(2\delta + \zeta)\right], \qquad (7)
\end{aligned}$$

where $\overline{\sigma}_h^0$ is the polarization average convoluted cross section, $\delta \equiv \tan^{-1}\frac{b}{a}$, P_T (P_L) is the degree of transverse (longitudinal) polarization, and ζ is the angle of the μ^+ transverse polarization relative to that of the μ^- as measured using the direction of the μ^-'s momentum as the $\hat{z}$ axis. It is clear from the form of this expression that one requires transverse polarization ($P_T \neq 0$). Longitudinal polarization only provides sensitivity only to the combination $\overline{\sigma}_h^0 \propto a^2 + b^2$ only. The CP-violation is contained in the $\sin\zeta$ term. One can measure the coefficients in the $\cos\zeta$ and $\sin\zeta$ terms, $\frac{a^2-b^2}{a^2+b^2}$ and $\frac{-2ab}{a^2+b^2}$, by running at fixed $\zeta = 0, \pi/2, \pi, 3\pi/2$ and measuring the asymmetries (taking $P_T^+ = P_T^- \equiv P_T$ and $P_L^\pm = 0$)

$$\begin{aligned}
\mathcal{A}_I &\equiv \frac{\overline{\sigma}_h(\zeta=0) - \overline{\sigma}_h(\zeta=\pi)}{\overline{\sigma}_h(\zeta=0) + \overline{\sigma}_h(\zeta=\pi)} = P_T^2\frac{a^2 - b^2}{a^2 + b^2} = P_T^2\cos 2\delta\,, \\
\mathcal{A}_{II} &\equiv \frac{\overline{\sigma}_h(\zeta=\pi/2) - \overline{\sigma}_h(\zeta=-\pi/2)}{\overline{\sigma}_h(\zeta=\pi/2) + \overline{\sigma}_h(\zeta=-\pi/2)} = -P_T^2\frac{2ab}{a^2 + b^2} = -P_T^2\sin 2\delta\,.
\end{aligned}$$

Provided the overall production rate and the background are well-known, the fractional error in these asymmetries can be approximated as $\frac{\delta A}{A} \propto P_T^2\sqrt{L}$. So the importance of the transverse polarization is seen to be critically important, and one could give up some luminosity in exchange for achieving the highest possible transverse polarization. A detailed study[21] has concluded that using such beams will provide a good CP determination provided one has sufficient luminosity.

7 Heavy Higgs Bosons

Supersymmetric models are characterized by Higgs sectors that contain multiple physical Higgs bosons. Often the lightest of these Higgs bosons is a Standard Model-like state with a mass close to the Z boson mass. In the MSSM, there are two complex Higgs doublet fields which after symmetry breaking and the Higgs mechanism there remain a light, neutral h and two (usually) heavier H and A bosons. As one change the parameters of the model so that

[3]The properties of the Higgs bosons in the MSSM is discussed further in the section on heavy Higgs bosons.

the heavy H and A become heavier, they decouple, and one is left with the light h whose properties become very much like that of the Standard Model Higgs boson. Therefore a like scenario is that the h boson will be studied at a future linear collider via the Higgs-strahlung process $e^+e^- \to Zh$ while the analogous processes for the heavier states, $e^+e^- \to ZH$ and $e^+e^- \to ZA$, will not be viable because the H and A are too heavy and/or the cross section is too small because the couplings of these Higgs bosons to the gauge bosons are suppressed. A linear collider operating in the $\gamma\gamma$ mode[22, 23, 24, 25, 26, 27, 28] can produce the heavy Higgs bosons H and A in the s-channel. However the energy spread for this option is much larger than it would be for a muon collider, and a large luminosity (~ 200 fb^{-1}) would be required.

The most efficient procedure is the operate the muon collider at maximum energy and produce the H and A (often as overlapping resonances) via the radiative return mechanism[6]. By looking for a peak in the $b\bar{b}$ final state, the H and A can be discovered and, once discovered, the machine $\sqrt{s}$ can be set to M_A or M_H and factory-like precision studies pursued. Note that the A and H are typically broad enough that $R = 0.1\%$ would be adequate to maximize their s-channel production rates. If we have a muon collider, we would have precise control and knowledge of the collision energy. In supersymmetric extensions of the Standard Model the combination of gauge invariance and supersymmetry often implies that there are relationships between the masses of the Higgs bosons in the theory. For example in the (MSSM) there is a mass-squared sum rule given by

$$M_h^2 + M_H^2 = M_A^2 + M_Z^2 + \Delta \tag{8}$$

where Δ is a calculable radiative correction[29, 30] that depends on the details of the supersymmetric spectrum. Since a muon collider can measure the Higgs boson masses very accurately and the Z-boson mass is known precisely from elsewhere[31] (from s-channel production at the LEP experiment), one can experimentally determine the value of Δ to high precision. Theoretically one knows what value to expect for a given supersymmetric spectrum for Δ, and a possible crosscheck of experiment with theory is possible. Radiative corrections have in the past given us indirect information of the value of the top quark mass before it was discovered and is at the present time providing constraints on the mass of the as yet undiscovered Higgs boson. The measurement of Δ would represent the extension of measuring radiative corrections to an extended Higgs sector.

Solving the sum rule for the mass-difference of the heavy Higgs bosons, one gets

$$M_A - M_H = \frac{M_h^2 - M_Z^2 - \Delta}{M_A + M_H} . \tag{9}$$

In the decoupling limit, $M_A \to \infty$ and $M_H \to \infty$, the mass difference becomes small and eventually the Higgs resonance peaks will overlap.

The mass of the light Higgs boson h can be measured to less than an MeV in the s-channel. The mass-difference between the heavy Higgs bosons H and A can also be measured in the s-channel, but here the ultimate precision that can be obtained depends sensitively on the parameter space of the MSSM. As the masses become larger (and approach the decoupling limit) or as $\tan\beta$ increases, the heavy Higgs states become more degenerate. Excluding the case where the mass-difference is smaller than the Higgs widths themselves (in which case the resonance peaks overlap and cannot be distinguished), a reasonable expectation[32] is that the mass-difference can be measured to a few tens of MeV with a modest muon collider integrated luminosity of 0.1 fb^{-1}. So the radiative correction can be experimentally determined by measuring Higgs boson masses in the s-channel to something of order 10 GeV^2 using Eq. (9).

It will also be interesting to measure and compare the branching ratios of these supersymmetric Higgs boson and compare them to theoretical predictions. The H and A decay into $b\bar{b}$ more often than into $t\bar{t}$ for $\tan\beta > 5$. For some regions of parameter space both types of decay can be seen. If this turns out to be the case, one can use the ratio of the decay rates to measure $\tan\beta$. Decays into supersymmetric decays might also occur if they are kinematically possible. These additional decay modes would place constraints on the parameters describing the masses and couplings of the supersymmetric particles and could as a result narrow down the space of the soft supersymmetry breaking parameters.

In this context chargino (supersymmetric partners to the electrically charged bosons of the Standard Model) production via Higgs exchange has been studied recently[33]. For values of the parameters where the H and A resonances are still separable ($\tan\beta < 8$ and $M_A < 400$ GeV), the Higgs-chargino couplings can be measured to a few percent. In another study[34] the interference of the CP-even and the CP-odd amplitudes of neutral Higgs boson exchange in the pair production of neutralinos (supersymmetric partners to the neutral bosons of the Standard Model). Using longitudinally polarized muons beams and exploiting the neutralino decays $\tilde{\chi}_i^0 \to \ell^{\pm}\tilde{\ell}^{\mp}$ where $\ell = e, \mu, \tau$, it was found that large measurable asymmetries result for nearly degenerate H and A and intermediate values of M_A and $\tan\beta$. This measurement could be useful for extracting the couplings of the Higgs bosons to neutralinos.

8 Nonminimal Higgs Sectors

The Higgs sector of the Standard Model should properly be viewed as a simple parameterization of the mechanism of electroweak symmetry breaking. An elementary scalar field that is an electroweak doublet is sufficient to describe all the masses and physical effects seen in experiments to date. The true nature of the physics giving rise to electroweak symmetry breaking may be more complicated than this simple description. For example, in supersymmetric models such as the MSSM already discussed, the Higgs sector includes additional states. This can occur even in the absence of supersymmetry. More generally the Higgs sector does not have to involve elementary scalars at all. The Standard Model-like Higgs boson may arise from a more complicated theory. So while the standard interpretations of the electroweak precision data suggest there is a light Standard Model-like Higgs boson, alternative Higgs sector models can be constructed that are consistent with the data. This has important implications for a muon collider since the envisioned study mode is the production of the Higgs boson in the s-channel which requires the energy of the collider to be set equal to the Higgs boson mass.

One example of an extended Higgs sector that does not have a light Standard Model-like Higgs boson is a two-Higgs-doublet model with a heavy (~ 1 TeV) Higgs boson with large couplings to vector bosons. The large $\Delta S > 0$ and $\Delta T < 0$ coming from the heavy Higgs with large VV coupling is compensated by an even larger $\Delta T > 0$ coming from a small ($\Delta M \sim 1$ GeV is sufficient) mass splitting between the $H^{\pm}$ and the other heavy neutral Higgs boson. This type of Higgs boson would be discovered at the LHC and appear to be a heavy Standard Model Higgs boson because of the unsuppressed couplings to the $W^{\pm}$ and Z gauge bosons. The combination of the direct detection of this Higgs boson and the electroweak precision data would indicate the presence of an exotic Higgs sector. A characteristic feature of these models is that there exists a lighter (< 500 GeV) Higgs boson with suppressed couplings to the gauge bosons which would escape detection at the LHC and a linear collider[35]. A muon collider could be key to discovering such a Higgs boson.

The Higgs sector in the so-called type II two Higgs doublet extension to the Standard Model has been studied recently in Ref. [36]. This study found that in these models the mode $\mu^+\mu^- \rightarrow H^{\pm}W^{\mp}$ is a very promising channel in which to discover the charged Higgs boson.

9 Higgs Threshold Measurement

Operating a collider near the production threshold is a well-established method for measuring the masses, widths, and spins of the produced particles. For example at LEP II the $e^+e^- \rightarrow W^+W^-$ threshold was exploited to accurately determine the W mass. At an electron-positron or muon collider one can choose energies close to the Zh production threshold to determine properties of the Higgs boson. The rapid rise in the cross section near the threshold is sensitive to the Higgs mass. These measurement require tens of inverse femtobarns to provide a measurement of the Higgs mass to less than 100 MeV[37].

In the ideal case where the normalization of the measured Zh cross section as a function of $\sqrt{s}$ can be precisely predicted, including efficiencies and systematic effects, sensitivity to the Standard Model Higgs boson mass is maximized by a single measurement of the cross section at $\sqrt{s} = M_Z + M_h + 0.5$ GeV, just above the real particle threshold. With a $\sim \pm 180$ MeV measurement of m_h from initial running or from a linear collider; $\sqrt{s}$ can be set quite close to this optimal point. As an example of the precision that might be achieved, suppose $m_h = 100$ GeV and backgrounds are neglected. The Zh cross section is 120 fb and is rising at a rate of 0.05 fb/MeV. With $L = 50$ fb^{-1} and including an overall (b-tagging, geometric and event identification) efficiency of 40%, this yields 2.4×10^3 events, or a measurement of the cross section to about 2%. From the slope of the cross section one concludes that a m_h measurement with accuracy of roughly 50 MeV is possible.

In practice, there will be systematic errors associated with experimental efficiencies as well as for theoretical predictions of the Zh cross section and h branching ratio(s) that will be very difficult to reduce below the 1% level. The ratio of the cross section measured at $\sqrt{s}$ well-above threshold in the initial H discovery to that measured right at threshold is thus the key to determining m_h.

With sufficient luminosity it is possible to determine the Higgs boson mass to a high and very valuable level of precision by measuring the $\ell^+\ell^- \rightarrow Zh \rightarrow Zb\bar{b}$ cross section just above threshold and normalizing to a second measurement either well above threshold or near the Zh cross section peak. One simultaneously determines $\sigma(Zh)\mathrm{BF}(h \rightarrow b\bar{b})$ at a level of accuracy that could distinguish between the Standard Model Higgs sector and its many possible extensions.

The shape of the production cross section can be used in addition to determine the spin and charge conjugation and parity (CP) properties of the Higgs boson[38]. These threshold measurements are of interest for the planned linear colliders and would be possible at a muon collider if hundreds of inverse femtobarns of luminosity were available.

10 Summary

The current schedule has the LHC beginning to take data in 2007 and will likely reveal the path the particle physics will take over the next few decades at least. The indirect experimental clues indicate that a likely scenario is that a Higgs boson with mass < 250 GeV will be discovered. This would provide a strong motivation for construction a electron-positron linear collider as the next step to study the properties of the Higgs boson and to explore the Higgs sector (and any other new physics) more completely.

A muon collider, which presumably might follow the operation of a linear collider, could play a crucial role in the further development of particle physics. First, the demonstration that a muon collider is actually feasible would open up the possibility of extending the energy frontier much further than is being contemplated with other technologies. Second, the development of muon beams can have, as a possible offshoot, generation of high intensity neutrino beams (from decaying muons) that could yield valuable information about neutrinos from long baseline experiments. Finally the unique role of the muon collider in producing Higgs bosons in the s-channel offers the possibility that the Higgs sector, which is probably more complicated that is described by the Standard Model Higgs sector, might be studied in detail. The muon collider can produce not only a light Standard Model-like Higgs boson in the s-channel, but can also produce and disentangle heavier Higgs bosons such as the H and A of supersymmetric models.

Lepton colliders can produce Higgs bosons near threshold giving important information on its mass and spin. This would require operating a muon collider (or linear collider) with energy near $M_Z + M_h$ and would require large integrated luminosities of around 100 fb^{-1}.

Muon colliders may provide the only method for finding and separating two heavy, highly degenerate Higgs bosons of the MSSM. The s-channel production mechanism together with the possibility of polarizing the muon beams may prove essential to separating the CP-even state from the CP-odd state especially if the resonance peaks are overlapping.

Finally muon collider offer exciting possibilities in addition to their role as Higgs factories. The prospect of extending the energy frontier of accelerators should keep physicists interested in muon colliders for many decades to come.

References

[1] C. M. Ankenbrandt *et al.*, "Status of Muon Collider Research and Development and Future Plans," *Phys. Rev. ST Accel. Beams* **2**, 081001 (1999) *[arXiv:physics/9901022]*.

[2] R. Raja and A. Tollestrup, "Calibrating the energy of a 50-GeV x 50-GeV muon collider using spin precession," *Phys. Rev. D* **58**, 013005 (1998) *[arXiv:hep-ex/9801004]*.

[3] D. Ayres *et al.* [Neutrino Factory and Muon Collider Collaboration], "Expression of interest for R & D towards a neutrino factory based on a storage ring and a muon collider," *arXiv:physics/9911009*.

[4] N. Holtkamp *et al.*, "A feasibility study of a neutrino source based on a muon storage ring," *SLAC-REPRINT-2000-054*

[5] A. Blondel *et al.*, "ECFA/CERN studies of a European neutrino factory complex," *CERN-2004-002*.

[6] V. D. Barger, M. S. Berger, J. F. Gunion and T. Han, "Higgs boson physics in the s-channel at $\mu^+\mu^-$ colliders," *Phys. Rept.* **286**, 1 (1997) *[arXiv:hep-ph/9602415]*.

[7] V. D. Barger, M. S. Berger, J. F. Gunion and T. Han, "s-channel Higgs boson production at a muon muon collider," *Phys. Rev. Lett.* **75**, 1462 (1995) *[arXiv:hep-ph/9504330]*.

[8] V. D. Barger, M. Berger, J. F. Gunion and T. Han, "Physics of Higgs factories," in *Proc. of the APS/DPF/DPB Summer Study on the Future of Particle Physics (Snowmass 2001)* ed. N. Graf, eConf **C010630**, E110 (2001) *[arXiv:hep-ph/0110340]*.

[9] M. S. Berger, "Higgs bosons at muon colliders," in *Proc. of the APS/DPF/DPB Summer Study on the Future of Particle Physics (Snowmass 2001)* ed. N. Graf, eConf **C010630**, P114 (2001) *[arXiv:hep-ph/0110390]*.

[10] M. Battaglia and K. Desch, "Precision studies of the Higgs boson profile at the e^+e^- linear collider," *arXiv:hep-ph/0101165*.

[11] R. Barate *et al.* [ALEPH Collaboration], "Observation of an excess in the search for the standard model Higgs boson at ALEPH," *Phys. Lett. B* **495**, 1 (2000) *[arXiv:hep-ex/0011045]*.

[12] P. Abreu *et al.* [DELPHI Collaboration], "Search for the standard model Higgs boson at LEP in the year 2000," *Phys. Lett. B* **499**, 23 (2001) *[arXiv:hep-ex/0102036]*.

[13] M. Acciarri *et al.* [L3 Collaboration], "Higgs candidates in e^+e^- interactions at $\sqrt{s} =$ 206.6-GeV," *Phys. Lett. B* **495**, 18 (2000) *[arXiv:hep-ex/0011043]*.

[14] G. Abbiendi *et al.* [OPAL Collaboration], "Search for the standard model Higgs boson in e^+e^- collisions at $\sqrt{s}$ =192-GeV - 209-GeV," *Phys. Lett. B* **499**, 38 (2001) *[arXiv:hep-ex/0101014]*.

[15] A. N. Okpara, "LEP-wide combination of the search for the standard model Higgs boson," *arXiv:hep-ph/0105151*.

[16] G. G. Hanson, "Towards a Higgs factory / muon collider," *Nucl. Instrum. Meth. A* **503**, 96 (2003).

[17] A. Caldwell, "Muon collider and Higgs factory," *J. Phys. G* **29**, 1569 (2003).

[18] R. Godang, S. Bracker, M. Cavaglia, L. Cremaldi and D. Cline, "Resolution of nearly mass degenerate Higgs bosons and production of black hole systems of known mass at a muon collider," *arXiv:hep-ph/0411248*.

[19] B. Autin, A. Blondel and J. R. Ellis, "Prospective study of muon storage rings at CERN," *CERN-99-02*

[20] V. D. Barger, T. Han and C. G. Zhou, "Higgs boson decays to tau pairs in the s-channel at a muon collider," *Phys. Lett. B* **480**, 140 (2000) *[arXiv:hep-ph/0002042]*.

[21] B. Grzadkowski, J. F. Gunion and J. Pliszka, "How valuable is polarization at a muon collider? A test case: Determining the CP nature of a Higgs boson," *Nucl. Phys. B* **583**, 49 (2000) *[arXiv:hep-ph/0003091]*.

[22] G. Jikia, "Z boson pair production via photon fusion at a high-energy photon linear collider," *Nucl. Phys. B* **405**, 24 (1993).

[23] M. S. Berger, "The W boson loop background to $H \to ZZ$ at photon-photon colliders," *Phys. Rev. D* **48**, 5121 (1993) *[arXiv:hep-ph/9307259]*.

[24] D. A. Dicus and C. Kao, "Production of Z boson pairs at photon linear colliders," *Phys. Rev. D* **49**, 1265 (1994) *[arXiv:hep-ph/9308330]*.

[25] G. J. Gounaris, P. I. Porfyriadis and F. M. Renard, "The heavy neutral Higgs signature in the $\gamma\gamma \to ZZ$ process," *Eur. Phys. J. C* **19**, 57 (2001) *[arXiv:hep-ph/0010006]*.

[26] M. S. Berger, "Test of an equivalence theorem at one loop," *arXiv:hep-ph/9207275*.

[27] M. M. Muhlleitner, M. Kramer, M. Spira and P. M. Zerwas, "Production of MSSM Higgs bosons in photon photon collisions," *Phys. Lett. B* **508**, 311 (2001) *[arXiv:hep-ph/0101083]*.

[28] D. M. Asner, J. B. Gronberg and J. F. Gunion, "Detecting and studying Higgs bosons in two-photon collisions at a linear collider," *Phys. Rev. D* **67**, 035009 (2003) *[arXiv:hep-ph/0110320]*.

[29] M. S. Berger, "Radiative Corrections To Higgs Boson Mass Sum Rules In The Minimal Supersymmetric Extension To The Standard Model," *Phys. Rev. D* **41**, 225 (1990).

[30] V. D. Barger, M. S. Berger, A. L. Stange and R. J. N. Phillips, "Supersymmetric Higgs boson hadroproduction and decays including radiative corrections," *Phys. Rev. D* **45**, 4128 (1992).

[31] D. E. Groom *et al.* [Particle Data Group Collaboration], "Review of particle physics," *Eur. Phys. J. C* **15**, 1 (2000).

[32] M. S. Berger, "Higgs sector radiative corrections and s-channel production," *Phys. Rev. Lett.* **87**, 131801 (2001) *[arXiv:hep-ph/0105128]*.

[33] H. Fraas, F. Franke, G. Moortgat-Pick, F. von der Pahlen and A. Wagner, "Precision measurements of Higgs chargino couplings in chargino pair production at a muon collider," *Eur. Phys. J. C* **29**, 587 (2003) *[arXiv:hep-ph/0303044]*.

[34] H. Fraas, F. von der Pahlen and C. Sachse, "Interference of Higgs boson resonances in $\mu^+\mu^- \rightarrow \tilde{\chi}_i^0\tilde{\chi}_j^0$ with longitudinal beam polarization," *Eur. Phys. J. C* **37**, 495 (2004) *[arXiv:hep-ph/0407057]*.

[35] P. Chankowski *et al.*, "Do precision electroweak constraints guarantee e^+e^- collider discovery of at least one Higgs boson of a two-Higgs-doublet model?" *Phys. Lett. B* **496** (2000) 195 *[hep-ph/0009271]*.

[36] C. A. Marin, "Higgs production at a muon collider in the two Higgs doublet model of type II," *arXiv:hep-ph/0405021*.

[37] V. D. Barger, M. S. Berger, J. F. Gunion and T. Han, "Precision Higgs boson mass determination at lepton colliders," *Phys. Rev. Lett.* **78**, 3991 (1997) *[arXiv:hep-ph/9612279]*.

[38] D. J. Miller, S. Y. Choi, B. Eberle, M. M. Muhlleitner and P. M. Zerwas, "Measuring the spin of the Higgs boson," *Phys. Lett. B* **505**, 149 (2001) *[arXiv:hep-ph/0102023]*.

In: Astrophysics and Condensed Matter
Editor: Thomas G. Hardwell, pp. 175-193
ISBN: 978-1-60456-332-0

Chapter 6

CRYSTAL IDENTIFICATION AND CLASSIFICATION

Rositsa Jordanova, Victor Lamzin and Matthew R. Groves*
EMBL Outstation Hamburg, Notkestrasse 85
22603 Hamburg, Germany

Abstract

This article will provide a review of the current methodologies in use for the identification and classification of protein crystallization experiments, covering computer-aided identification, the use of cross-polarization, UV-stimulated fluorescence and the addition of fluorescent markers. The current state of the field will be reviewed and potential new directions will be explored.

1.1. Introduction: The Need for Improved Protocols for Crystal Identification and Classification

The ability to determine three-dimensional, high resolution structures of biologically relevant macromolecules has had a major impact on modern biology. The insights gained into protein structure on an atomic level have permitted detailed analysis of the biochemical roles played by proteins in biological processes and provide important information for the development of novel pharmaceuticals for human health. X-ray diffraction remains the dominant method for the determination of three dimensions structures, not only due to the resolution of the models produced, but also due to the fact that models of assembly subunits can be used as entry points for the analysis of large complexes using lower resolution methods such as electron microscopy (EM) and small angle X-ray scattering (SAXS). While the role of X-ray diffraction techniques has been central, macromolecular crystallography is still utterly dependent upon the production of macromolecular crystals.

The value of high resolution structures has been globally recognized, and a number of structural genomics consortia (SGC) have been established. In addition to populating the protein databank (www.rcsb.org, Berman *et al.*, 2000) these consortia have also been at the

* E-mail address: groves@embl-hamburg.de. Phone: +49-40-89902-161. (Corresponding Author)

forefront of novel developments in the structure determination pipeline which are aimed at increasing the efficiency with which structures are produced. Whilst these developments are primarily designed to optimize the handling of multiple targets within an SGC they represent additional benefit to structural biologists in that these novel methods can equally be applied to single "high value" targets as they are to the large, multiple target lists of SGC.

As this review is predominantly concerned with recent developments in the analysis of crystallization experiments no summary of recent advances in the production of material for structural analysis will be provided. However, a recent issue of *Acta Crystallographia* (**D62**, Part 10) was dedicated entirely to reports on the advances made within the SPINE (Structural Proteomics In Northern Europe) SGC and would provide interested readers with a excellent overview of other developments within the field.

1.2. (High Throughput) Crystallization

One of the most significant effects of the funding of SGC worldwide has been the recent rapid development of facilities that are able to automate the process of crystallization screening. Traditional methods to determine conditions that support the crystallization of proteins were centred around manually performed vapour diffusion experiments (McPherson, 1999), with significant time being required to manually perform the screen and, more importantly, significant sample volumes (1-2 μl per experiment). As some authors have suggested that ~400 independent conditions represents an optimal screening of chemical space for crystallization conditions (e.g. Rupp, 2003), and sample concentrations are typically in the range of 10 μg/μl, large amounts of highly pure samples (~10 mg) were required for an adequate screening of each biological sample. This effectively limited the number of systems that were amenable to research using structural biology to those which were readily available in such high quantities, either directly from source or using recombinant methods.

The advent of high throughput crystallization robotics has significantly reduced the sample requirements. Typical experimental volumes are now ~100 nl for individual vapour diffusion experiments and microfluidics permit screening of 96 conditions with as little as 1.4 μl of sample (e.g. Fluidigm, http://www.fluidigm.com/). Whilst initially dedicated to an SGC or individual research institutions, increasingly these facilities also now accept samples from non-associated researchers (e.g. the HTP facility at EMBL, Hamburg, Müller-Dieckmann (2006)), allowing the benefits of high throughput methods to be applied to a much wider range of structural projects.

The increased availability of such high throughput facilities, coupled with the reduction in required sample, has resulted in the generation of a huge number of crystallization experiments. High-throughput robotics can typically generate 1000s of crystallization experiments per day, with each experiment requiring repeated evaluation over a time period that can extend to weeks. As these experiments are typically visualized automatically, there is a pressing need to develop methodologies for the automated identification and classification of crystallization experiment images. This need is driven not only by the needs of the individual project to correctly identify protein crystals, but also to provide a scoring of the images such that conditions more likely to produce protein crystals can be identified.

Although the goal of any crystallization screening is the identification of high quality crystals, typically only a small subset of conditions will result in protein crystals. While results from individual samples may vary widely, an estimate of 0.4% of screening conditions which result in successful initial crystallization "hits" is provided by a recent publication (Berry *et al.*, 2006). Even when no initial "hits" are determined, the behaviour of the protein sample in the screen as a whole (i.e. pH ranges under which heavy precipitation is observed, etc.) will provide valuable experimental information. This information may be used in subsequent experiments to refining parameters in the search for conditions under which a particular sample may crystallize. Classically, the results of crystallization images may be segregated into the following categories:

1. Clear Drop
2. Light Precipitation
3. Heavy Precipitation
4. Microcrystalline Precipitate
5. Phase Separation
6. Cluster of Crystals
7. Single Crystals

However, interpretation of the crystallization images can often lead into the situation in which different researchers would classify the images into different classes (e.g. Wilson & Berry, 2005). This is primarily due to the lack of clear definitions of what constitutes "light" precipitation (sample at the edge of solubility) and "heavy" precipitation – which is usually interpreted as a sign that the sample becomes denatured under these conditions. The situation is further complicated in that the drop classification may change with time and also that diffraction quality crystals may spontaneously appear from any of these drop classifications. Finally, any single crystallization experiment may result in the presence of more than a single phenomenon (e.g. light precipitation and phase separation). A number of excellent resources exist to aid the researcher in the classification of the crystallization experiment, of which the web site written by Therese Bergfors is possibly the best-known (The Pictorial Library of Crystallization Drop Phenomena: http://alpha2.bmc.uu.se/terese/crystallization/library.html).

The difficulties in a reproducible drop classification amongst disparate researchers and the generation of large numbers of crystallization images makes an urgent case for the development of novel methodologies to aid researchers in the identification and classification (or "scoring") of crystallization images. Broadly, these developments may be segregated into two general classes: computational and physical. The computation solutions proposed are designed to use the images obtained from crystallization robotics and apply computer algorithms to classify the results of the crystallization experiment. Physical methods may be defined as developments to use the physical properties of biological macromolecules (polarization, fluorescence, etc.) in order to identify (and thus classify) phenomena within the crystallization experiment.

However, the identification of a crystalline sample is merely one step in a structural determination pipeline. Subsequent transfer of the sample to a suitable X-ray source is still carried out manually and manual intervention is still required to centre the sample on an X-ray source for initial sample characterization and, ultimately, data collection. Given the increasing number of samples under structural analysis it is not surprising that many of these techniques are additionally being applied to the identification of samples within their cryo-supports on X-ray sources, in order to automate the process of X-ray data collection.

The characterization of the diffraction properties of a sample is an essential step in the optimization of the crystallization conditions and information gained from any X-ray diffraction is used to guide further optimization experiments of each individual sample.

2.1. Computational Classification and Identification

Automated analysis of protein crystallization images is largely performed using image analysis of crystallization images produced by high-throughput robotics. As such each software solution tends to be developed for a specific imaging system and differences in lighting, pixel resolution and depth of focal plane strongly influence the performance of each software package when applied to a "non-native" imaging system. The general method of computational classification relies strongly upon the detection of textures (pixel intensity variance) within a crystallization image, in addition to the detection of image discontinuities; which are then used to define boundaries of objects within the crystallization image. The objects identified can then be classified using a number of attributes, such as size, shape and texture. Features which are strongly associated with protein crystals, such as long straight boundaries and regions of low variance within the defined objects can also be evaluated.

Figure 1 provides an illustrative example of some of the processes behind automated image classification as implemented in the software package ALICE (Analysis of Images from Crystallization Experiments, Wilson (2002)). In the example chosen an image of crystallization experiment (1a) is processed and the gradient between neighbouring pixels is evaluated (1b), with the darker points representing areas which possess the steepest gradients. This "gradient image" displays a significant improvement in the quality of the image, as lighting effects due to the curvature of the drop are largely removed, while other features present in the image (such as the smaller spherical objects) are retained.

Automated analysis of the gradient image can then be performed to detect the presence of straight lines (as indicated in 1c), or a threshold can be applied to the image with the aim of enclosing individual objects (as indicated in 1d). Once objects (or regions within the image) have been identified, these areas can be evaluated in terms of a number of difference geometrical parameters (e.g. object surface area to perimeter ratios and variation in intensity within the object for both the original image and the gradient image).

a) Collected image b) Gradient image c) Line detection d) Object detection

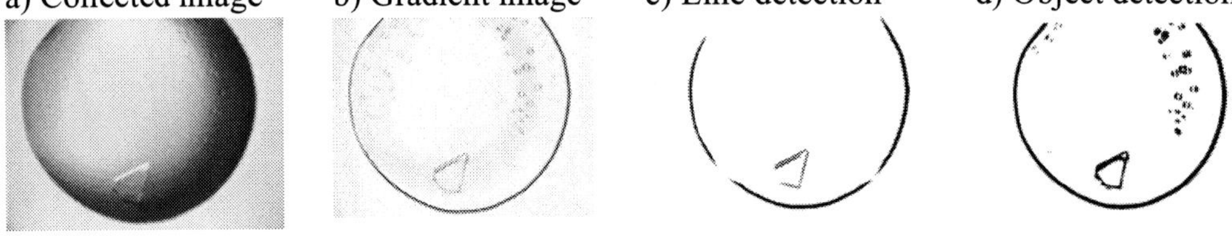

Figure 1. An Example of Automated Image Analysis.

An illustrative example of some of the processes used for automated image classification of crystallization drops. The image collected under white light (1a) is processed to produce a "gradient" image (1b), in which the value of each pixel is based upon the gradient in intensity to the neighbouring pixels. This "gradient" image can be analyzed for the presence of lines of similar gradient values (1c) or a threshold defined to define objects (1d). Images are taken from Wilson (2002) © IUCr.

Whilst this kind of image processing can result in a high success rate of identification of proteins crystals, crystallization images which contain more features can be more difficult to classify, particularly to distinguish between images of crystallization experiments which contain microcrystalline precipitate and those which contain non-microcrystalline precipitate (Figure 2). The presence of microcrystalline precipitate is a clear indication of a condition that, with sufficient optimization, may eventually produce well ordered protein crystals, suitable for diffraction analysis. The example presented in Figure 2a shows a condition which has resulted in microcrystalline precipitate, whereas the example in Figure 2c has resulted in a finer, non-microcrystalline precipitate. From the respective image analysis (2b and 2d) it is clear that shape information alone is insufficient to distinguish between these two cases, although the variation in intensity, both in the original image and in the "gradient" image may be used for classification, as the gradient magnitude varies more across crystalline objects.

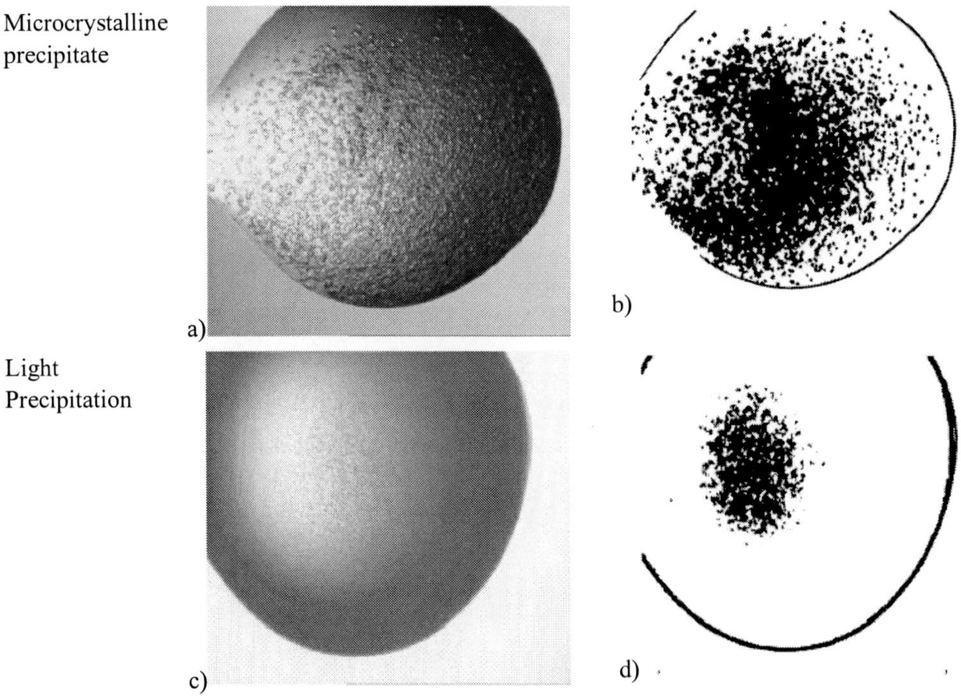

Figure 2. Gradient Image Analysis applied to Images Containing Microcrystalline and Light Precipitation

An illustrative example of the difficulties is distinguishing between microcrystalline precipitate (2a) and light precipitation (2c) using automated image classification of crystallization drops. The image collected under white light (2a & 2c) is processed to produce a "gradient" image (2b & 2d), in which the value of each pixel is based upon the gradient in intensity to the neighbouring pixels. The resulting "gradient" images contain no clearly distinguishing features. Images are taken from Wilson (2002) © IUCr.

2.2. Automated Fourier Transform Processing of Crystallization Drop Images

In the examples above the image processing has been carried out in the spatial domain, with no account taken of the relationship between objects or their spatial arrangement. An extension to the above methods is to perform a Fourier transform (FT) analysis of the images (Bern, 2004, Walker *et al.*, 2007) to more clearly identify images that contain periodic and directional structures over relatively large areas. The resulting FT analysis shows identifiable "spokes" (Figure 3) which are a clear indication of the presence of regularly shaped objects (Figure 3a & 3b). Unfortunately, the presence of regularly shaped artefacts, such as scratches on the crystallization plate surface, also produce such "spokes" and can thus result in the generation of false positives. The field of view also needs to be taken into consideration, as frequent imaging of each crystallization plate can often result in the spreading of a drop, such that it more closely borders the sides of the crystallization well. In this case the strong contribution from the linear edges of the crystallization well, or scratches on the crystallization plate (e.g. Figure 3e) may also contribute to the formation of such "spokes".

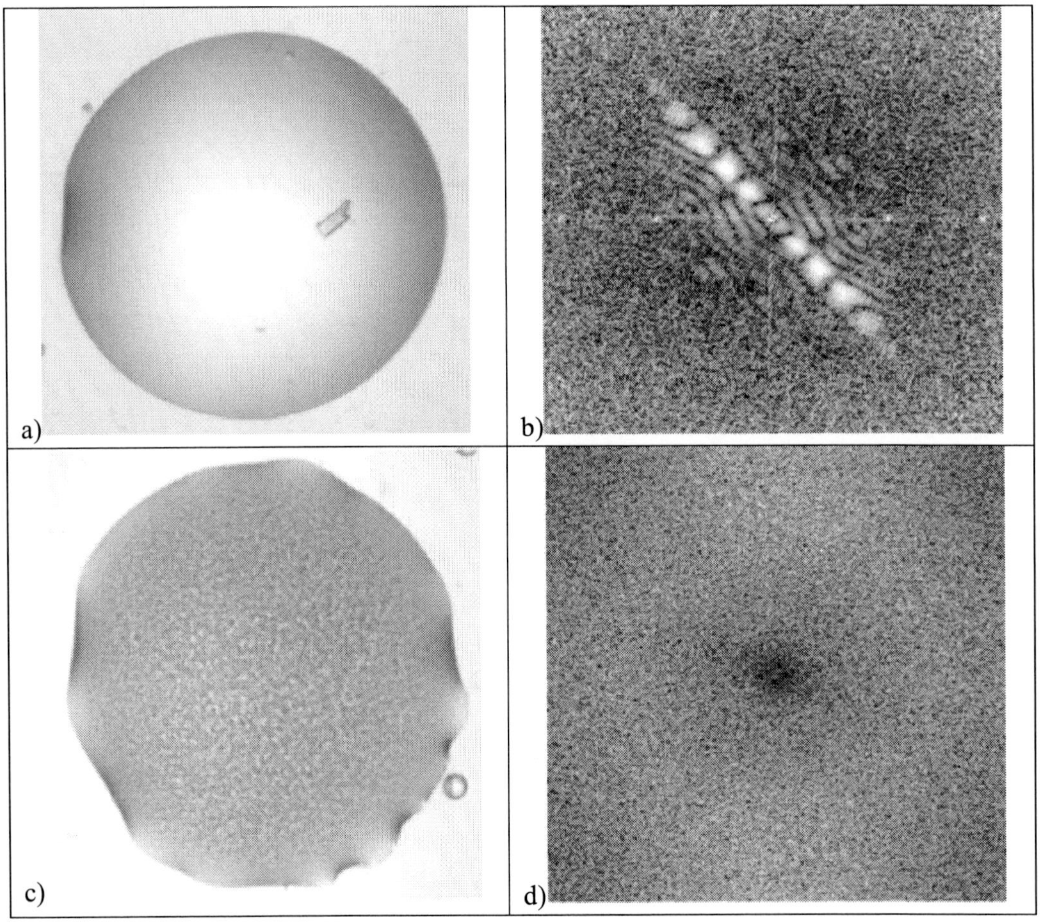

Figure 3. Continued on next page.

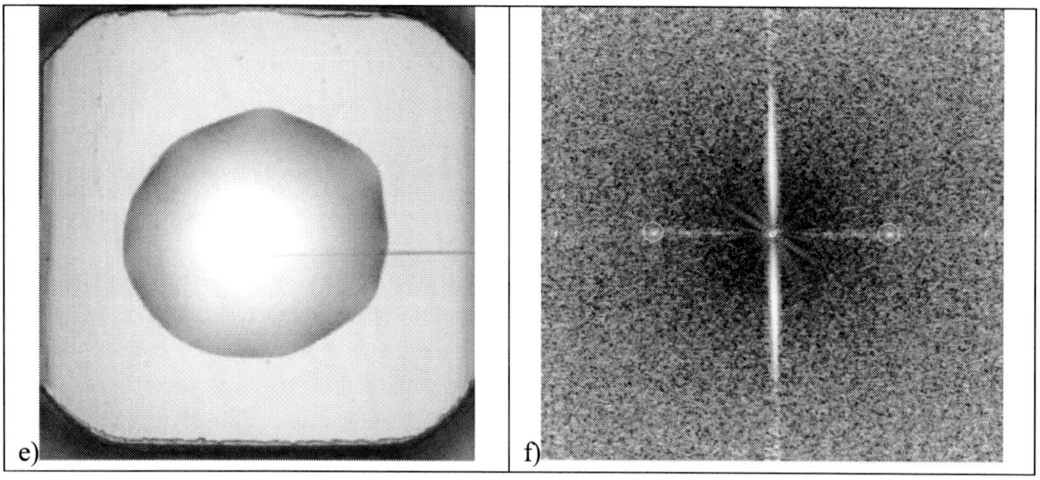

Figure 3. The Use of the Fourier transform to Analyse Crystallization Images.

An illustrative example of the use of the Fourier transform in the automated analysis of crystallization experiments. A single crystal (3a) has a regular and periodic structure that results in "spokes" in the Fourier transform (FT) of this image (3b). Such spokes are not visible in the FT of a drop containing only precipitation (3c) as shown by the equivalent transform (3d). Linear artefacts within the crystallization image (such as the scratch visible in 3e) also result in "spokes" in the FT image (3f). Images taken from Walker *et al.*, (2007) © IUCr.

In summary, the automation of crystal imaging has reached a highly advanced level, in which authors report the successful identification of protein crystals in up to 90% of cases within their image libraries. However, distinguishing microcrystalline precipitate from non-structured precipitate remains challenging, especially when the overall size of microcrystalline precipitate approaches the pixel size of the imaging cameras (~3 μm as a general rule of thumb) and thus, under standard illumination conditions, remains indistinguishable from light precipitate. While these samples are generally too small for data collection, there are a number of sources under construction (e.g. PETRA-III at DESY) which have the potential to collect meaningful data from such samples, through the generation of a μm scale beam. Therefore, it remains imperative to investigate additional methods to improve the contrast of protein crystals above the remaining background of the image, such that these small microcrystals can be successfully identified.

A number of different methods have been established to use the properties of biomolecules directly, such as UV fluorescence, rotation of polarized light (birefringence), etc. Additionally, markers have been added to samples prior to crystallization which also provide a protein (or nucleotide) specific marker (fluorescent dyes). The general aim of these methods is to improve the contrast of the crystallization images and thus improve the use of automated image analysis software in identifying small microcrystals as well as reducing the number of false negatives produced by the methods described above. We have separated these methods into those which are intrinsic to biomolecules and those which require modification of either the sample or the crystallization reagents.

3.1. Intrinsic Properties: UV Fluorescence

The amino acids tryptophan (Trp), tyrosine (Tyr) and phenylalanine (Phe) possess natural intrinsic fluorescence and are commonly used as markers in kinetic assays in which substrate binding perturbs the hydrophobic environment of these residues. However, the dominant fluorophore of proteins is the indole group of the tryptophan. Excitation of the fluorescence is possible over the range of 260-300 nm, with a peak of absorption at 280nm. Emission occurs in the wavelength range of 300 – 450 nm, with peak emission at 320 – 355 nm for Trp, 310 nm for Tyr (Lakowicz, 1999; Permyakov, 1993). A number of authors (Judge *et al.*, 2004; Lund *et al.*, 2005; Dierks *et al.*, 2007) have demonstrated that this property can be used in order to increase the contrast available within crystallization drop images to aid the identification of small microcrystalline objects. As each molecule within the sample under investigation possesses the same number of tryptophans, UV excited fluorescence has the possibility to act as a concentration-dependent marker of the protein state within the crystallization image. The crystalline state represents a highly concentrated and ordered arrangement of proteins and thus should provide the greatest source of fluorescence within a crystallization image. Precipitation is widely accepted to represent a lower concentration state of biomolecules within solution, essentially due to the lack of order in the arrangement of molecules within the precipitate leading to large enclosed solvent cavities.

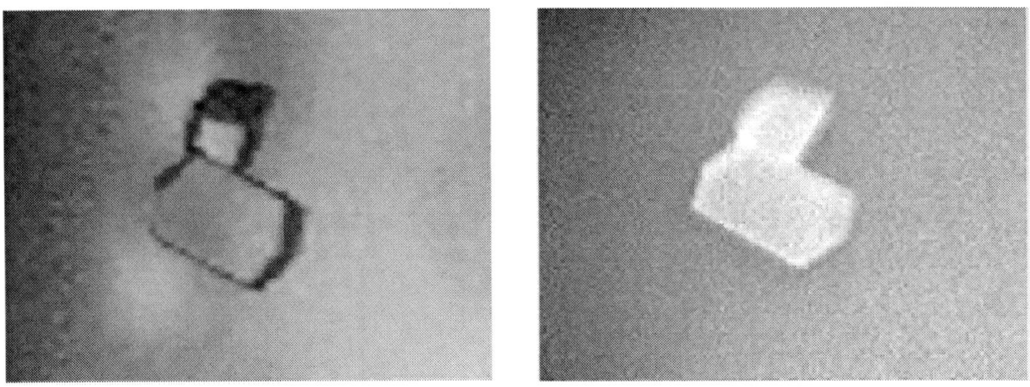

Visible Light Illumination   Ultraviolet Fluorescence

Figure 3. Intrinsic Fluorescence Imaging of Protein Crystals.

A comparison of the signals obtained using the intrinsic fluorescence properties of hen egg white lysozyme (4.7 % tryptophan content). When the intensity of the fluorescence signal is measured on a scale of 0 -255 the fluorescence signal obtained from the protein crystals is 30-60 units above that obtained from the background. Corresponding images of salt crystals are contrasted by less than 30 units. Images taken from Judge *et al.* 2004 © IUCr.

The use of the fluorescence removes signals in the image which arise from common artifacts of crystallization experiment images, such as scratches in the crystallization plate and other non-proteinaceous contributions. These contributions hamper effective identification of crystals in images that are collected using visible light microscopy. No meaningful study has been performed to assess the level of improvement of automated crystal identification using UV excited fluorescence, although Judge *et al.* (2004) demonstrate that the signals obtained from such images are such that protein crystals are contrasted by a factor of ~2 over salt crystals in the case of lysozyme (4.7% tryptophan content). However, the direct dependence of the fluorescence signal upon local protein concentration also implies that other crystallization experiment phenomena will impact on the visualization of small protein microcrystals in drops that contain precipitate or other phenomena. An illustrative example is given in Figure 4.

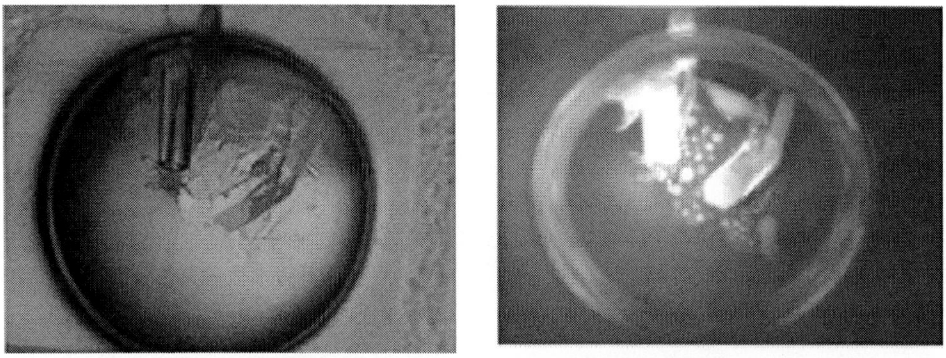

Visible Light Illumination Ultraviolet Fluorescence

Figure 4: Native Fluorescence Signals from Non-crystalline Phenomena.

A comparison of the signal strength obtained under native fluorescence conditions for protein crystals and other non-crystalline phenomena. While the protein crystals (PTP1B; 1.9 % tryptophan content) are clearly contrasted above the background when compared with the visible light illumination, a significant fluorescence signal is also seen from the round, non-crystalline objects. Images taken from Judge *et al.*, 2004. © IUCr.

While the protein crystals presented in Figure 4 are clearly visible, a significant signal is also obtained from other non-crystalline phenomena within the drop. The contrast between these two phenomena suggests that native fluorescence images may not contain sufficient signal to distinguish between small microcrystalline objects and other crystallization phenomena.

The use of UV excited fluorescence has further downstream applications, with a number of authors demonstrating that native fluorescence can be used as a method to improve crystal recognition once mounted upon an X-ray source (e.g. Pohl *et al.*, 2004; Vernede *et al.*, 2006). The improved crystal recognition can then be used to aid software solutions for automated centering and data collection. Pohl *et al.*, (2004) demonstrated that excitation wavelengths of 280 nm markedly improve crystal detection within a cryo-loop and can significantly reduce the computational processing time required for automated centering. Vernede *et al.*, (2006) extended the method by the use of a UV laser at 266 nm to allow fluorescence information to be collected from other amino acids within a protein (tyrosine and phenylalanine). The possibility to use additional amino acids to obtain a fluorescence signal removes the dependence upon the relatively rare tryptophan residue and extends the applicability of UV excited fluorescence to a wider range of biomolecules. The authors additionally demonstrated that exposure times of less than 1 second resulted in no significant damage to the protein crystals under investigation and that exposures of up to 60 seconds were required to induce structural damage to a protein under cryo-conditions (i.e. stabilization at 100 K; Vernede *et al.*, 2006). However, the effect of UV induced damage to protein crystals under non-cryo conditions can reasonably be expected to be significantly increased, potentially reducing the applicability of this method for the identification of protein crystals within a crystallization experiment.

3.2. Cross Polarization Analysis

The second intrinsic property of biomolecules that has been demonstrated to be useful for the analysis of crystallization experiments is birefringence, which exploits the anisotropic natural of crystalline materials to refract light. For birefringence to be produced the refractive index of light passing through the sample must vary with the polarization of the incoming light.

Under birefringent conditions crystalline material is clearly visible as light patches. The polarizing and analyzing filters are rotated simultaneously, whilst integrating the amount of transmitted light using a CCD camera. In this image 70 different crystallization experiments are simultaneously examined for crystal growth, in less than 1 min. Image taken from Bodenstaff *et al.*, 2002 © IUCr.

Figure 5. Birefringent Analysis of Multiple Crystallization Experiments

Thus, linearly polarized light wave becomes resolved into a faster and a slower traveling wave when passing through a birefringent sample. These waves will exhibit a phase difference (δ), which depends upon the magnitude of the plano-birefringence (Δn_p) the thickness of the sample (L) and the wavelength of the light (λ), given by the relation: $\delta = (2\pi/\lambda)\ \Delta n_p Z$. Protein crystals may be grouped into three categories: isotropic crystals, uniaxial crystals and biaxial crystals which, apart from a few exceptions, correspond to cubic, trigonal, tetragonal or hexagonal point-group symmetry or orthorhombic, monoclinic and triclinic point-group symmetry, respectively.

A number of authors have demonstrated the use of birefringence in the analysis of crystallization trails (Bodenstaff *et al.*, 2002; Echalier *et al.,* 2004). Bodenstaff *et al.* demonstrated that multiple crystallization experiments can be visualized simultaneously, with crystal material as small as 10 μm clearly identifiable (Figure 5). Additionally, Echalier *et al.* showed that the use of birefringence may be used to classify microcrystalline precipitate (Figure 6). The use of birefringence to distinguish between crystalline and non-crystalline precipitate has many implications. In cases where an initial screening of a biomolecules does not result in clear crystal growth, birefringence can be used to determine conditions which produce crystalline precipitate and which, after subsequent optimization, may result in the identification of crystallization conditions that would otherwise be missed.

Owen and Garman (2005) have extended the use of birefringence to examine isolated protein crystals. Using similar equipment to that of the groups described above they analyzed a single crystal within a cryo-mount and determined that birefringence could be used to identify optically homogenous areas within a single protein crystal. Their findings indicate that diffraction quality is strongly correlated with optical homogeneity of the sample.

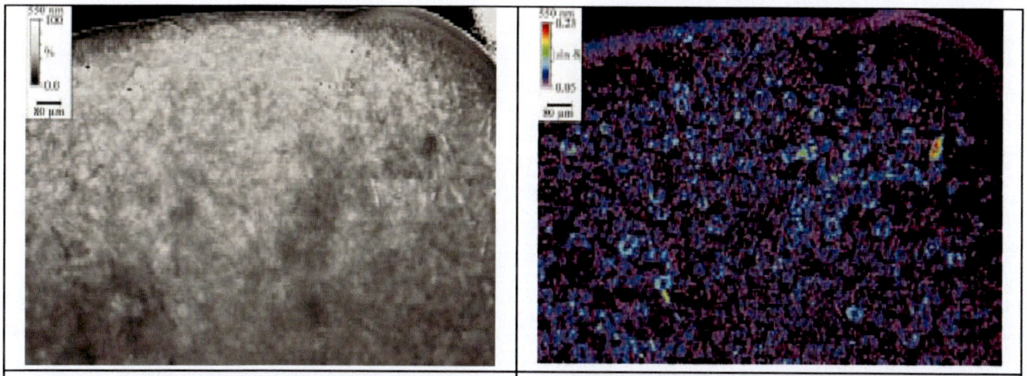

Visible light Illumination	Magnitude of the Phase shift (sin δ)

Figure 6. An example of the use of Birefringence in the Identification of Crystalline Precipitate.

A comparison of images collected under visible light illumination and under polarizing microscopy. A careful examination of the figure on the left will reveal the presence of small needle-like clusters. The identical drop is imaged using polarizing microscopy in the right hand image with the magnitude of the phase shift (expressed as sin (δ); blue: 0.05; red: 0.28), with enhanced signals available for the protein crystals. Images taken from Echalier *et al.*, (2004) © IUCr.

While birefringence is a powerful technique there are certain limitations inherent in the technique. The use of standard pressed plastic crystallization plates is prohibited as these are not constructed of an optically inert media, although alternative plates made of epoxy resin are under development. Echalier *et al* state that even plates that are marketed as possessing low birefringence still create enough background to mask the birefringence of crystalline precipitate, severely reducing the possibilities within current high-throughput facilities. Additionally, birefringence will result in false negatives when used to analyze certain types of crystal: crystals with cubic symmetry are not birefringent, whilst crystals of some of the lower symmetry space groups are also not birefringent in certain orientations. Finally, birefringence is not exclusively a property of protein crystals and many organic and inorganic materials that are present within crystallization screens can also form birefringent crystals, resulting in the generation of false positives.

4.1. Extrinsic Properties: Sample Staining Using a Coloured Dye

An initial attempt to improve the contrast between protein crystals and the remainder of the crystallization drop background was marked under the trademark Izit Crystal Dye™ by Hampton Research (Aliso Viejo, California; Cosenza *et al.*, 2003). The product consists of a solution of methylene blue trihydrate at an undefined concentration. The principle behind the increase in contrast is that the dye has a low molecular weight of ~320 Da and can penetrate the solvent channels that are present in protein crystals, but absent in salt crystals. This results in a preferential staining of proteins present within the crystals, which take up the blue colour of the dye. Thus, the major function of this dye was to distinguish between salt and protein crystals resulting from initial sparse matrix screening experiments.

As methylene blue has a (limited) affinity for proteins, this also results in a concentration dependent staining, in which protein precipitate is stained at a lower level than protein crystals. However, as the results are still evaluated under visible light illumination the presence of protein crystals can still be obscured by other artefacts within the drops, such as protein skin or precipitation and sufficient care has to be taken to titrate the amounts of dye present within the drop such that not all binding sites are saturated. In the event of over saturation the distinction between precipitation and crystals can be lost. This titration effectively requires prior knowledge of the phenomena present within the drop, reducing the applicability of this method for an automated evaluation of crystallization images.

Additionally, in cases in which the microcrystals are sufficiently small, insufficient contrast may be generated by the addition of the dye to distinguish crystalline material from other drop phenomena. According to the information supplied on the Hampton Research website (available at : http://www.hamptonresearch.com/) 1-3% of all macromolecules do not absorb sufficient amounts of dye for efficient detection; although in 1-3% of cases, the addition of Izit Crystal Dye™ is reported to have induced the formation of protein crystals.

While the addition of a dye to a crystallization image provides a simple method to increase the contrast available in crystallization experiment imaging; insufficient research has been performed to assess whether the addition of Izit Crystal Dye™ to drops significantly reduces the formation of crystals. More recently, the addition of some dyes (see below) has been demonstrated not to effect crystallization at concentrations below 0.1 mM. However, the addition of Izit Crystal Dye™, at concentrations that are sufficient to increase contrast of the crystal, is generally accepted to impact strongly on the diffraction properties of the crystal (unpublished data). As a result the dye has fallen out of use to some extent, although the use of protein or DNA specific markers for protein crystallization drop classification has been investigated and extended by other researchers, as described below.

4.2. Covalent Modification of the Sample

A recent publication by Forsythe *et al.,* (2006) expanded upon the use of UV fluorescence for the identification of protein crystals. The authors proposed that the covalent attachment of the fluorescent dye carboxyrhodamine to proteins could be used a method to improve the contrast of protein crystals in high throughput crystallization experiments. A number of well characterized samples were used (thaumatin, lysozyme and ribonuclease) and the effects of different labelling ratios, up to 10 % of the sample, were determined by comparing crystallization rates and diffraction properties. In the initial experiments the protein were labelled only on the N-terminal amide, but the authors also investigated the effects of randomly labelling surface exposed lysine amides. In both cases, the covalent attachment of carboxyrhodamine resulted in the anticipated increase in fluorescence signal of the samples. A significant enhancement in the degree of contrast of protein crystals above the background was observed, when compared with that available from native UV fluorescence alone (Figures 7). Subsequent crystallization condition sparse matrix screening experiments using the proteins canavalin, α-lactoglobulins A and B and chymotrypsinogen (all estimated to be 0.5 % labelled) demonstrated the effectiveness of this approach in identifying protein crystals, even when obscured by precipitate. Excitation of the dye was measured through a low bandpass filter at a wavelength of 470 nm, with emission determined through a 500 nm high bandpass filter.

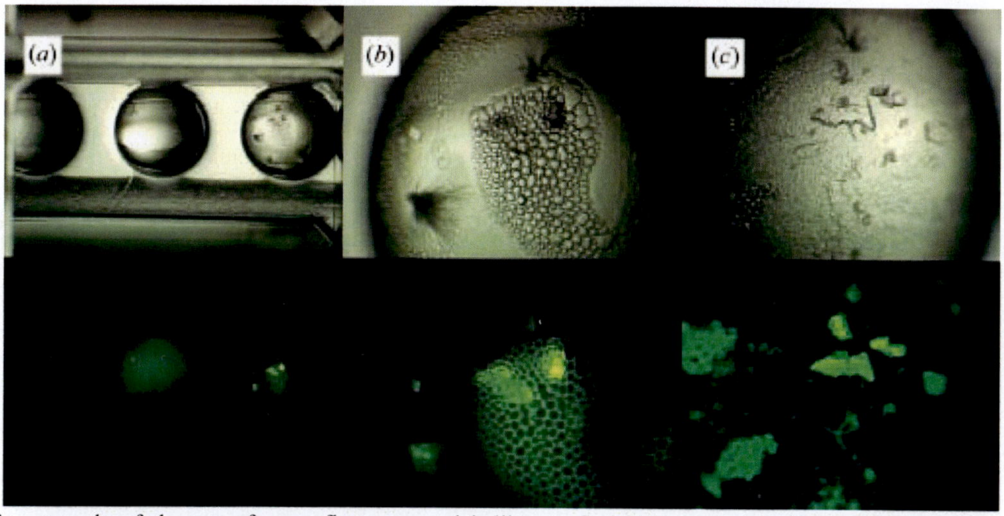

An example of the use of trace fluorescence labelling to find hidden crystals. The upper row shows crystallization experiment of chymotrypsinogen B as imaged using white light, while the lower row shows the corresponding image under fluorescent illumination. (a) Details a low-magnification image the crystallization wells, (b) shows a higher magnification image of the well on the right hand side of (a). Note that the crystals obscured by other proteinaceous phenomena remain clearly detectable in the fluorescence images. Subsequently, the foam layer obscuring the crystal was removed and the well re-imaged such that the crystals are more clearly visible under white light illumination (c). Note that the crystals are greatly contrasted above the other crystallization phenomena (precipitation and skin). Image taken from Forsythe *et al.*, (2006) © IUCr.

Figure 7. Trace Labelling of Proteins to Enhance the Fluorescence Signal

The published protocol for N-terminal attachment of carboxyrhodamine requires the transfer of the protein into 25 mM HEPES buffered at pH 7.5 and concentration to 2 - 10 mg/ml. A volume of carboxyrhodamine succinimidyl ester (CSE) dissolved in dimethyformamide (DMF) is then added to the protein sample and the reaction is left to progress of 0.5 -1 hour in a dark room at room temperature, before the reaction was stopped through the addition of ethanolamine as source of primary amines. The relative volumes of DMF solubilized CSE were chosen to yield the final labelling concentration with an example given of 9.8 μl of DMF solubilized CSE and 10 μl of ethanolamine being required to label 0.5 % of a 10 ml lysozyme solution at 5 mg/ml. (resulting in a solution containing 0.2 % organic solvent, with 5-10 % organic solvent content being generally accepted as a boundary for protein stability). The preparation of lysozyme samples that are randomly labelled on lysine side chains was prepared by dilution of the protein solution into a 0.1 M sodium borate buffer at pH 9.5 immediately prior to carrying out the derivatization reaction above. Finally, a gel filtration step was performed to remove residual labelling components. The authors established that no significant effect on the rate of formation of lysozyme, thaumatin or ribonuclease crystals in a sparse matrix screen (Hampton Research, Aliso Viejo, California) could be discerned when labelling concentrations were lower than 10 % of the sample, indeed a slight increase of the number of conditions that produced crystals could be observed for the labelled samples. Additionally, the authors investigated the diffraction properties and demonstrated that a 1 % labelling ratio produced no significant effect on the quality of the data collected.

These results led the authors to recommend the use of trace labelling in high throughput research, although there are a few minor points that need to be taken into account. Firstly, the transfer of the samples at high concentration into a salt free buffer described in the protocol above may not be suitable for a subset of proteins under analysis in high throughput platforms. This should not be considered to be a major drawback as the salt concentration in the buffer is not predicted to affect the reaction, but the buffer should not contain a source of primary amines (e.g. ammonium ions, or Tris buffer, etc.). Thus, this protocol could be modified to meet the specific requirements and features of different proteins (e.g. pI, number of lysine residues, salt concentration required to maintain solubility, etc.). However, this would require a pre-screening step for each sample under investigation, with a concomitant increase in the amount of manual intervention required for each sample.

Such conditions, coupled with the addition of 0.1 % DMF in the example above, may lead to the loss of a subset of samples through precipitation before optimal conditions are identified. Additionally, the general rule in purification of samples for crystal screening is to achieve as high a purity level as possible and a significant subset of researchers may feel that the inhomogeneity of the sample would adversely affect the crystallization properties of the sample. A counter argument to this would be to cite the growing use of protocols for the methylation of lysine residues, which have been shown to improve the crystallization properties of a number of different proteins (e.g. Walter *et al.*, 2006). The increase in fluorescence signals obtained markedly improves the possibility of correctly identifying protein microcrystals, even when occluded by other drop phenomena. This alone warrants the serious consideration of the technique as an additional method. While further data is still required on the effects of labelling on protein crystal diffraction properties the authors point out that the dye may be omitted at all subsequent steps, once the crystallization condition has been identified.

4.3. Non-covalent Modification of the Sample

In an alternative approach to the covalent linking of dyes to protein molecules a number of authors (Kettenberger *et al.,* (2005), Groves *et al.*, (2007)) have investigated the potential of adding fluorescent dyes to crystallization experiments. The advantage in this approach is that no additional handling of the sample is required, with the fluorescent dyes being added to the crystallization buffers. Kettenberger *et al.* demonstrated that the addition of a non-specific dye (SYBR-gold, Invitrogen) to crystals containing DNA or RNA molecules resulted in a strong fluorescence signal. In parallel experiments crystals which contained only protein components showed 30 – 50 fold less fluorescence signal, when examined under fluorescence conditions (Figure 8). In this example the authors added both 1:10000 dilution of the commercially available dye solution to preformed crystals of the 12 subunit Pol II to distinguish between DNA/protein complex crystals and protein crystals.

In the experiment described above the protein/DNA crystals are incubated in mother liquor solution containing trace amounts of SYBR-Gold (Invitrogen), a cyanine dye that binds to any DNA or RNA, irrespective whether it is single- or double- stranded. Upon binding of SYBR-gold to nucleic acids a dramatically enhancement of the fluorescence yield of SYBR-Gold is obtained whereas, as anticipated, proteins have essentially no effect (Tuma et al., 1999). However, no experiments were performed to establish the effect of SYBR-gold upon crystal growth or subsequent diffraction properties.

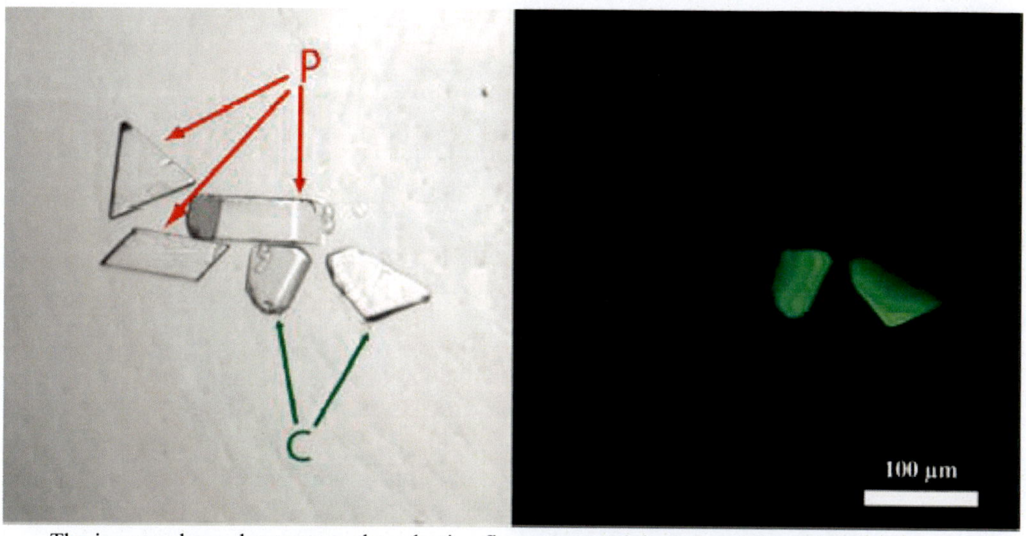

The images above demonstrate the selective fluorescent staining of nucleic acid-containing crystals. Protein crystals in the free form (P) and in the form of a nucleic acid complex (C) are shown. On the left, a transmission light microscopic image shows the positions of all crystals within the drop. On the right, the fluorescence images show selective staining of nucleic acid complex crystals after incubation with SYBR-Gold. Filters providing an excitation wavelength range of 450–490 nm and an emission wavelength range >515 nm were used. The P crystals are composed of complete 12-subunit Pol II in the free form and the crystals labelled "C" are the same complex in the presence of an 11-mer DNA duplex carrying a 15-nucleotide overhang, which was annealed to an 11-mer RNA (Kettenberger et al., 2004). Images taken from Kettenberger *et al.* (2006) © IUCr.

Figure 8. An Example of the Use of Non covalent Dyes as a Method to Identify DNA Crystals.

The authors also investigated the possibility of tracking the binding of a fluorescently labelled (Cy3) protein through the increase of the fluorescent signal when Pol II crystal were soaked in a solution containing the 20 kDa transcription elongation factor TFIIS, previously covalently labelled with Cy3. A significant increase in the fluorescence signal of the Pol II crystals was observed when Pol II crystals were incubated in a solution containing 2 mg/ml TFIIS for 4 hours. This observation raises the possibility of studying larger macromolecular assemblies for complexes that are insufficiently stable to allow purification, through the growth of smaller stable complexes and the use of fluorescence microscopy to monitor subsequent soaking experiments.

A parallel experiment was performed by the authors of this review, in which the effect of the addition of a non-covalent dye (1,8-ANS, Molecular Probes) to crystallization screening solutions was examined (Groves *et al.*, 2007). Using a sample set of 10 proteins (lysozyme, proteinase K, glucose isomerase, insulin, xylanase, trypsin, Rv0153, Rv2234, ppm1p and tap42p) the crystallization frequency under sparse matrix conditions was compared under a variety of different dye concentrations. The published results indicate that at concentrations of dye lower than 1 mM no effect on the crystallization rate could be determined. The addition of the dye had the anticipated effect of increasing the fluorescence properties of the macromolecules under investigation, with a significant signal available at dye concentrations as low as 0.9 nM. At such low dye concentrations the implication is that the weak affinity between the dye and protein macromolecules is sufficient to sequester the dye to protein crystals, resulting in an additional enhancement of the fluorescence signals obtained from protein crystals above the background. The fluorescence images showed sufficient contrast that lysozyme crystals were clearly identifiable, even in the presence of heavy protein precipitation (Figure 9g). Additionally, small protein microcrystals (cf. Figure 9d), effectively undetected under visible light illumination are clearly distinguishable. As the dye is added immediately prior to crystallization screening, no significant increase in handling is obtained, although an increased tendency to precipitate was observed for samples in which the dye was present at concentrations in excess of 1 mM.

X-ray diffraction studies were carried out on one of the samples that demonstrated the strongest increase in fluorescence, ppm1p. The high resolution data (1.6 Å) data did not reveal to presence of a single, well order dye molecule at any location on the surface of the refined protein structure (Groves *et al.*, 2007). This finding could support the hypothesis that the interaction between the dye and protein is non-specific and transitory. A transitory interaction (or high k_{off}) would be predicted to have limited effect on the crystallization. While the general effect of the presence of the dye on data collection has yet to be more rigorously evaluated, with data available from only a limited number of systems, the dye can be omitted from any subsequent optimization screens once initial hits have been identified.

The use of 1,8-ANS was made based upon the particular properties of this dye, which is one of the oldest commercially available dyes, specifically that it has been used to demonstrate the presence of a wide variety of proteins. The modern trend in fluorescence microscopy has been to produce dyes that are more specific, but such products would potentially limit the efficacy of this technique, in which the non-specificity of the dye is highly advantageous. The use of more modern dyes may indeed provide additional benefits, but with an increase in cost and handling.

This procedure shows the same benefits in sample identification as the covalent modification of proteins demonstrated by Forsythe *et al.*, (2006) with the added advantage that no additional pre-screening steps are required, and is therefore easier to integrate within a HTP platform. As observed for the covalent modifications and the use of intrinsic UV fluorescence the enhanced fluorescence signal obtained from the crystals can also be used in downstream processes, such as the centring of macromolecular crystals on an X-ray source and, potentially, automated sample handling from the crystallization drop to the X-ray source. The disadvantages to the technique are similar to those mentioned above, in terms of the addition of an impurity to the sample prior to screening and the potential effect on protein structure. It should also be noted that the nature of the protein under investigation plays a crucial role in the degree of enhancement of fluorescence obtained, given that the signal obtained from the dye requires hydrophobic residues to be exposed for fluorescence conditions to be met. As the size of macromolecular complexes under investigation increases, the likelihood of such surfaces being available increases, although small and highly charged proteins may show insufficient interaction with the dye to generate a detectable signal.

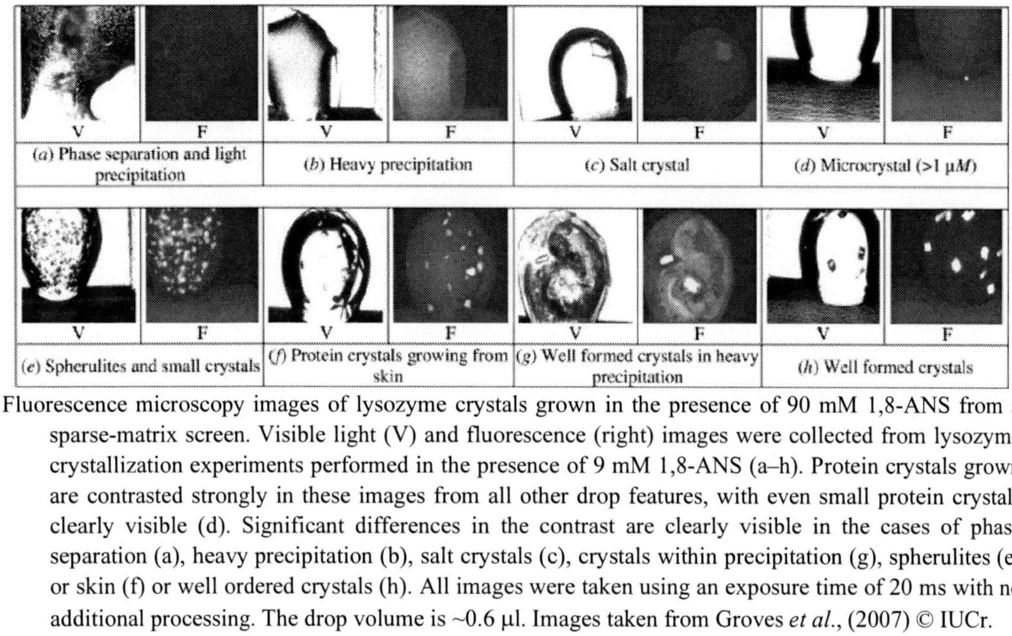

Fluorescence microscopy images of lysozyme crystals grown in the presence of 90 mM 1,8-ANS from a sparse-matrix screen. Visible light (V) and fluorescence (right) images were collected from lysozyme crystallization experiments performed in the presence of 9 mM 1,8-ANS (a–h). Protein crystals grown are contrasted strongly in these images from all other drop features, with even small protein crystals clearly visible (d). Significant differences in the contrast are clearly visible in the cases of phase separation (a), heavy precipitation (b), salt crystals (c), crystals within precipitation (g), spherulites (e) or skin (f) or well ordered crystals (h). All images were taken using an exposure time of 20 ms with no additional processing. The drop volume is ~0.6 μl. Images taken from Groves *et al.*, (2007) © IUCr.

Figure 9. An Example of the Use of Non-covalent Dyes as a Method to Identify Protein Crystals

5. Conclusion

In the preceding sections we have reviewed current methods that are available to rapidly identify conditions under which protein crystals are formed. No true integration of all of the above methods has yet been established, although we have demonstrated that the performance of crystal recognition software (e.g. XREC, Pothineni, *et al.*, 2006; Groves *et al.*, 2007) is improved through the use of fluorescence microscopy by a factor of over 50 %. Commercial systems (e.g. Inspect-X, Emerald Biosystems) have recently been brought onto the market which integrate UV, cross-polarization and visible light microscopy, but insufficient data is currently available to establish whether modifications are still necessary to the material used to produce crystallization plates. As noted above, current crystallization plates are not suitable for analysis under birefringent conditions and a number of the plates currently in use by structural biologists worldwide show significant background under UV illumination. A concerted effort is required both by academia and industry to provide a solution for this integration, in addition to investigating additional approaches to macromolecular crystal identification.

As high throughput crystallization robotic advance to requiring ever lower sample volumes it is foreseen that the detection of smaller protein microcrystals will become more pressing. The limitations in visible light microscopy are well balanced by developments in fluorescence imaging and birefringence which are, to an extent, less sensitive to the absolute crystal size. Indeed, the biophysical approaches above can be used to categorize the protein solution state (precipitation, phase separation, etc.) in conditions that are chemically far from those under which the final diffraction quality crystals may be produced. Thus, methods that provide information on the spatial and concentration distribution of macromolecules in crystallization experiments could become an integral and essential part of structural biology in the near future.

References

Berman, H.M., Westbrook, J., Feng, Z., Gilliland, G., Bhat, T.N., Weissig, H., Shindyalov, I.N. & Bourne, P.E. (2000) *Nucleic Acids Research*, **28** pp. 235-242.

Bern, M., Goldberg, D., Stevens, R. C. & Kuhn, P. (2004). *J. Appl. Cryst.* **37**, 279±287.

Lakowicz, J. R. (1999). Principles of Fluorescence Spectroscopy, 2nd ed. New York: Kluwer Academic/Plenum Publishers.

Berry, I.M. et al., (2006) *Acta Cryst.* **D62**, 1137-1149

Bodenstaff, E. R., Hoedemaeker, F. J., Kuil, M. E., De Vrind, H. P. & Abrahams, J. P. (2002). *Acta Cryst.* **D58**, 1901±1906.

Cosenza, L., Bray, T. L., DeLucas, L. J., Gester, T. & Hamrick, D. T. (2003). *Use of Dye to Distinguish Salt and Protein Crystals Under Microcrystallization Conditions.* WO 2003012430. University of Alabama Birmingham Research Foundation, USA.

Dierks, K., Meyer, A., Einspar, H. & Betzel, C. (2007) *Acta Cryst.* **A63**, s20

Echalier, A., Glazer, R. L., Fülöp, V. & Geday, M.A. (2004) *Acta Cryst.* (2004). D60, 696-702

Forsythe, E., Achari, A & Pussey, M. (2006) *Acta Cryst.* **D 62**, 339-346

Geerlof, A. et al. (2006). *Acta Cryst.* **D62**, 1125–1136.

Judge, R.A., Swift, K. & González, C. (2005) *Acta Cryst.* **D61**, 60-66

Kettenberger, H. & Cramer, P. (2006) *Acta Cryst.* (2006). **D62**, 146-150

Müller-Dieckmann, J. (2006) *Acta Cryst* **D 62** 1446-1452.

McPherson, A. (1999) Crystallization of Biological Macromolecules, Cold Spring Harbor Laboratory Press, ISBN: 0-87969-527-7.

Owen, R. L. & Garman, E. (2005). *Acta Cryst.* **D61**, 130–140.

Kettenberger, H., Armache, K. J. & Cramer, P. (2004). *Mol. Cell*, **16**, 955–965.

Permyakov, E. A. (1993). Luminescent Spectroscopy of Proteins. Boca Raton, FL, USA: CRC Press.

Pohl, E., Ristau, U., Gehrmann, T., Jahn, D., Robrahn, B., Malthan, D., Dobler H. & Hermes

C. (2004) *J. Synchrotron Rad.* **11**, 372-377
Pothineni, S. B., Strutz, T. & Lamzin, V. S. (2006). *Acta Cryst.* **D62**, 1358–1368.
Rupp, B. (2003) *J. Struct. Biol.* **142**, 162-169
Tuma, R.S., Beaudet, M.P., Jin, X., Jones, L.J., Cheung, C.Y., Yue, S. & Singer, V.L. (1999) *Anal. Biochem*, **268**, 278-288.
Vernede, X., Lavault, B., Ohana, J., Nurizzo, D., Joly, J., Jacquamet, L., Felisaz, F., Cipriani, F & Bourgeois, D. (2006) *Acta Cryst.* (2006). D62, 253-261.
Voet, D. & Voet, J. G. (1995). *Biochemistry*, 2nd ed. New York: John Wiley & Sons.
Walter, T.S. et al. (2006) *Structure*, **14**, 1617-1622
Walker, C.G., Foadi, J. & Wilson, J. (2007) J. *Appl. Cryst.* **40**, 418-426
Wilson, J. (2002). *Acta Cryst.* **D58**, 1907±1914.
Wilson, J. & Berry, I.M. (2005) *J. Appl. Cryst.* **38**, 493-500

In: Astrophysics and Condensed Matter
Editor: Thomas G. Hardwell, pp. 195-214
ISBN: 978-1-60456-332-0

Chapter 7

MØLLER ENERGY-MOMENTUM COMPLEX FOR A CHARGED BLACK HOLE SOLUTION IN HETEROTIC STRING THEORY

I. Radinschi*
Department of Physics, "Gh. Asachi" Technical University,
Iasi, 700050, Romania

Abstract

It is well-known that one of the most challenging problems of general relativity, which has not yet yield a satisfactory solving is the energy-momentum localization. For giving an answer to the problem of energy and momentum localization, many attempts have been made in the past but this remains an important issue to be settled. The difficulty relies in the lack of a generally accepted expression for the energy density. Since Einstein has given his important expression for the energy-momentum complex, the localization of energy is also connected to the use of other energy-momentum complexes, including the prescriptions of Landau-Lifshitz, Papapetrou, Bergmann-Thomson, Weinberg and Møller. The Einstein, Landau-Lifshitz, Papapetrou, Bergmann-Thomson and Weinberg energy-momentum complexes give meaningful results only if the calculations are restricted to quasi-Cartesian coordinates. The energy-momentum complex of Møller gives the possibility to determine the expressions for energy in any coordinate system. In his recent paper, Lessner formulated his viewpoint that the Møller definition is a powerful concept of energy and momentum in general relativity. The concept of energy-momentum complex was reconsider by Virbhadra and his collaborators, and since then many interesting studies were done in this area. Chang, Nester and Chen showed that the energy-momentum complexes are actually quasi-local and legitimate expressions for the energy-momentum. They pointed out the existence of a direct relationship between energy-momentum complexes and quasi-local expressions because every energy-momentum complex is connected to a legitimate Hamiltonian boundary term.

We evaluate the energy distribution of a charged black hole solution in heterotic string theory in the Møller prescription. This solution was given by A. Sen and is an exact classical solution in the low-energy effective field theory, which describes a

*E-mail address: radinschi@yahoo.com

black hole carrying a finite amount of charge and angular momentum. We consider that the angular momentum vanishes and we have two metrics ds^2 and ds_E^2. The first metric describes the transformed solution and the second is the Einstein metric, which is given by $ds_E^2 = e^{-\Phi'} ds^2$.

For both metrics we obtained meaningful results for the energy distribution in the Møller prescription. The energy distribution depends on the mass m and the arbitrary number α. If we consider only the first term in the expressions for the energy distribution, we obtain for the energy the expression $E = m$ in the both cases. For the Einstein metric we perform some calculations and the case $\alpha = 0$ corresponds to the value $E = M$, the energy is equal to the mass M of the black hole.

Our result sustain the Lessner opinion that the Møller energy-momentum complex is a powerful tool for evaluating the energy and momentum in general relativity, and the very important Cooperstock hypothesis which states that "the energy and momentum are confined to the regions of non-vanishing energy-momentum tensor of the matter and all non-gravitational fields". This result, and many others obtained by several authors support the idea that the Møller energy-momentum complex yield acceptable results for different space-times.

Keywords: Møller energy-momentum complex, charged black hole solution, heterotic string theory

PACS: 04. 20 Dw, 04. 70. Bw, 11. 17.+y, 97. 60. Lf

1. Introduction

It is a problem of actuality that one of the most interesting problems of general relativity the energy-momentum localization has not yet yielded a satisfactory solving. Numerous attempts at constructing an energy-momentum density have been made but none of them yield a generally accepted expression. There are many methods used for the energy and momentum localization, and we notice the superenergy-tensors, the pseudotensorial method and here the energy-momentum complxes are included, the quasi-local quantities and the tele-parallel theory of gravitation. We mention the works of Bel [1]-[2] and Robinson [3] and, influenced by their interesting results the papers of many others physicist [4]. Referring to the quasi-local approach very important are the works on quasi-local mass of Brown and York [5] and Hayward [6]. Dealing with this method implies to determine the effective energy of a source by measurements on a two-surface. In this case the masses are evaluated over a two-surface as opposed to an integral spanning over a three-surface of a local density as in the case for pseudocomplexes. Furthermore, there are a lot of works which have been done to construct superenergy tensors [4], [7].

Concerning the pseudotensorial definitions, which imply to work with energy-momentum complexes we point out that there are various prescriptions including those of Einstein [8]-[9], Landau-Lifshitz [10], Papapetrou [11], Bergmann-Thomson [12], Weinberg [13] and Møller [14]. There are doubts that these prescriptions could give acceptable expressions for energy for a given space-time and the problem consists in a) most of the energy-momentum complexes (ELLPW) and Bergmann-Thomson (except the Møller energy-momentum complex) are non-covariant, coordinate dependent expressions and yields meaningful results only the calculations are carried out in quasi-Cartesian coordinates, and b) this is in good agreement with the equivalence principle, which states that gravity cannot be detected at a point.

In the recent years the issue of the energy-momentum localization by using the energy-momentum complexes was re-opened. Many researchers studied different space-times and interesting results were obtained [15]-[18], demonstrating that these definitions are powerful concepts for energy-momentum localization in general relativity. For many geometries the (ELLPW), Bergmann-Thompson and Møller prescriptions yield good results, and this is also available in the case of 2 and 3 dimensional space-times. The concept of energy-momentum complex was reconsider by Virbhadra and his collaborators [15] and since then reliable results were obtained in this area [16]-[18]. Although the energy-momentum complexes of Einstein, Landau-Lifshitz, Papapetrou, Bergmann-Thomson and Weinberg are coordinate dependent they can give a reasonable result if the calculations are performed in quasi-Cartesian coordinates. On the other hand, the Møller energy-momentum complex allows performing the calculations in any coordinate system. Some interesting results sustain this conclusion, that the pseudotensorial prescriptions are powerful concepts for energy-momentum localization. Aguirregabiria, Chamorro and Virbhadra [17] obtained that several energy-momentum complexes "coincide" for any metric of the Kerr-Schild class. Furthermore, we point out the work of Virbhadra [18] where he established that several energy-momentum complexes (ELLPW) comply with the quasi-local mass definition of Penrose for a general non-static spherically symmetric metric of the Kerr-Schild class. He determined the expression for energy distribution in the case of this general metric. He concluded that though the energy-momentum complexes are non-tensors under general coordinate transformations, the local conservation laws with them hold in all coordinate systems. Furthermore, Chang, Nester and Chen [19] showed that the energy-momentum complexes are actually quasi-local and legitimate expression for the energy-momentum, and established the existence of a direct relationship between energy-momentum complexes and quasi-local expressions because every energy-momentum complexes is strongly connected to a legitimate Hamiltonian boundary term.

Møller [14] elaborated an expression which enables one to calculate the energy distribution in any coordinate system, not only in quasi-Cartesian coordinates. The energy and energy current density components of the Møller energy-momentum complex can be transformed as a four-vector density with respect to the group of purely spatial transformations. It seems that this definition, which is coordinate independent gave a good solving to the problem of energy and momentum localization. But, Møller himself noticed some drawback of his prescription [20]. He showed that the total energy-momentum components P_i does not transform like a four-vector under Lorentz transformations. So, the expression obtained for the energy density is not the best candidate for describing the energy distribution. Later, Kovacs [21] sustained that the expression given by Møller for his energy-momentum complex is suitable for the evaluation of energy. Further, Novotny [21] agreed that Møller's conclusion about his energy-momentum components P_i is right, and this expression indeed does not transform like a four-vector under Lorentz transformations. But Lessner, in his important work [22] noticed that the problem lies with the interpretation of the result from a special relativistic point of view instead of a general relativistic one. In conclusion, the Møller prescription can be considered as a powerful tool for evaluating the energy distribution of a given space-time. The advantage in using the Møller energy-momentum complex is that there are no constraints to carry out the calculations in quasi-Cartesian coordinates, so we can evaluate the energy and momentum components in any coordinate system.

A plethora of meaningful results recently obtained support the conclusion given by Lessner [22] in his recent paper about the Møller definition in general relativity. Virbhadra [15] calculated the energy distribution in the Kerr-Newmann space-time and his result is very interesting. The Møller definition provides the existence of an amount of energy in the exterior of the Kerr-Newmann black hole. In the particular case of the Reissner-Nordström space-time the energy is the same as that proposed by Komar and evaluated by Cohen and de Felice [23]. Further, the definition of the effective gravitational mass given by Kulkarani, Chellathurai and Dadhich [24] has used by the last two to compute the effective mass of a Kerr-Newmann black hole and a Kerr black hole embedded in a magnetic field. This formalism yields the effective gravitational mass of the Reissner-Nordström space-time that a neutral test particle "feels" in this field.

Xulu [25] calculated the energy distribution in the Kerr-Newman space-time using the Møller prescription. He gave important conclusions about the expressions for energy distribution obtained in the Møller prescription for the Schwarzschild, Vaydia and Janis-Newmann-Winicour space-times [26].

I-Ching Yang et. al. [27] obtained applying the Møller energy-momentum complex the energy distributions of the Garfinkle, Horowitz and Strominger (GHS) solution and of the dyonic dilaton black hole solution. They pointed out that the energy distributions are shared both by the interior and the exterior of these charged dilaton black holes. The presumption that the Møller energy-momentum complex is a 3-dimensional scalar under purely spatial transformation in these energy distributions was verified.

Very interesting are the results obtained by Vagenas [28]-[29]. He used the Møller and Einstein energy-momentum complexes [28] in order to perform the calculations for the energy distributions associated with the two-dimensional Schwarzschild and Reissner-Nordström black hole backgrounds. He concluded that while the Møller definition provides meaningful physical results, the Einstein energy-momentum complex fails to do so in these gravitational backgrounds. The explicit expression for the energy of a two-dimensional Schwarzschild black hole was found and the result was extracted in three different coordinate systems (gauges). The conclusion was that in all three gauges the result is the same. This result support the opinion that the Møller prescription allows to carry out the calculations in any coordinate system and gives meaningful outcomes. In another work, he calculated the effective gravitational mass as well as the energy and momentum distributions of a radiating charged particle in the Einstein universe with the Møller energy-momentum complex [29]. The Einstein universe is known as a generalization of Bonnor and Vaidya space-time and several space-times are limiting cases of it. The results obtained in the case of the Reissner-Nordström black hole background are the same with those derived by de la Cruz and Israel, Cohen and Gautreau, Cohen and de Felice using the Komar mass and by other authors [23].

Gad [30] studied the energy distribution associated with a stringy charged black hole in the Møller prescription and obtained an acceptable result. The energy distribution presents a dependence on the mass of the black hole, electric charge and the asymptotic value of the dilaton field. Furthermore, he calculated the energy and momentum distributions associated with a Gödel-type space-time in the Møller prescription [31]. This result is also an interesting one. The author investigated the energy and momentum associated with this space-time in cylindrical coordinates. He evaluated the energy and momentum density components of the Gödel-type space-time as well as of the homogeneous Gödel-type space-time. A conclusion was that the Møller definition does not support the Cooperstock hypothesis for the homogeneous space-time.

The present author, using the Møller formalism computed the energy distribution of a conformal scalar dyon (CSD) black hole [32]. This black hole solution was developed by Virbhadra [33] and is an exact solution of Einstein–Maxwell conformal scalar field equations which is characterized by the scalar charge, electric charge and magnetic charge. This solution is also the magnetic generalization of the Bekenstein solution [34]. The total gravitational energy of the CSD black hole obtained in the limit $r \to \infty$ is equal to the CSD charge. This result is in total agreement with those obtained by Virbhadra in the Weinberg prescription [35] and by the author in the Tolman prescription [32].

The subject of another paper is the evaluation of the energy distribution of an anisotropic model of universe based on the Bianchi type I metric by using the Møller energy-momentum complex [36]. We obtained a result which agrees with those of Banerjee and Sen [37], Xulu [38] and ourself [39]-[40]. The total energy density vanishes everywhere because the energy contributions from the matter and field inside an arbitrary two-surface, in the case of the anisotropic model based in the Bianchi type I metric, cancel each other. Our result supports the viewpoint of Tryon [41] which assumed that "the Universe appeared from nowhere about 10^{10} years ago and the conventional laws of physics need not have been violated at the time of the creation of the Universe. According with his Big Bang model, the Universe is homogenous, isotropic and closed and consists of matter and anti-matter equally." Furthermore, the net energy of the Universe may be equal to zero. Our result supports the calculations of Rosen [42] which are in good concordance with the studies of Tryon. We studied the Bianchi I space-time because by using an adequate transformation of coordinates we are allowed to reach the line element which describes the spatially Friedmann–Robertson–Walker space-time. For an universe described by a line element that can be reduce to the spatially flat FRW metric the total energy vanishes [43].

Another work is dedicated to the evaluation of the energy of an anisotropic model of universe based on the Bianchi type VI0 metric by using the Møller energy-momentum complex [44]. The energy due to the matter plus field is determined to be equal to zero. We obtain in the Møller prescription the same result for the total energy of a Bianchi type VI0 universe as those obtained in the Tolman, Landau and Lifshitz, Papapetrou and Bergmann prescriptions [44]-[45].

The subject of other work, obtained by using the Møller prescription, is the one that yield the expression for energy distribution [46] of a charged regular black hole solution given by E. Ayón-Beato and A. Garcia (ABG) [47]. This is a solution to the coupled system of the Einstein field and equations of the nonlinear electrodynamics. This solution describes a regular black hole with mass M and electric charge q and avoids thus the singularity problem. Further, the metric at large distances has the same behaviour as the Reissner-Nordström solution. The usual singularity of the RN solution, at $r = 0$, has been smoothed out and now it simply corresponds to the origin of the spherical coordinates. The Møller formalism provides in the expression for energy a term which agrees with the Komar [48] prescription.

In collaboration with I-Ching Yang we calculated the energy of the Melvin magnetic universe [49]. The energy distribution depends on the magnetic field.

In another paper we elaborated a study about the difference of energy between the Einstein and Møller prescriptions [50]. We compare the difference of energy between the Einstein and Møller prescriptions with the energy density for some black hole solutions and found out a special connection between the aforementioned two quantities. We carried out these calculations for the Schwarzschild, Kerr, Reissner-Nordström, charged regular black hole solution given by E. Ayón-Beato and A. Garcia (ABG) [47] and for the static spherically symmetric nonsingular space-time given by Dymnikova. Further study is needed to understand the difference of energy between the Einstein and Møller energy-momentum complexes for other classes of black hole solutions.

A result which also point out the importance of the pseudotensorial prescriptions and particularly of the Møller prescription is that connected to the evaluation of the energy distribution of a static spherically symmetric nonsingular black hole [51]. For this black hole solution we noticed that total energy is given by the mass of the black hole, which is the same as the ADM mass for this metric. This result is the same in the Papapetrou, Weinberg and Møller prescriptions. The Cooperstock hypothesis [52] which states that the energy and momentum are confined to the regions of non-vanishing energy-momentum tensor of the matter and all non-gravitational fields is also supported.

The subject of another paper is the study of the energy distribution of the gamma metric in the Møller prescription [53]. The result is the same as obtained by Virbhadra [54] in the Weinberg prescription.

Other work is focused on the evaluation of the energy of some $2+1$ dimensional black hole solutions [55] with the Møller energy-momentum complex. We study the cases of the non-rotating and uncharged black hole, charged black hole, the solution coupling to a static scalar field and the static and circularly symmetric exact solution of the Einstein-massless scalar equation. Our conclusion is that the Møller definition provides meaningful physical results for these four aforementioned $2+1$ dimensional black hole solutions.

In this chapter we evaluate the energy distribution of a charged black hole solution in heterotic string theory applying the Møller prescription. This black hole solution was given by A. Sen [56]. We study the energy associated with this solution because we think it can yield us an interesting result. For this purpose, we consider the case when the angular momentum vanishes. We have two cases. In the first case, the metric that we use describes a rotating charged black hole solution generated by starting from a rotating black hole solution carrying no charge (the Kerr solution) with the help of the twisting procedure [57], [64], [65]. The second case is that of the Einstein metric, which is obtained from the first metric by a transformation which implies to use the dilaton field. The Møller energy-momentum complex gives us consistent results for these two geometries. Through the chapter we use geometrized units ($G=1, c=1$) and follow the convention that Latin and indices run from 0 to 3 and Greek indices run from 1 to 3.

2. Energy in the Møller Prescription

An interesting problem that have arisen in recent years is the construction of classical soliton [58]-[59] and black hole [60]-[61] solutions in string theory. We refer to the solutions in low dimensions [62] and to the solutions which correspond to solvable conformal field theories [63]. The importance of the study of black holes in string theory is explained in the following way. Because string theory is expected to provide us with a finite and clearly defined theory of quantum gravity, the answer of many questions related to the black hole evaporation could be solved in the context of string theory. To do this, is required to construct black hole solutions in string theory. The string theory possess extended symmetries. There are two classes of symmetries which were analyzed in the supergravity theories. The first class of symmetries [64] gives us the possibility to generate charged solutions starting from a neutral one. This is the method used by Sen [56] to obtain a charged black hole solution. The second class of symmetries allows us to construct solutions carrying magnetic charge starting from solutions carrying electric charge [65]. Further, in an important paper A. Sen [66] described a method to construct the most general black hole solution in one particular four dimensional string theory, the heterotic string theory compactified on a six dimensional torus. In another paper, A. Sen [67] investigated the possibility that some of the extremal black hole solutions in string theory which have the same quantum numbers as the Bogomol'nyi saturated elementary string states can be identified to elementary string excitations. Together with G. Horowitz [68] he constructed and study the electrically charged, rotating black hole solution in heterotic string theory compactified on a $(10-D)$ dimensional torus. Furthermore, Horne and Horowitz has shown that an arbitrarily small amount of angular momentum change the properties of extremal charged black holes coupled to a dilaton and gave an exact rotating black hole string solution [69]. These studies and other important works [70] support the importance of the black hole solutions in string theory.

An interesting and important problem connected to these black hole solutions is their energy-momentum localization. We consider that a good manner to evaluate the energy associated with a black hole solution in string theory in the one which used the Møller energy-momentum complex [14]. This is because of the aforementioned important results obtained in the Møller prescription and of the Lessner opinion [22] and Cooperstock very important hypothesis [52].

We begin with the presentation of the procedure that allows us obtaining a black hole solution in string theory. As we pointed out above, it is well-known that in the string theory there is the possibility to construct rotating charge-neutral black hole solutions [57]. These solutions are identical to the Kerr solution [57] of ordinary Einstein gravity with the dilaton taking a constant value. Rotating charged black hole solutions in these theories have been studied [67] in the limit of small angular momentum.

The solution constructed by A. Sen [56] is an exact classical solution in the low-energy effective field theory, which describes a black hole carrying a finite amount of charge and angular momentum. Furthermore, the solution given by Sen [56] is different from that obtained by Horne and Horowitz [69] because it involves the antisymmetric tensor field in a nontrivial way. The method he used is the twisting procedure [57]. As we pointed out, this method generates inequivalent classical solutions starting from a given classical solution of string theory. As a general method (see above), with the twisting procedure we can generate charged black hole solutions with the help of a charge-neutral solution. Sen [56] generated the rotating charged black hole solution by starting from a rotating black hole solution carrying no charge (the Kerr solution).

The standard Kerr solution [57], which is the most general static black hole solution of the Einstein equation, is given by

$$ds^2_{Kerr} = \frac{r^2 + a^2\cos^2\theta - 2\,m\,r}{r^2 + a^2\cos^2\theta}\,dt^2 - \frac{r^2 + a^2\cos^2\theta}{r^2 + a^2 - 2\,m\,r}\,dr^2 -$$

$$-(r^2 + a^2\cos^2\theta)\,d\theta^2 - \frac{\sin^2\theta}{r^2 + a^2\cos^2\theta}((r^2 + a^2)(r^2 + a^2\cos^2\theta) + \tag{1}$$

$$+2\,m\,r\,a^2\sin^2\theta)\,d\varphi^2 - \frac{4\,m\,r\,a\sin^2\theta}{r^2 + a^2\cos^2\theta}\,dt\,d\varphi,$$

where

$$a = \frac{J}{m}. \tag{2}$$

In (2) J is the angular momentum. One has

$$A_i = 0,\ B_{i\,k} = 0,\ \Phi = 0, \tag{3}$$

where A_i is the Maxwell field, Φ is the dilaton field and $B_{i\,k}$ is the antisymmetric tensor gauge field.

Now, the next step is to generate from the neutral solution the transformed solution. Using the twisting procedure the transformed solution, which describes a rotating charged black hole, is obtained from the Kerr black hole solution. The metric that described it is given by

$$ds^2 = \frac{(r^2 + a^2\cos^2\theta - 2\,m\,r)(r^2 + a^2\cos^2\theta)}{(r^2 + a^2\cos^2\theta + 2\,m\,r\sinh^2\frac{\alpha}{2})^2}\,dt^2 -$$

$$-\frac{r^2 + a^2\cos^2\theta}{r^2 + a^2 - 2\,m\,r}\,dr^2 - (r^2 + a^2\cos^2\theta)\,d\theta^2 - ((r^2 + a^2)\times$$

$$\times(r^2 + a^2\cos^2\theta) + 2\,m\,r\,a^2\sin^2\theta + 4\,m\,r(r^2 + a^2)\sinh^2\frac{\alpha}{2} + \tag{4}$$

$$+4\,m^2\,r^2\sinh^4\frac{\alpha}{2}) \times \frac{(r^2 + a^2\cos^2\theta)\sin^2\theta}{(r^2 + a^2\cos^2\theta + 2\,m\,r\sinh^2\frac{\alpha}{2})^2}\,d\varphi^2 +$$

$$+\frac{4\,m\,r\,a\cosh^2\frac{\alpha}{2}(r^2 + a^2\cos^2\theta)\sin^2\theta}{(r^2 + a^2\cos^2\theta + 2\,m\,r\sinh^2\frac{\alpha}{2})^2}\,dt\,d\varphi,$$

where for the dilaton field we have the expression

$$\Phi' = -\ln\frac{r^2 + a^2\cos^2\theta + 2\,m\,r\sinh^2\frac{\alpha}{2}}{r^2 + a^2\cos^2\theta} \tag{5}$$

and with

$$A'_\phi = -\frac{2\,m\,r\,\sinh\alpha\,\sin^2\theta}{r^2 + a^2\cos^2\theta + 2\,m\,r\,\sinh^2\frac{\alpha}{2}} \tag{6}$$

and

$$A'_t = \frac{2\,m\,r\,\sinh\alpha}{r^2 + a^2\cos^2\theta + 2\,m\,r\,\sinh^2\frac{\alpha}{2}}, \tag{7}$$

which are the only nonvanishing components of the Maxwell field A_i. Further, the non-zero component of the antisymmetric tensor gauge field $B_{i\,k}$ is given by

$$B'_{t\,\phi} = \frac{2\,m\,r\,a\,\sinh^2\frac{\alpha}{2}\sin^2\theta}{r^2 + a^2\cos^2\theta + 2\,m\,r\,\sinh^2\frac{\alpha}{2}} \tag{8}$$

Now, the Einstein metric is obtained with the relationship $ds^2_E = e^{-\Phi'}\,ds^2$, with Φ' given by (5) and one obtains

$$\begin{aligned}
ds^2_E = {} & \frac{r^2 + a^2\cos^2\theta - 2\,m\,r}{r^2 + a^2\cos^2\theta + 2\,m\,r\,\sinh^2\frac{\alpha}{2}}\,dt^2 - \\
& -\frac{r^2 + a^2\cos^2\theta + 2\,m\,r\,\sinh^2\frac{\alpha}{2}}{r^2 + a^2 \ - 2\,m\,r}\,dr^2 - \\
& -(r^2 + a^2\cos^2\theta + 2\,m\,r\,\sinh^2\tfrac{\alpha}{2})\,d\theta^2 + \\
& +\frac{4\,m\,r\,a\,\cosh^2\frac{\alpha}{2}\sin^2\theta}{r^2 + a^2\cos^2\theta + 2\,m\,r\,\sinh^2\frac{\alpha}{2}}\,dt\,d\varphi - \\
& -((r^2 + a^2)(r^2 + a^2\cos^2\theta) + +2\,m\,r\,a^2\sin^2\theta + \\
& +4\,m\,r(r^2 + a^2)\sinh^2\tfrac{\alpha}{2} + \\
& +4\,m^2\,r^2\sinh^4\tfrac{\alpha}{2}) \times \frac{\sin^2\theta}{r^2 + a^2\cos^2\theta + 2\,m\,r\,\sinh^2\frac{\alpha}{2}}\,d\varphi^2.
\end{aligned} \tag{9}$$

This metric describes a black hole solution with mass M and charge Q given by

$$\begin{aligned}
M &= \frac{m}{2}(1 + \cosh\,\alpha), \\
Q &= \frac{m\,\sinh\alpha}{\sqrt{2}}.
\end{aligned} \tag{10}$$

In these relations α is an arbitrary number. Further, for this black hole solution we have for the angular momentum J and the magnetic dipole momentum μ the expressions

$$\begin{aligned}
J &= \frac{m\,a}{2}\,(1 + \cosh\alpha), \\
\mu &= \frac{m\,a\,\sin\alpha}{\sqrt{2}}.
\end{aligned} \tag{11}$$

Now, for our purpose, we consider the case when the angular momentum vanishes. We have two cases.

The first one is the case when we obtain the metric from (4) with $a = 0$ and we obtain

$$ds^2 = \frac{(r^2 - 2\,m\,r)\,r^2}{(r^2 + 2\,m\,r\,\sinh^2\frac{\alpha}{2})^2}\,dt^2 - \frac{r^2}{r^2 - 2\,m\,r}\,dr^2 - r^2\,d\theta^2 - r^2\,\sin^2\theta\,d\varphi^2. \quad (12)$$

In the second case, the Einstein metric ($ds_E^2 = e^{-\Phi'}\,ds^2$, with Φ' given by (5)) is given by

$$\begin{aligned} ds_E^2 = \frac{r^2 - 2\,m\,r}{r^2 + 2\,m\,r\,\sinh^2\frac{\alpha}{2}}\,dt^2 - \frac{r^2 + 2\,m\,r\,\sinh^2\frac{\alpha}{2}}{r^2 - 2\,m\,r}\,dr^2 - \\ -(r^2 + 2\,m\,r\,\sinh^2\tfrac{\alpha}{2})\,d\theta^2 - \\ -(r^2 + 2\,m\,r\,\sinh^2\tfrac{\alpha}{2})\,\sin^2\theta\,d\varphi^2. \end{aligned} \quad (13)$$

In the following we present the definition of the Møller energy-momentum complex. Its good properties indicates it as a powerful tool for energy-momentum localization.

As we pointed out above, Møller gave an expression which allows performing the calculations in any coordinate system, not only in quasi-Cartesian coordinates. Furthermore, Møller argued that his expression enables one to obtain the same values for the total energy and momentum as the Einstein energy-momentum complex for a closed system.

The Møller energy-momentum complex ${M_i}^k$ [14] has the expression

$${M_i}^k = \frac{1}{8\,\pi}\,{\chi_i}^{k\,l}{}_{,l}, \quad (14)$$

where

$${\chi_i}^{k\,l} = \sqrt{-g}\left(\frac{\partial g_{i\,n}}{\partial x^m} - \frac{\partial g_{i\,m}}{\partial x^n}\right)\,g^{k\,m}\,g^{n\,l}. \quad (15)$$

The quantity ${\chi_i}^{k\,l}$ is the Møller superpotential and satisfies the antisymmetric property

$${\chi_i}^{k\,l} = -{\chi_i}^{l\,k}. \quad (16)$$

${M_0}^0$ is the energy density and ${M_\alpha}^0$ are the momentum density components.

The Møller energy-momentum complex ${M_i}^k$ satisfies the local conservations laws

$$\frac{\partial {M_i}^k}{\partial x^k} = 0. \quad (17)$$

The energy of the physical system in a four-dimensional background is given by

$$\begin{aligned} E = \iiint {M_0}^0\,dx^1\,dx^2\,dx^3 = \\ = \frac{1}{8\,\pi}\iiint \frac{\partial {\chi_0}^{0l}}{\partial x^l}\,dx^1\,dx^2\,dx^3. \end{aligned} \quad (18)$$

Using the Gauss theorem we compute the energy contained in a sphere of radius r and one obtains

$$E = \frac{1}{8\pi} \oint \chi_0^{\ 0l} \sin\theta \, d\theta \, d\varphi. \tag{19}$$

After performing some calculations, for the metric described by (12) the non-zero $\chi_i^{\ kl}$ components of the Møller energy-momentum complex are given by

$$\begin{aligned}
\chi_2^{\ 2\,1} &= -2\,\frac{r\,\sin\theta(-r+2\,m)}{r+2\,m\,\sinh^2\frac{\alpha}{2}},\\
\chi_3^{\ 3\,1} &= -2\,\frac{r\,\sin\theta(-r+2\,m)}{r+2\,m\,\sinh^2\frac{\alpha}{2}},\\
\chi_3^{\ 3\,2} &= 2\,\frac{r\,\cos\theta}{r+2\,m\,\sinh^2\frac{\alpha}{2}},\\
\chi_0^{\ 0\,1} &= -2\,\frac{m\,r\,\sin\theta\,(-2\,r\,\sinh^2\frac{\alpha}{2}-r+2\,m\,\sinh^2\frac{\alpha}{2})}{(r+2\,m\,\sinh^2\frac{\alpha}{2})^2}.
\end{aligned} \tag{20}$$

From (20) it results that the $\chi_i^{\ k\,l}$ components of the Møller energy-momentum complex depend on the mass m, arbitrary number α and r coordinate. The only component involved in the expression for the energy is the $\chi_0^{\ 0\,1}$ component. For the evaluation of the energy associated with the black hole solution described by the metric given by (12), we use (18) and (20) and after some calculations and applying the Gauss theorem, we obtain for the energy distribution the expression

$$E = m\,(1 + \frac{\alpha^2}{2} + \frac{\alpha^4}{24} + O(\alpha^6)). \tag{21}$$

From (21) it results that the energy distribution depends on the mass m and arbitrary number α. In the particular case, when we put in the expression for the energy the value $\alpha = 0$ we obtain for the energy distribution

$$E = m. \tag{22}$$

We plot the expression for the energy taking into account only the first three terms from (21), as in the equation below

$$E = m\,(1 + \frac{\alpha^2}{2} + \frac{\alpha^4}{24}). \tag{23}$$

In the Fig.1 we give the graph of the energy. The energy distribution E is plotted on Y-axis against α on X-axis for $m = 1$.

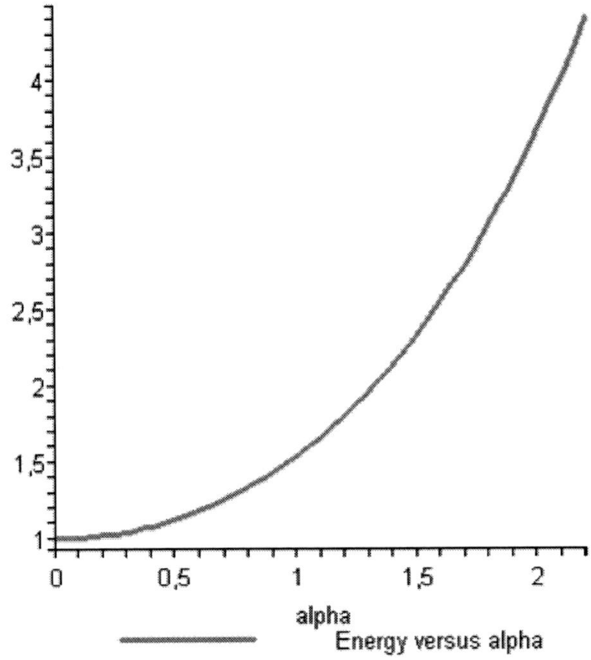

Figure 1.

In the case of the Einstein metric, which is given by (13) for the non-vanishing components $\chi_i^{\,k\,l}$ of the Møller energy-momentum complex we obtain the expressions

$$\chi_2^{2\,1} = -2\,\frac{\sin\theta(-r+2\,m)(r+\,m\,\sinh^2\frac{\alpha}{2})}{r+2\,m\,\sinh^2\frac{\alpha}{2}},$$

$$\chi_3^{3\,1} = -2\,\frac{\sin\theta(-r+2\,m)(r+\,m\,\sinh^2\frac{\alpha}{2})}{r+2\,m\,\sinh^2\frac{\alpha}{2}},$$

$$\chi_3^{3\,2} = 2\,\cos\theta, \tag{24}$$

$$\chi_0^{0\,1} = 2\,\frac{m\,r\,\sin\theta(1+\sinh^2\frac{\alpha}{2})}{r+2\,m\,\sinh^2\frac{\alpha}{2}}.$$

In this case the components of the Møller energy-momentum complex depend on the mass m, arbitrary number α and r coordinate. Because of the equations (10) the χ_i^{kl} components also depend on the mass M and charge Q of the black hole. Plugging (24) in (18), after a series of calculations and applying the Gauss theorem, we evaluate the integral over the surface of a sphere with radius r and obtain the energy distribution which is given by

$$E = m\,(1+\frac{\alpha^2}{4}+\frac{\alpha^4}{48}+O(\alpha^6)). \tag{25}$$

Taking into account (10) we obtain for the energy distribution the expression

$$E = \frac{2\,M}{1+\cosh\alpha}(1+\frac{\alpha^2}{4}+\frac{\alpha^4}{48}+O(\alpha^6)). \tag{26}$$

In the Fig.2 we make the graph for the expression for the energy given by (26). The energy distribution E is plotted on Y-axis against α on X-axis, in the case $M = 1$ and we take into account only the first three terms from (26). Further, from this figure we observe the difference between the energy distribution in this case and in the one described by the equation (21) and represented in the Fig.1.

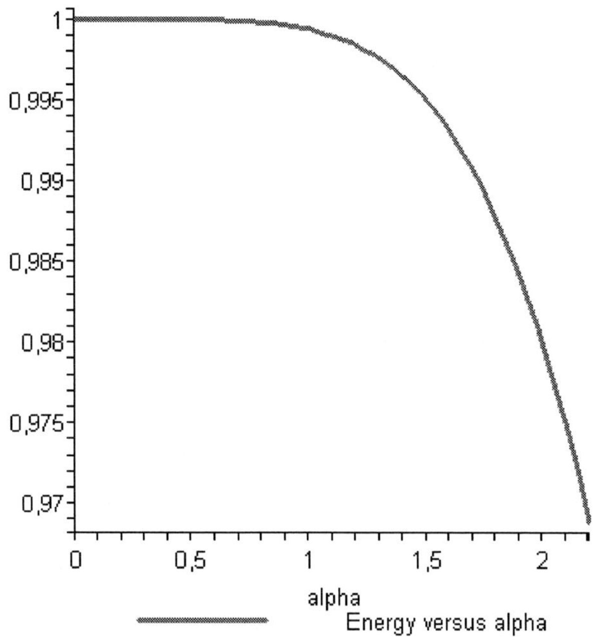

Figure 2.

For the space-time described by the metric given by the equation (12), we obtain that the second and the third terms in the expression for the energy are twice the value than those obtained in the case of the Einstein metric, which is given by the equation (13). The first terms are the same in the both expressions for the energy distribution, which are given by the equations (21) and (25).

Furthermore, in the case of the Einstein metric if the arbitrary number α vanishes, $\alpha = 0$ and using the eqations (10) the energy is given by

$$E = M. \tag{27}$$

This result is the same as for the Schwarzschild solution and from the equations (10) we obtain

$$Q = \frac{m\,\sinh\alpha}{\sqrt{2}} = 0. \tag{28}$$

For the aforementioned two metrics described by the equations (12) and (13), in the approximation of a vanishing angular momentum $a = 0$, we obtain for the energy associated with these black hole solutions meaningful results. If we consider for the arbitrary number α the value $\alpha = 0$, the expressions for the energy distribution are equal to m. For the Einstein metric we perform some calculations and in the case $\alpha = 0$ we obtain that the energy associated with this space-time is equal to the mass M of the black hole.

3. Discussion

The issue of the localization of energy continues to be an unsolved one since Einstein has given his important conclusion of the special theory of relativity that mass is equivalent to energy. Misner et al [71] noticed that to look for a local energy-momentum means that is looking for the right answer to the wrong question. They also concluded that the energy is localizable only for spherical systems. Cooperstock and Sarracino [72] pointed out that if the energy is localizable in spherical systems then it is also localizable in any space-time. Bondi [73] gave his opinion that "a nonlocalizable form of energy is not admissible in general relativity, because any form of energy contributes to gravitation and so its location can in principle be found".

Even the method of localization of energy by using several energy-momentum complexes has many adepts there was also much criticism related to the use of these prescriptions for energy and momentum localization. The main lack of the pseudotensorial method is that in many cases is restricted to quasi-Cartesian coordinates. On the other hand, it is important to point out that a new concept has been arised, that of quasi-local energy. To date, a large number of definitions for quasi-local mass have been proposed [6]-[7]. But the quasi-local mass definitions have also some weakness [74].

Virbhadra and his collaborators [15] re-opened the issue of the energy-momentum complexes and obtained meaningful results for many space-times. In his recent paper Virbhadra [18] emphasized that though the energy-momentum complexes are non-tensors under general coordinate transformations, the local conservation laws with them hold in all coordinate systems. He concluded that several energy-momentum complexes (ELLPW) comply with the quasi-local mass definition of Penrose for a general non-static spherically symmetric metric of the Kerr-Schild class. He obtained the expression of the energy distribution for this general metric. After these works, other authors used different energy-momentum prescriptions to investigate the energy distributions for many space-times. They demonstrated that for many given geometries several energy-momentum complexes yield the same and acceptable result [16], [25]-[31]. Chang, Nester and Chen [19] showed that the energy-momentum complexes are actually quasi-local and legitimate expression for the energy-momentum, and established the existence of a direct relationship between energy-momentum complexes and quasi-local expressions because every energy-momentum complexes is strongly connected to a legitimate Hamiltonian boundary term. Furthermore, important studies have been done about the new idea of quasi-local approach for energy-momentum complexes [19], [75] and a large class of new pseudotensors connected to the positivity in small regions have been constructed [76]. In this light, the quasi-local quantities are associated with a closed 2-surface (L. B. Szabados, [75] and http://relativity.livingreviews.org/Articles/lrr-2004-4/). The Hamiltonian boundary term determines the quasi-local quantities for finite regions and the special quasi-local energy-momentum boundary term expressions correspond each of them to a physically distinct and geometrically clear boundary condition [77].

As we pointed out above in Introduction, most of energy-momentum complexes are dependent on the coordinate system. Only the Møller energy-momentum complex allows performing the calculations in any coordinate system. Another argument which sustains the use of the Møller prescription is the Lessner [22] conclusion that the Møller energy-momentum complex is a powerful representation of energy and momentum in general relativity. The viewpoint of Lessner [22] is that "The energy-momentum four-vector can only transform according to special relativity only if it is transformed to a reference system with an everywhere constant velocity. This cannot be achieved by a global Lorentz transformation". All these conclusions made us to decide to apply the Møller energy-momentum complex for evaluating the energy distribution of some important space-times [32], [36], [44], [46], [49], [50], [51], [53] and [55]. In all these cases we obtained good results.

In this chapter we choose to study the energy associated with the black hole solution given by A. Sen [56] because we found this a very interesting issue. Furthermore, we use the Møller prescription to compute the energy distribution of this charged black hole solution in heterotic string theory [78]. We consider the case when the angular momentum vanishes. We have two cases. Firstly, the metric that we use describes a rotating charged black hole solution generates by starting from a rotating black hole solution carrying no charge (the Kerr solution) with the help of the twisting procedure [57], [64], [65]. Secondly, we study the Einstein metric, which is obtained from the first metric by a transformation which implies to use the dilaton field. We conclude that the Møller energy-momentum complex yields consistent results for both metrics.

For the metrics given by (12) and (13) we obtained meaningful results for the energy distribution in the Møller prescription. The first term in the expressions for the energy distribution given by (21) and (25) is the same and equal with the mass m. In the case of the space-time described by the metric given by (12), the Møller energy-momentum complex gives for the other terms involved in the expression for the energy twice the value than those obtained in the case of the Einstein metric.

The expressions for the energy given by the equations (21) and (25) can be interpreted as the effective gravitational masses that a neutral test particle "feels" in the fields described by (12) and (13). In both cases, the energy distribution depends on the mass m and the arbitrary number a. In the case $\alpha = 0$ we obtain that $E = m$ for the aforementioned space-times. Furthermore, because of the dependence of the physical parameters M and Q (see equations (10)) on m and α , the energy distribution given by (25) depends on the mass M and charge Q of the black hole. Using the equations (10), for the Einstein metric in the case $\alpha = 0$ the expression for the energy is equal to the mass M of the black hole and is given by $E = M$.

These results sustain that the Møller energy-momentum complex is a powerful tool for obtaining the energy distribution in a given space-time. Further, they support the Lessner opinion [22] and the Cooperstock very important hypothesis [52].

Acknowledgments

This work was supported by Grant GR 80/23.05.2007.

References

[1] L. Bel, *C. R. Acad. Sci. Paris,* **247**, 1094 (1958); L. Bel, *PhD Thesis (CDU et SEDES Paris 5e)* (1960).

[2] L. Bel, *Commun. Mayh. Phys.* **138**, 59 (1962).

[3] I. Robinson, *Class. Quant. Grav.* **14**, A331 (1997).

[4] M. A. G. Bonilla and J. M. M. Senovilla, *Phys. Rev. Lett.* **11**, 783 (1997); M. A. G. Bonilla and J. M. M. Senovilla, *Gen. Rel. Grav.* **29**, 91 (1997); J. M. M. Senovilla, *gr-qc/9901019*, to be published in *Gravitation and Relativity in General*, eds. A. Molina, J. Martin, E. Ruiz and F. Atrio (World Scientific, 1999); P. Teyssandier, *gr-qc/9905080;* J. Garecki, *Annalen Phys.* **10**, 911 (2001); M. P. Dabrowski and J. Garecki, *Class. Quant. Grav.* **19**, 1 (2002); R. Lazkog, J. M. M. Senovilla and R. Vera, *Class. Quant. Grav.* **20**, 4135 (2003).

[5] J. D. Brown and J. W. York, *Phys. Rev.* **D47**, 1407 (1993).

[6] S. A. Hayward, *Phys. Rev.* **D49**, 831 (1994).

[7] J. M. M. Senovilla, *Class. Quant. Grav.* **17**, 2799 (2000).

[8] A. Einstein, *Preuss. Akad. Wiss. Berlin* **47**, 778 (1915); Addendum-ibid. **47**, 799 (1915); A. Trautman, in *Gravitation: an Introduction to Current Research*, ed. L. Witten (Wiley, New York, 1962, p. 169).

[9] R. P. Wallner, *Acta Physica Austriaca* **52**, 121 (1980).

[10] L. D. Landau and E. M. Lifshitz, *The Classical Theory of Fields* (Pergamon Press, 1987, p. 280).

[11] A. Papapetrou, *Proc. R. Irish. Acad.* **A52**, 11 (1948).

[12] P. G. Bergmann and R. Thomson, *Phys. Rev.* **89,** 400 (1953).

[13] S. Weinberg, *Gravitation and Cosmology: Principles and Applications of General Theory of Relativity* (John Wiley and Sons, Inc., New York, 1972, p. 165).

[14] C. Møller, *Ann. Phys. (NY)* **4,** 347 (1958).

[15] K. S. Virbhadra, *Phys. Rev.* **D41,** 1086 (1990); S. Virbhadra, *Phys. Rev.* **D42**, 2919 (1990); N. Rosen and K. S. Virbhadra, *Gen. Rel. Grav.* **25**, 429 (1993); K. S. Virbhadra and J. C. Parikh, *Phys. Lett.* **B331,** 302 (1994); K. S. Virbhadra, *Pramana - J. Phys.* **45**, 215 (1995).

[16] A. Chamorro and K. S. Virbhadra, *Pramana - J. Phys.* **45**, 181 (1995); A. Chamorro and K. S. Virbhadra, *Int. J. Mod. Phys.* **D5**, 251 (1997); S. S. Xulu, *Int. J. Theor. Phys.* **37**, 1773 (1998); S. S. Xulu, *Int. J. Mod. Phys.* **D7**, 773 (1998); E. C. Vagenas, *Int. J. Mod. Phys.* **A18**, 5949 (2003); E. C. Vagenas, *Int. J. Mod. Phys.* **A18**, 5781 (2003); E.

C. Vagenas, *Mod. Phys. Lett.* **A19**, 213 (2004); Th. Grammenos, *Mod. Phys. Lett.* **A20**, 1741 (2005); I. Radinschi, *Acta Phys. Slovaca* **49(5)**, 789 (1999); I. Radinschi, *Mod. Phys. Lett.* **A15**, Nos. 11&12, 803 (2000); I. Radinschi, *Fizika* **B9(4)**, 203 (2000); I. Radinschi, *Acta Phys. Slovaca* **50(4)**, 609 (2000); I. Radinschi, *Nuovo Cimento* **B115**, N, 501 (2000); I. Radinschi, *Acta Phys. Hung. NS-H* **12**, 47 (2000); I. Radinschi, *Mod. Phys. Lett.* **A16**(10), 673 (2001); T. Bringley, *Mod. Phys. Lett.* **A17**, 157 (2002); M. Súkeník and J. Sima, *gr-qc/0101026*; M. Sharif, *Int. J. Mod. Phys.* **D13**, 1019 (2004).

[17] J. M. Aguirregabiria, A. Chamorro and K. S. Virbhadra, *Gen. Rel. Grav.* **28,** 1393 (1996).

[18] K. S. Virbhadra, Phys. Rev. **D60,** 104041 (1999).

[19] Chia-Chen Chang, J. M. Nester and Chiang-Mei Chen, *Phys. Rev. Lett.* **83**, 1897 (1999).

[20] C. Møller, *Ann. Phys. (NY)* **12,** 118 (1961).

[21] D. Kovacs, *Gen. Rel. Grav.* **17**, 927 (1985); J. Novotny, *Gen. Rel. Grav.* **19**, 1043, (1987).

[22] G. Lessner, Gen. Relativ. Gravit. **28**, 527 (1996).

[23] V. de la Cruz and W. Israel, *Nuovo Cimento* **51**, 744 (1967); J. M. Cohen and R. Gautreau, *Phys. Rev.* **D19,** 2273 (1979); J. M. Cohen and F. de Felice, *J. Math. Phys.* **25**, 992 (1984); M. Ahmed and S. M. Hossain, *Prog. Theor. Phys.* **93**, 901 (1996); I-Ching Yang, Ching-Tzung Yeh, Rue-Ron Hsu and Chin-Rong Lee, *Int. J. Mod. Phys.* **D6,** 349 (1997); I. Radinschi, *Mod. Phys. Lett.* **A16(10)**, 673 (2001).

[24] R. Kulkarni, V. Chellathurai and N. Dadhich, *Class. Quant. Grav.* **5**, 1443 (1998).

[25] S. S. Xulu, *Mod. Phys. Lett.* **A15**, 1511 (2000); *PhD Thesis, hep-th/0308070*

[26] S.S. Xulu, *Astrophys. Space. Sci.* **23**, (2003).

[27] I-Ching Yang, Ching-Tzung Yeh, Rue-Ron Hsu and Chin-Rong Lee, *Int. J. Mod. Phys.* **D6,** 349 (1997); I-Ching Yang, Wei-Fui Lin and Rue-Ron Hsu, *Chin. J. Phys.* **37**, 113 (1999).

[28] E. C. Vagenas, *Int. J. Mod. Phys.* **A18**, 5781 (2003).

[29] E. C. Vagenas, *Mod. Phys. Lett.* **A19**, 213 (2004).

[30] Ragab M. Gad, *Astrophys. Space Sci.* **295**, 459 (2005).

[31] Ragab M. Gad, *Astrophys. Space Sci.* **293**, 453 (2004).

[32] I. Radinschi, *Mod. Phys. Lett.* **A15 (35)**, 2171 (2000).

[33] K. S. Virbhadra and J. C. Parikh, *Phys. Lett.* **B331,** 302 (1994).

[34] J. S. Bekenstein, *Ann. Phys. (N.Y)* **91,** 75 (1975).

[35] K. S. Virbhadra and J. C. Parikh, *Phys. Lett.* **B131,** 302 (1994)

[36] I. Radinschi, *Fizika* **B9(4)**, 203 (2000).

[37] N. Banerjee and S. Sen, *Pramana-J. Phys.* **49**, 609 (1997).

[38] S.S. Xulu, *Int. J. Theor. Phys.* **39**, 1153 (2000).

[39] I. Radinschi, *Acta Phys. HN-S* **12**, 47 (2000).

[40] I. Radinschi, *Acta Phys. Slovaca* **50(4)**, 609 (2000).

[41] E. P. Tryon, *Nature* **246**, 396 (1973).

[42] N. Rosen, *Gen. Rel. Grav.* **26**, 319 (1994).

[43] V. B. Johri, D. Kalligas, G. P. Sing and C. W. F. Everitt, *Gen. Rel. Grav.* **27**, 313 (1995).

[44] I. Radinschi, *Chin. J. Phys.* **39(3)**, 231 (2001).

[45] I. Radinschi, *Chin. J. Phys.* **39(5)**, 1 (2001).

[46] I. Radinschi, *Mod. Phys. Lett.* **A16(10)**, 673 (2001).

[47] E. Ayón-Beato and A. Garcia, *Phys. Lett.* **B464**, 25 (1999).

[48] A. Komar, *Phys. Rev.* **113**, 934 (1959).

[49] I-Ching Yang and I. Radinschi, *Fizika* **B14**, 311 (2005).

[50] I-Ching Yang and I. Radinschi, *Chin. J. Phys.* **42**, 40 (2004).

[51] I-Ching Yang and I. Radinschi, *Chin. J. Phys.* **41**, 326 (2003).

[52] F. I. Cooperstock, *Mod. Phys. Lett.* **A14**, 1531 (1999).

[53] I-Ching Yang and I. Radinschi, *Mod. Phys. Lett.* **A17**, 1159 (2002).

[54] K. S. Virbhadra, *gr-qc/9606004*

[55] I-Ching Yang and I. Radinschi, *gr-qc/0309130*, to appear in *AIP*

[56] A. Sen, *Phys. Rev. Lett.* **69(7)**, 1006 (1992).

[57] S. Ferrara, J. Scherk and B. Zumino, *Nucl. Phys.* **B121**, 393 (1977); E. Cremmer, J. Scherk and S. Ferrara, *Phys. Lett.* **68B**, 234 (1977); **74B**, 61 (1978); E. Cremmer and J. Scherk, *Nucl. Phys.* **B127**, 259 (1977); E. Cremmer and B. Julia, *Nucl. Phys.* **B159**, 141 (1979); M. De Roo, *Nucl. Phys.* **B255**, 515 (1985); *Phys. Lett.* **156B**, 331 (1985); E. Bergshoef, I. G. Koh and E. Sezgin, *Phys. Lett.* **155B**, 331 (1985); M. De Roo and P. Wagemans, *Nucl. Phys.* **B262**, 646 (1985); L. Castellani, A. Ceresole, S. Ferrara, R. D'Auria, P. Fre and E. Maina, *Nucl. Phys.* **B268**, 317 (1986); *Phys. Lett.* **161B**, 91 (1985); S. Cecotti, S. Ferrara and L. Girardello, *Nucl. Phys.* **B308**, 436 (1988); M.

Duff, *Nucl. Phys.* **B335**, 610 (1990); J. H. Horne, G. T. Horowitz and A. Steif, *Phys. Rev. Lett.* **68**, 568 (1992); M. Rocek and E. Verlinde, *Report No. IASSNS-HEP-91-68*; P. Horava, M. Rocek *Report No. EFI-91-57*; A. Giveon and M. Rocek, *Report No. IASSNS-HEP-91-84*; S. Kar, S. Khastagir and A. Kumar, *Report No. IP-BBSR-91-51*; J. Panvel, *Report No. LTH 282*; *Nucl. Phys.* **B393**, 95 (1993); R. Adler, M. Bazin and M. Schiffer, *Introduction to General Relativity* (McGraw Hill, New York, 1975); K. Thorne, R. Price and D. Macdonald, *Black Holes: The Membrane Paradigm* (Yale Univ. Press, New Haven, 1986); J. H. Horne and G. T. Horowitz, *Phys. Rev.* **D46**, 1340 (1992); S. F. Hassan and A. Sen, *Report No. TIFR/TH/91-40*.

[58] A. Dabhlokar, G. Gibbons, J. Harvey and F. R. Ruiz, *Nucl. Phys.* **B340**, 33 (1990); A. Dabhlokar, and J. Harvey, *Phys. Rev. Lett.* **63**, 719 (1989); A. Strominger, *Nucl. Phys.* **B343**, 167 (1990); C. G. Callan, J. Harvey and A. Strominger, *Nucl. Phys.* **B359**, 611 (1991); *Nucl. Phys.* **B367**, 60 (1991); M. Duff and J. Lu, *Phys. Rev. Lett.* **66**, 1402 (1991); *Nucl. Phys.* **B354**, 141 (1991).

[59] R. Khuri, *Phys. Lett.* **B294**, 325 (1992); R. Khuri, *Nucl. Phys.* **B387**, 315 (1992); A.Sen, *hep-th/9210050*, TIFR-TH-92-57.

[60] G. Gibbons, *Nucl. Phys.* **B207**, 337 (1982); G. Gibbons and K. Maeda, *Nucl. Phys.* **B298**, 741 (1988); D. Garfinkle, G. T. Horowitz and A. Strominger, *Phys. Rev.* **D43**, 3140 (1991); G. T. Horowitz and A. Strominger, *Nucl. Phys.* **B360**, 197 (1991); S. Giddings and A. Strominger, *Phys. Rev. Lett.* **67**, 2930 (1991).

[61] R. Kallosh, *Phys. Lett.* **B282**, 80 (1992); R. Kallosh, A. Linde, T. Ortin, A. Peet and A. Van Proeyen, *Phys. Rev.* **D46**, 5278 (1992).

[62] G. Mandal, A. Sengupta and S. Wadia, *Mod. Phys. Lett.* **A 6**, 1685 (1991).

[63] E. Witten, *Phys. Rev.* **D44**, 314 (1991); C. Nappi and E. Witten, *Phys. Lett.* **B293**, 309 (1992).

[64] G. Veneziano, *Phys. Lett.* **B 265**, 287 (1991); K. Meissner and G. Veneziano, *Phys. Lett.* **B 267**, 33 (1991); *Mod. Phys. Lett.* **A 6**, 3397 (1991); M. Gasperini, J. Maharana and G. Veneziano, *Phys. Lett.* **B 272**, 277 (1991); M. Gasperini and G. Veneziano, *CERN Report No. CERN-TH-6321-91*; A. Sen, *Phys. Lett.* **B 271**, 295 (1991); A. Sen, *Phys. Lett.* **274**, 34 (1991); S. F. Hassan and A. Sen, *Nucl. Phys.* **B375**, 103 (1992); J. Panvel, *Phys. Lett.* **B 284**, 50 (1992).

[65] A. Shapere, S. Trivedi and F. Wilczek, *Mod. Phys. Lett.* **A**, **6**, 2677 (1991); A. Sen, *Nucl. Phys.* **B404**, 109 (1993); T. Ortin, *Phys. Rev.* **D47**, 3136 (1993).

[66] A. Sen, *Nucl. Phys.* **B440,** 421 (1995).

[67] A. Sen, *Mod. Phys. Lett.* **A10**, 2081 (1995).

[68] G. T. Horowitz and A. Sen, *Phys. Rev.* **D53**, 808 (1996).

[69] J. H. Horne and G. T. Horowitz, *Phys. Rev.* **D46**, 1340 (1992).

[70] C. V. Johnson, N. Kaloper, R. Khuri and R. C. Myers, *Phys. Lett.* **B368**, 71 (1996); C. V. Johnson, and R. C. Myers, *Phys. Rev.* **D52**, 2294 (1995).

[71] C. W. Misner, K. S. Thorne and J. A. Wheeler, *Gravitation, W. H. Freeman and Co., NY*, 603 (1973).

[72] F. I. Cooperstock and R. S. Sarracino, *J. Phys. A: Math. Gen.* **11**, 877 (1978).

[73] H. Bondi, *Proc. R. Soc. London* **A427**, 249 (1990).

[74] G. Bergqvist, *Class. Quantum Gravit.* **9**, 1753 (1992).

[75] Chiang-Mei Chen and J. M. Nester, *Class. Quant. Grav.* **16**, 1279 (1999); Chiang-Mei Chen and J. M. Nester, *Grav. Cosmol.* **6**, 257 (2000); Chiang-Mei Chen, James M. Nester and Roh Suan Tung, *Phys. Lett.* **A203**, 5 (1995); L. B. Szabados, *Living. Rev. Relativity* **7**, 4 (2004).

[76] Lau Loi So, James M. Nester and Hsin Chen, gr-qc/0605150, to appear in Proceedings of the 7th International Conference on Gravitation and Astrophysics; S. Deser, J.S. Franklin and D. Seminara, *Class. Quant. Grav.* **16**, 2815 (1999).

[77] James M. Nester *Class. Quant. Grav.* **21**, S261 (2004); Chiang-Mei Chen, James M. Nester and Roh-Suan Tung, *Phys. Rev.* **D72**, 104020 (2005).

[78] I. Radinschi, *gr-qc/0302047*

In: Astrophysics and Condensed Matter
Editor: Thomas G. Hardwell, pp. 215-234
ISBN: 978-1-60456-332-0

Chapter 8

CHEMICAL BONDING THEORY OF SINGLE CRYSTAL GROWTH: A CRYSTALLOGRAPHIC VIEWPOINT

Congting Sun and Dongfeng Xue*
State Key Laboratory of Fine Chemicals, Department of Materials Science and Chemical Engineering, School of Chemical Engineering, Dalian University of Technology, 158 Zhongshan Road, Dalian 116012, China

Abstract

Starting from the crystallographic viewpoint, the chemical bonding theory of single crystal growth was proposed and introduced into the morphology prediction and control of inorganic crystals and powders, which can also help us to further understand the process of single crystal growth. In this chapter, the inorganic crystals within seven crystal systems were respectively selected to quantitatively predict their ideal morphology. The predicted morphologies by calculating the vertical growth rate of selected planes, can agree well with experimental observations. We took $Mg(OH)_2$ as an example to comprehensively analyze the effect of both crystal structure and reaction conditions on the powder morphology. Our results indicate that the current theory can lead to better understanding of the single crystal growth and the morphology control of various powders, which can guide the control of crystal and powder morphology from a kinetic viewpoint.

Key words: Chemical bonding theory, crystal growth, crystallization, morphology control

1. Introduction

Controlling the shape, size, and structure of inorganic crystals and powders is of fundamental importance due to the strong correlation between these parameters and physical/ chemical properties [1-4]. Crystal morphology affects the efficiency of downstream process (such as filtering, washing, and drying), and influences material properties such as bulk density and mechanical strength which play a major role in storage and handling [5]. To give an effective

* E-mail address: dfxue@chem.dlut.edu.cn. (Corresponding author.)

guide to control the morphology and size of crystals, prediction of crystal morphology has been an active research field [5-14]. Essentially, the shape of crystal highly depends on the ratio between the areas of crystallographic facets that enclose the single crystal, and the dominant crystal surface growth forms {*hkl*} are the reflection of their relative growth rates. As the morphology of single crystals is determined by both the crystal structure (the internal factors), the crystal growth conditions, and the processes of that growth (the external factors), we can see the fundamentality that understand how the morphology of a single crystal is determined through the interconnection between the internal and external factors. Therefore, it is necessary to introduce the crystallographic and growth theories into crystal synthesis (i.e., from both thermodynamic and kinetic viewpoints) to comprehensively understand crystal growth, to predict crystal morphology and further to guide the shape control of materials. In the actual crystal growth process, the different growth conditions and processes result in the discrepancy of morphology among the crystals, viz. the diversity of crystal morphologies can be found for a single material in different reaction conditions. In order to effectively control the morphology and size of the materials and to obtain the crystals for the industrial processes, the ideal morphology of crystals should be predicted primarily through analyzing the special crystal structure and the reaction condition in detail.

There are several theoretical methods available for the calculation of the morphological importance. The oldest is the empirical Bravais, Friedel, Donnay and Harker (BFDH) law, which assumes that the most energetically stable, and slowest growing faces are the ones with the highest density of material and the largest spacing between adjacent layers of material [11, 16]. This theory is applicable for rather isotropic crystal structures, but it is of limited use for more complex crystal structures that may have a large anisotropy in interactions between molecules. The periodic bond chain theory [11, 16] which is based on the existence of periodic bond chains (PBCs) and connected nets in a crystal structure was subsequently introduced by Hartman and Perdok. In this theory, PBCs are periodically repeated chains of strong intermolecular interactions. This concept was later used by Hartman and Bennema to formulate the attachment energy law, which states that the growth rate of a crystalline face is proportional to its attachment energy E_{hkl}^{att}, where the attachment energy E_{hkl}^{att} is defined as the energy released when one additional growth slice of thickness d_{hkl} is attached to the crystal face identified by the Miller indices *hkl*. However, recent study shows that this theory breaks down for complex connected net orientation topologies, in terms of connected nets having a roughing temperature equal to zero [8]. Along with the rapid development of the computer software and techniques, Monte Carlo [6, 8, 9] and molecular dynamics methods have been applied to study the most favored interactions between the crystal habit surfaces and specific solvents in order to establish the modification to the crystal shape in terms of relative growth rates. Recently, an approach that combines the kinetic Monte Carlo simulations and atomistic molecular dynamics simulations, has been suggested by Piana and Gale to study urea crystal growth from aqueous and methanol solution [6]. The molecular dynamics simulation techniques have been also adopted by Vlieg et al. to explain solvent-effected crystal habits, particularly developing surface modeling techniques to take account of the influence of surface steps. In addition, the attachment energy has been replaced by the step energy in their model [8, 9].

From the above context, crystals with different crystal structures usually take geometrical external forms following the symmetries involved in their respective structures, but they may also take very different external forms if the growth conditions are dissimilar. In order to obtain the actual single crystals with different morphologies and sizes, various techniques for crystal growth have been devised and developed with the development of fabrication technology. Methods for growing bulk single crystals are mainly divided into the following approaches: melting growth, solution growth, hydrothermal growth, assisted solvent growth, etc. The majority of crystals are produced by pulling from the melt, and crystal dimensions have reached 10 cm for InP, $LiTaO_3$, sapphire and other oxide crystals, 15 cm for $LiNbO_3$ and GaAs, and 30 to 40 cm for silicon and for halide scintillator crystals with melt dimensions up to 1m. The other methods for growth of bulk crystals also develop rapidly, for example, Hydrothermal growth of large high-quality low-dislocation-density quartz by VNIISIMS/Russia and Sawyer/USA, hydrothermal growth of large ZnO and emerald crystals, high-temperature solution growth of perovskites and garnets by top-seeded solution growth TSSG by Linz *et al.* and by the accelerated crucible rotation technique (ACRT) by Scheel and by Tolksdorf, TSSG growth of KTP and CLBO (the latter by Sasaki and Mori), aqueous solution growth of KDP (Sasaki, Zaitseva), Bridgman–Stockbarger growth of alkali halides and of CaF_2, float-zone growth of TiO_2 rutile by Shindo *et al.*, skull-melting growth of zirconia by Wenckus and Osiko, and Czochralski growth of numerous oxide compounds [17], arc-fusion growth of large MgO single crystal [18, 19].

On the other hand, microcrystals (powders) due to their size effect have received steadily growing interests as a result of their peculiar and fascinating properties, and applications superior to their bulk counterparts. In some case, shape control allows one to tailor the property with a greater versatility than can be achieved otherwise. Among various strategies for controlled synthesis, the “soft chemistry” rout, which is based on a solution process, is an effective method for the chemical synthesis of micro- and nanostructured materials with well-controlled shapes, sizes and structures. To date, a variety of functional materials with desired shape, morphology, and crystal phase have been achieved by applying solution chemistry approaches [20-41].

Generally speaking, crystal growth is the process of the birth and development of a solid phase with a regular structure out of a disordered and irregular state, and thus it can be regarded as a first-order phase transition. Lots of efforts had been devoted to predict the crystal morphology, however, mainly for molecular crystals. In order to systematically study the shape-property relationship and fully explore potential applications of general inorganic materials, it is necessary to develop a generic model to predict the morphology of inorganic crystal within seven crystal systems. The chemical bonding theory [5, 12–15] of single crystal growth is thus introduced into the morphology prediction and control of inorganic crystals from crystallographic viewpoint in this chapter. In this theory, we assume that growth units formed by solute and solvent pass through the boundary layer, and enter the lattice sites by bonding to the atoms on the corresponding crystal surface in the crystallization process [14]. For any crystal the constituent chemical bonds at each particular crystallographic plane can be quantitatively studied on the basis of its detailed crystallographic structure in the view of thermodynamics. All theoretical relations were generally derived and formulated by the thermodynamics of the whole crystallizing system, which includes the crystallized phase, crystal-solution interface and the mother solution phase. The microscopic characteristics of chemical bonds such as the bond number and strength distributed within particular crystallographic volume can be quantitatively determined along any [*hkl*] direction, the corresponding vertical growth rate of these selected planes can thus be quantitatively obtained in terms of the bonding parameters. The morphology of the crystal can be predicted by means of the relative growth rate of the selected planes. In this work, we selected seven typical inorganic crystals within seven crystal systems as examples to show how this chemical bonding theory functions, and the ideal morphology of the samples has been quantitatively predicted, which is in a good agreement with the experimental one. We also took $Mg(OH)_2$ as an example to analyze the effect of both crystal structure and reaction conditions on the powder morphology. Our present results indicate that the chemical bonding theory of single crystal growth is applicable to seven crystal systems and can provide a comprehensive understanding of the single crystal growth and the morphology control of various powders, which can also guide the control of crystal and powder morphology from a kinetic viewpoint.

2. Chemical Bonding Theory of Single Crystal Growth and Theoretical Calculation

2.1. Chemical Bonding Theory of Single Crystal Growth

Starting from the thermodynamic viewpoint and combining with the crystallographic and growth theories, comprehensive understanding the crystallization process is of fundamental importance in crystal growth and design. Herein, the crystallization process is assumed to be divided into the following three phase zones: liquid phase zone, transitional phase zone and crystal phase zone (Figure 1) [13].

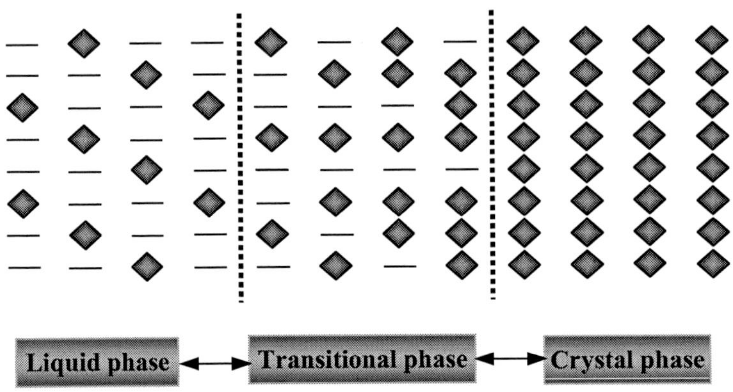

Figure 1. Schematic drawing of the three phase zones for the crystallization process.

As can be seen in Figure 1, growth units (shown as the rhombus) are first formed by bonding the constituent atoms of the crystal with each other in the liquid phase zone, and the bond strength is strong inside. These growth units can exist in the form of ions, ionic groups, complexes, and so on. The growth units then diffuse into the crystal phase zone and, further, gradually and selectively bond with the atoms on the corresponding crystal surfaces along the particular orientation (i.e., crystallographic orientation). In the whole process of crystal growth, the change of the growth units is not obvious due to the strong bond strength inside, and these growth units could be treated as stable growth groups. In other words, the crystallization process can also be regarded as the aggregated process of growth units from the liquid phase to the crystal phase.

The transitional phase zone is an interfacial phase, which is between the liquid phase zone and the crystal phase zone. In this zone, since the distance between the growth units becomes shorter, the arrangement of growth units is partially modulated. Growth units closed to the crystal phase tend to integrate and bond with each other in terms of the crystal intrinsic structure, which is attributed to the strong bonding gravitation of the corresponding crystallographic plane. On the other hand, the growth units closed to the liquid phase are dispersed in the solution and have a tendency to arrange in the form of that in the liquid phase zone.

Clearly, the bond strength inside the growth units is often stronger but has a little influence on the crystal growth. On the contrary, the weaker chemical bonds formed between growth units in the transitional phase zone, which can be affected easily by the external factors such as impurity, supersaturation, and pH value of the solution, are the determinant factors for the crystal growth rate and morphology. In addition, the crystal morphology is likewise depended on the intrinsic properties of the crystal, namely the crystal structure and the bond strength of the constituent chemical bonds. For any crystal the constituent chemical bonds at each particular crystallographic plane can be quantitatively studied on the basis of its detailed crystallographic structure in the view of thermodynamics. All theoretical relations were generally derived and formulated by the thermodynamics of the whole crystallizing system, which includes the crystallized phase, crystal-solution interface and the mother solution phase. The microscopic characteristics of chemical bonds such as the bond number and strength distributed within particular crystallographic volume can be quantitatively determined along any [*hkl*] direction, the corresponding vertical growth rate of these selected planes can thus be quantitatively obtained in terms of the bonding parameters.

The crystal growth is the process that the solid-liquid interface moves into the liquid, and at the same time the Gibbs free energy of the whole system decreases. Due to the existence of the driven force, the solid-liquid interface would develop into the solution by a distance of Δx along the crystallographic direction. This process leads to the decrease of the Gibbs free energy of the whole system, and the decreased proportion is defined as ΔG. Therefore, the work of driven force in above process can be expressed as:

$$W = f \cdot A \cdot \Delta x \tag{1}$$

where f is the driven force per unit area on the interface, and A is the area of the interface. On the other hand, the work of the driven force is equal to the decrease of the Gibbs free energy of the whole system, thus,

$$f \cdot A \cdot \Delta x = -\Delta G \tag{2}$$

where the minus represents that the Gibbs free energy is decreased.

For the nucleation in super-cooling melt, i.e., the arrangement of atoms within a small volume in the liquid phase transfers into that in solid phase, once overcome the obstacle of the nucleation barrier, the system will reach a state that the Gibbs free energy decreases. The above process involves that the metastable phase is converted into the stable phase and causes the Gibbs free energy of the whole system decreased. Since the released energy is related to the change of volume in phase transition, it is defined as "volume free energy" (ΔG_V).

Meanwhile, a new interface is formed along with the subsequent crystal growth. The atoms inside the bulk phase are symmetrically surrounded by the other atoms, but the ones on the interface suffer unsymmetrical force due to the unsaturated bonds. The force on the inside atoms is thus different from that on the surface atoms. The surface atoms will be departure from the equilibrium position of the regular arrangement, which leads to the Gibbs free energy of the whole system increased. This part of energy is defined as "interface free energy" (ΔG_S). Therefore, the total change of the Gibbs free energy in the whole system can be expressed as the following form:

$$\Delta G = \Delta G_S + \Delta G_V = A \cdot \gamma - \frac{V}{\Omega_S} \cdot \Delta g \tag{3}$$

where γ is the specific surface free energy, A is the area of the interface, V is the volume formed by the interface moving into the liquid phase, which can also be expressed as $V=A \cdot \Delta x$, Ω_S is the volume of an atom, and Δg is the chemical potential per mole. The first term in eq 1 expresses the free energy of formation of the new interface and the second term describes the difference in the chemical potential between the crystalline phase and the surrounding mother liquor.

By substituting eq 1 into eq 3, we then obtain the following relationship:

$$f \cdot A \cdot \Delta x = A \cdot \gamma - \frac{V}{\Omega_S} \cdot \Delta g \tag{4}$$

Meanwhile, under the thermodynamic condition, the driven force f can be expressed as:

$$f = \frac{\Delta g}{\Omega_S} \tag{5}$$

By substituting eq 5 into eq 4, eq 4 can be expressed as:

$$\frac{\Delta g}{\Omega_S} \cdot A \cdot \Delta x = A \cdot \gamma - \frac{V}{\Omega_S} \cdot \Delta g \tag{6}$$

Further, eq 6 may be transformed into the following equation

$$\frac{A \cdot \gamma}{2 \cdot A \cdot \Delta x} = \frac{\Delta g}{\Omega_S} \tag{7}$$

The rate of crystal growth can be expressed as the rate of displacement of a given crystal surface in the direction perpendicular to the face. Different crystallographic faces of a crystal, however, usually have different linear growth rates. Variations occur in the shape of crystals when individual faces grow at different rates, the overall crystal habit being determined by the slowest growing face. The rate of the crystal growth is proportioned to the driven force,

$$R \propto f \tag{8}$$

Herein, we define a coefficient k, and eq 8 becomes

$$R = k \cdot f \tag{9}$$

By substituting eq 5 into eq 9, we can then obtain

$$R = k \cdot \frac{\Delta g}{\Omega_S} \tag{10}$$

Furthermore, if we combine eq 7 and eq 10, the following formulate can be obtained

$$R = k \cdot \frac{A \cdot \gamma}{2 \cdot A \cdot \Delta x} \tag{11}$$

Meanwhile, the surface free energy γ can also be expressed as

$$\gamma = \frac{E}{A} \tag{12}$$

where E is the total bond energy between the adjacent net planes. By substituting eq 12 into eq 11, we can then obtain

$$R = k \cdot \frac{E}{2 \cdot A \cdot \Delta x} \tag{13}$$

According to the relationship between the bond energy E, and the bond valence S

$$E = \sum_n E_i = \sum_n J \cdot S_i = J \cdot \sum_n S_i \tag{14}$$

where E_i is the bond energy of individual bond and J is a constant for a given compound.
We can obtain the formulate by substituting eq 14 into eq 13

$$R = k \cdot \frac{J \cdot \sum_n S_i}{2 \cdot A \cdot \Delta x} = \frac{k \cdot J}{2} \cdot \frac{\sum_n S_i}{A \cdot \Delta x} = K \cdot \frac{\sum_n S_i}{A \cdot \Delta x} \tag{15}$$

Therefore, the chemical bonding theory of single crystal growth, which takes account of both crystal structure and growth conditions from the crystallographic viewpoint, can be described as the following formula:

$$R = K\frac{N \cdot S}{D \cdot A} \tag{16}$$

where R is the vertical growth rate of the selected plane, N is the bond number, S is the bond strength, D is the interplanar distance, A is the area of selected planes, and K is a constant for different planes in an equilibrium system. The "bond" here is an expression of not only thermodynamic equilibrium but also kinetic equilibrium. Therefore, both structure and reaction conditions have been taken into account in the chemical bonding theory of single crystal growth.

2.2. Theoretical Calculation

By using the eq 16, the vertical growth rate of selected planes of the crystal/powder can be quantitatively calculated. Combining the relative growth rates of the unique forms with their inherent multiplicities yields the overall crystal shape which can then, in turn, be predicted using the chemical bonding theory of single crystal growth.

Firstly, the crystal surfaces which are the most prominent and emerge finally in the process of crystallization and growth should be selected. These selected planes are always the low Miller index planes (h, k, $l < 5$) that have larger interplanar distance and weaker bond strength between the adjacent net planes. To reduce the number of the selected planes, the concept of periodic bond chains (PBCs) is introduced here, to calculate the crystal planes that have bigger possibility to be the crystal surfaces.

Then, bond valence model (BVM) is used to scale the bond strength concerning the oriented growth of crystals. The bond strength between atoms i and j is associated with the bond valence s_{ij}, which may be related to the bond length d_{ij} is by the following equation:

$$s_{ij} = \exp\left[\frac{d_0 - d_{ij}}{B}\right] \tag{17}$$

where the parameters d_0 and B are constants fitted empirically, d_{ij} is the bond length between the atoms i and j, bond valence s_{ij} is measured in valence units (v. u.), which formally corresponds to units of electronic charge. Since the ideal morphology of the crystal is the natural shape formed in an equilibrium system without influence from any external factors, i.e., in the bonding process, the weaker bonds formed between the growth units are not affected by the ambient condition (i.e., crystallizing environments), the bond strength can be expressed by that of the constituent chemical bonds.

Chart 1. Both predicted and experimental growth morphologies of selected samples within seven crystal systems

Crystal system	Typical crystal	Ideal morphology	Experimental morphology
Cubic	Cu_2O	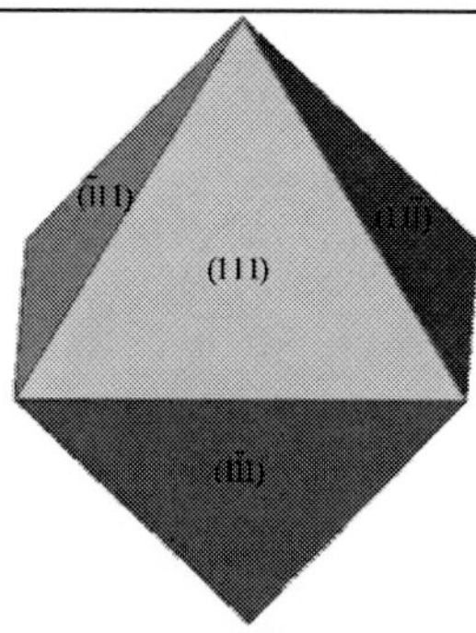	
Trigonal	$Mg(OH)_2$	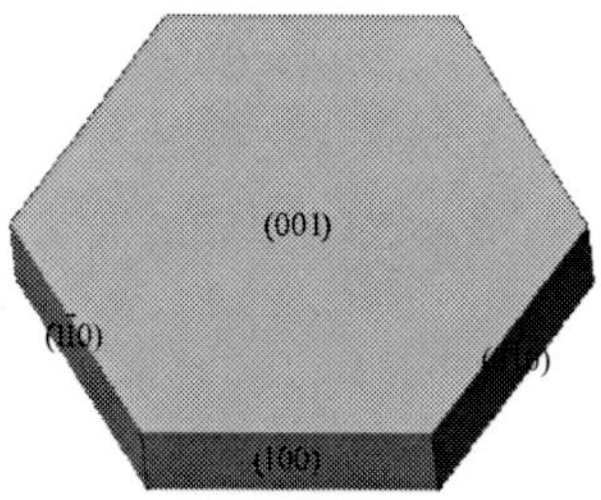	
Tetragonal	KDP (KH_2PO_4)	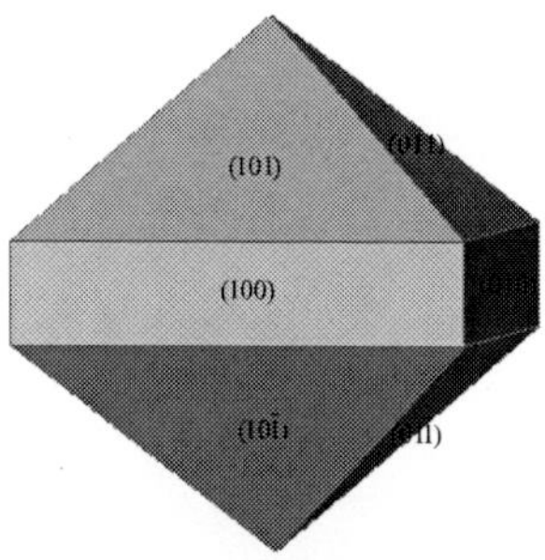	
Orthorhombic	$BaCO_3$	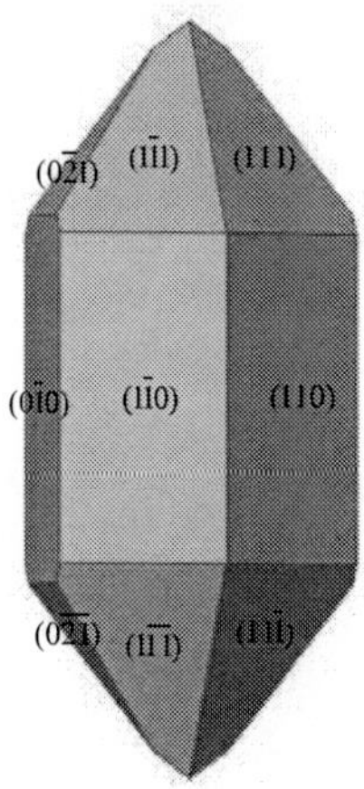	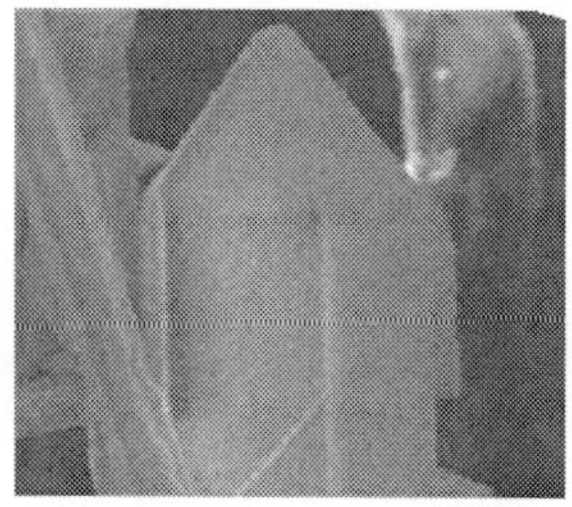

Chart 1. Continued

Crystal system	Typical crystal	Ideal morphology	Experimental morphology
Hexagonal	ZnO	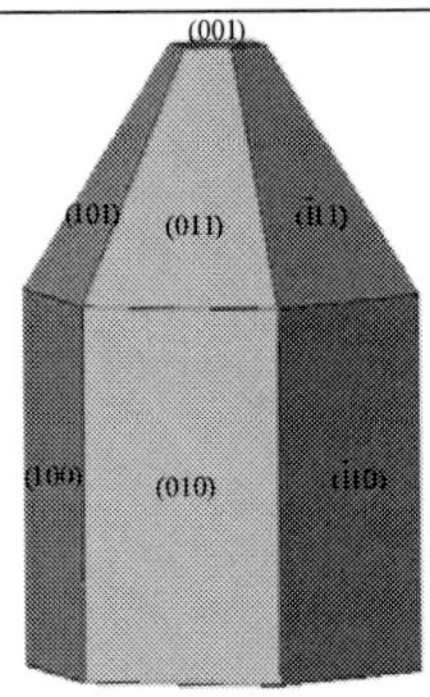	
Monoclinic	$MgCO_3 \cdot 3H_2O$	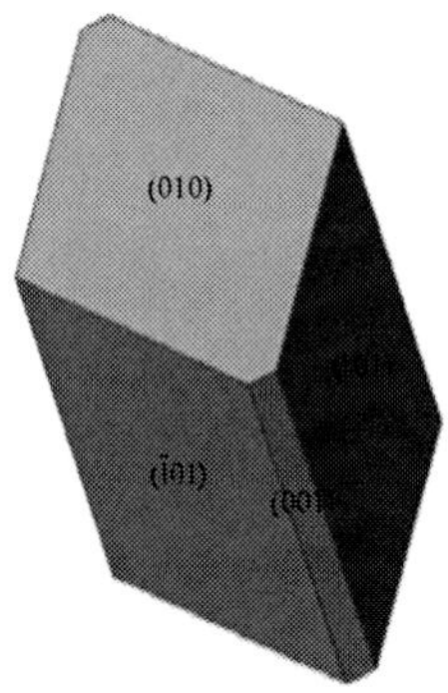	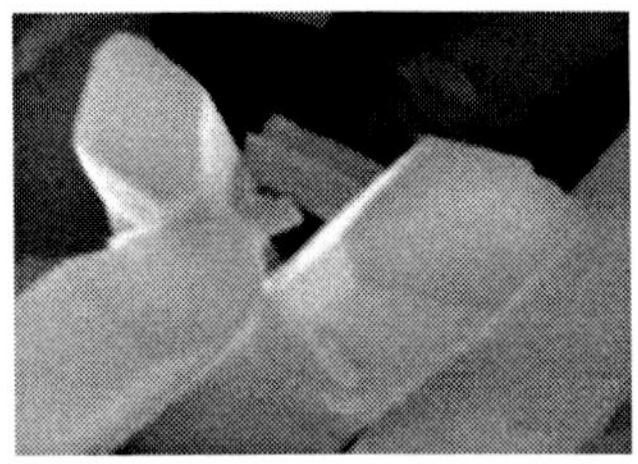
Triclinic	$CuSO_4 \cdot 5H_2O$	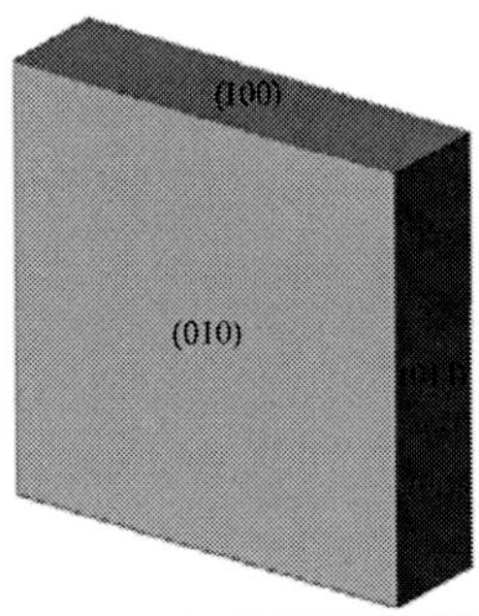	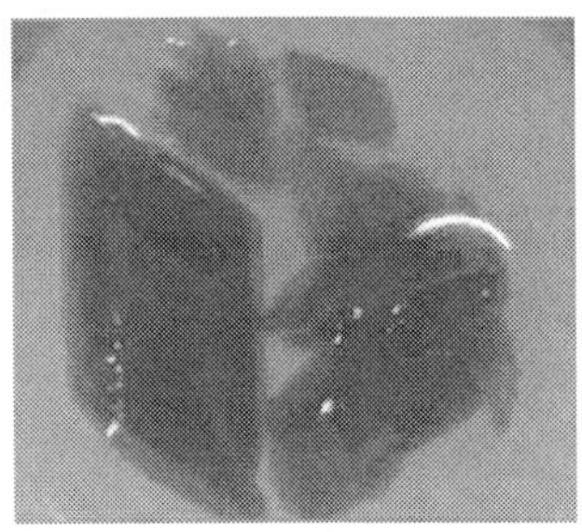

The obtained morphology is synergistically determined by intrinsic characteristics and detailed growth conditions. From the thermodynamic viewpoint, the crystal growth process is mainly determined by its intrinsic (e.g., crystallographic) characteristics. From the kinetic viewpoint, the growth process is on the other hand affected strongly by the external growth conditions. Therefore, to further understand the effect of kinetic control in the final morphology and size of the materials, the ideal morphology of crystal should be obtained primarily through analyzing the special crystal structure and the reaction condition in detail. In this section, using the chemical bonding theory of single crystal growth, typical inorganic crystals/powders within seven crystal systems were selected to predict their ideal morphologies, and the results are listed in Chart 1. Compared with the experimental growth morphologies of these crystals and powders, it shows that the ideal morphologies of the powders are in good agreement with the experimental ones.

3. Discussion

In the process of crystal growth, the configuration of growth units and the bonding process between them are easily affected by the growth conditions; therefore, the crystal morphology may be changed by the growth conditions consequently. A series of experimental approaches have thus been employed in order to obtain the diverse morphologies and sizes for a single material from the kinetic viewpoint. In theory, the chemical bonding theory of single crystal growth has been successfully applied to KDP, ADP, and $5Mg(OH)_2{\cdot}MgSO_4{\cdot}2H_2O$ (512MHSH), using which to analyze and explain the experimental phenomena from both the intrinsic structure and external conditions [5, 12-15], further to provide an effective guide for material synthesis. In this section, we also take $Mg(OH)_2$ as an example to discuss the phenomenon that various morphologies of $Mg(OH)_2$ can be obtained in different reaction conditions in detail.

3.1. Morphology Studies of KDP and ADP Crystallites Using Chemical Bonding Theory

It is well-known that potassium dihydrogen phosphate (KDP) and ammonium dihydrogen phosphate (ADP) possess good piezoelectric and nonlinear optical properties, which have drawn many attentions for their significantly theoretical interests and commercial applications in the optical modulation, frequency conversion, electrooptics, and so on. KDP and ADP crystallites have been successfully synthesized in ethanol–water solution in the previous work [12-15], and the size and morphology could be modified by adding different quantity of ethanol into the mother solution. KDP and ADP crystallize in the tetragonal system, space group $I\bar{4}2d$, with lattice parameters $a = b = 0.74528$ nm, $c = 0.69717$ nm for KDP and $a = b = 0.7502$ nm, $c = 0.7546$ nm for ADP. The unit cell structures are shown in Figure 2, and it can be seen that the distinction between KDP and ADP is their different constituent cations. Different constituent cations result in different bond strength, bond number and bond direction between the adjacent cations (K^+, NH_4^+) and anions ($H_2PO_4^-$).

When an amount of ethanol is added into the undersaturated solution, ethanol molecules may form large number of hydrogen bonds with water molecules and emit heat, which is equivalent to reduce solvent water from the undersaturated solution, and then the solution becomes instantaneously supersaturated, which leads to spontaneous nucleation and rapid growth. Meanwhile, ethanol molecules may interact with hydrogen bonds of crystal surfaces through hydroxyl, and the ethyl blocks the normal diffusion of growth units, which consequently suppresses the vertical growth of the crystal surfaces containing lots of hydrogen bonds.

From Chart 1, it is obvious that the ideal morphology of KDP is enclosed by {100} and {101} planes. Structurally, {100} planes are composed of positive K^+ and negative $H_2PO_4^-$ growth units alternatively, while {101} planes are terminated by positive K^+ in growing

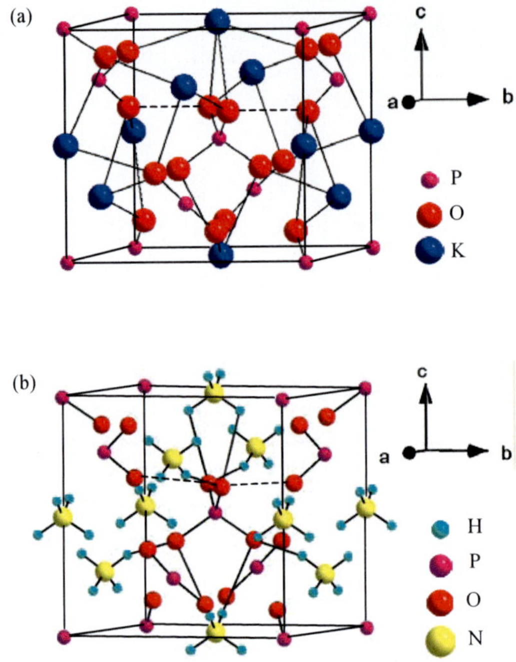

Figure 2. The tetragonal unit cell of (a) KDP and (b) ADP crystals. All atoms except hydrogen atoms of $H_2PO_4^-$ are drawn in color solid spheres, and the black dotted lines represent hydrogen bonds formed by two adjacent $H_2PO_4^-$ groups.

solution. Ethanol molecules will easily absorbed on the {100} planes by forming hydrogen bonds with the H^+ of $H_2PO_4^-$ groups; consequently, the normal growth rate of {100} planes decreases, while the growth rate of {101} planes has not obvious change due to the structure of interfaces, viz., the discrepancy of the rate between these two planes becomes larger. As a result, the facet {100} with a slower growth rate will be exposed more on the crystal surface. Therefore, the morphology of KDP becomes slender, (i.e., the aspect ratio increases) when ethanol is added into the reaction solution. And the effect of ethanol molecules on ADP is the same as on KDP. It is worth noting that use of a capping agent is an effective method to control the crystal shape. Therefore, it can be concluded that the effect of ethanol on crystal morphology and size may be classified into two aspects, extracting solvent water from solution and absorbing onto different crystal surfaces to suppress crystal growth. The former factor influences the volume of crystallites, while the later mainly affects the crystal morphology.

3.2. Morphology Studies of $Mg(OH)_2$ Crystallite Using Chemical Bonding Theory

As a flame-retardant filler in composite materials, magnesium hydrate ($Mg(OH)_2$) has shown many excellent functions in modern technology. It also has been widely used in wastewater treatment, solid–liquid separation, production of catalysts, etc. In recent years, it is found that $Mg(OH)_2$ can be used as starting product for controlling the synthesis of the micro- and nanoscale magnesium oxide (MgO), and the crystallite size, morphological features of

$Mg(OH)_2$ can be retained well. MgO has a number of attractive features desired for various applications in catalysis, refractory material industries, paint, and superconductors. Therefore, a variety of methods have been used to control the morphology and size of $Mg(OH)_2$ from the kinetic viewpoint. $Mg(OH)_2$ crystallizes in the trigonal system, space group $P\bar{3}m1$, with lattice parameters of $a = b = 0.3148$ nm, $c = 0.4772$ nm, and $\gamma = 120°$. The ideal morphology of $Mg(OH)_2$ has been quantitatively predicted by the chemical bonding theory of single crystal growth, and the normal growth rates of all selected planes are calculated and summarized in Chart 2. From Chart 2, it is obvious that the relative growth rate of {001} plane is the slowest. The detailed structure of every selected plane is also listed in Figure 3. The result of calculation indicates that {001} and {100} planes are predominant within all selected planes for $Mg(OH)_2$ powder, which lead to a natural morphology (shape) composed of hexagonal flake (shown in Chart 1). The diversity of crystal morphologies that can be found for a single material is testimony to the fact that the macroscopic shape is sensitive to the growth conditions [6]. Until now, different shapes of $Mg(OH)_2$ samples, such as the hexagonal lamella-like, wire-like, rod-like morphologies, etc. have been obtained via using different methods and under various conditions.

Chart 2. Relative growth rate of every selected plane of $Mg(OH)_2$ crystal

Plane	N_{Mg-O}	$N_{H\cdots O}$	D_{hkl} (nm)	R_{hkl}	R_{hkl}/R_{001}
001	0	2	0.4772	3.968	1.00
100	2	2	0.2726	20.402	5.14
010	2	2	0.2726	20.402	5.14
$1\bar{1}0$	2	2	0.2726	20.402	5.14
$0\bar{1}1$	4	2	0.2367	36.841	9.28
101	4	2	0.2367	36.841	9.28
111	4	2	0.1495	36.837	9.28

The strong influence of the growth solvent on crystal habit has been well-documented, and it is known that the underlying mechanism of incorporation involves face-specific, solvent-solute interactions resulting in different degrees of solvent binding that produces the inhibition of growth units bonding into the crystal on specific individual crystal habit planes [42]. By comparing the samples obtained in different solvent, it can be found that their morphologies are different. When $Mg(OH)_2$ powders grow in bidentate solvents, such as ethylenediamine or 1,6-diaminohexane, it is favorable to form a rod-like $Mg(OH)_2$, but in ammonia solution, it tends to obtain hexagonal lamella-like $Mg(OH)_2$. As the solvent molecule, ethylenediamine may also act as a bidentate ligand to form the relatively stable Mg^{2+} complex (see Figure 4(a)), and the growth units in this case can be considered as $[Mg(en)_2]^{2+}$ and OH^- instead of Mg^{2+} and OH^-. As shown in Figure 4, the atoms of this complex tend to exist in one plane due to the space structure of six-coordinated Mg^{2+}, which limits Mg^{2+} bonding with OH^- along the directions parallel to this plane to a certain extent. In the process of bonding into the crystal lattice, i.e. in the transitional phase zone, OH^- as the growth units favor to form bonds with the complex along the vertical direction of the plane which is shown in Figure 4(b), and this $[Mg(en)_2](OH)_2$ complex is subsequently bonding into the crystal lattice. Starting from the crystal structure of $Mg(OH)_2$, when the $Mg(OH)_2$ crystallites grow along [001] direction, the complexes ($[Mg(en)_2](OH)_2$) instead of Mg^{2+} as the growth units have the tendency to form the hydrogen bonds along this direction, which at the same time causes the coordinate bonds between Mg^{2+} and the N atoms of ethylenediamine to become weaker and break finally, and then the Mg^{2+} without ethylenediamine could form bonds with the OH^- existed in the solution, which makes it possible to grow along the other orientations. In this case, using ethylenediamine as the solvent is beneficial for growth of 1D $Mg(OH)_2$. Herein, ethylenediamine not only serves as the ligand solvent (chelating reagent) to slowly release Mg^{2+}, but also the "oriented" functional molecules which lead the growth of $Mg(OH)_2$ along [100] direction. On the other hand, while $Mg(OH)_2$ grows in the H_2O–$NH_3{\cdot}H_2O$ solution, the growth units can be considered as Mg^{2+} and OH^-, which approaches to the ideal crystallization process. From Chart 2, the rates of (001) and (100) planes are the slowest and the ideal morphology is determined by these two plane groups. Structurally, it can be found that both {001} and {100} planes are composed of Mg^{2+}, but the atom density of {001} planes is larger than that of {100} planes (Figure 3(a) and (b)). In the process of growth, the quantity of dissociative $NH_3{\cdot}H_2O$ molecules adsorbed on the {001} planes is more than the {100} planes, and the more the adsorbed quantity of $NH_3{\cdot}H_2O$, the stronger suppressed effect on the growth of the plane. According to the above discussion, the value of R_{100}/R_{001} in the actual reaction condition is larger than that in ideal condition, therefore, hexagonal lamella-like $Mg(OH)_2$ is beneficial to obtained in H_2O–$NH_3{\cdot}H_2O$ solution.

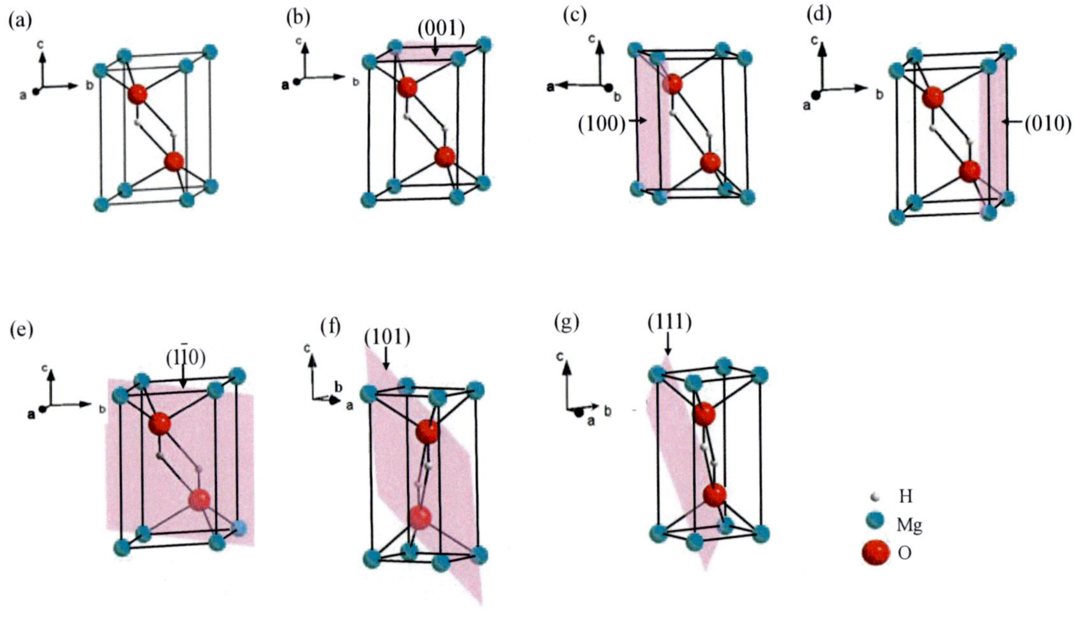

Figure 3. The unit cell of $Mg(OH)_2$ and the arrangement of atoms on every selected plane. (a) The structure of unit cell of $Mg(OH)_2$; (b) The arrangement of atoms on (001) plane; (c) The arrangement of atoms on (100) plane; (d) The arrangement of atoms on (010) plane; (e) The arrangement of atoms on (1-10) plane; (f) The arrangement of atoms on (101) plane; (g) The arrangement of atoms on (111) plane.

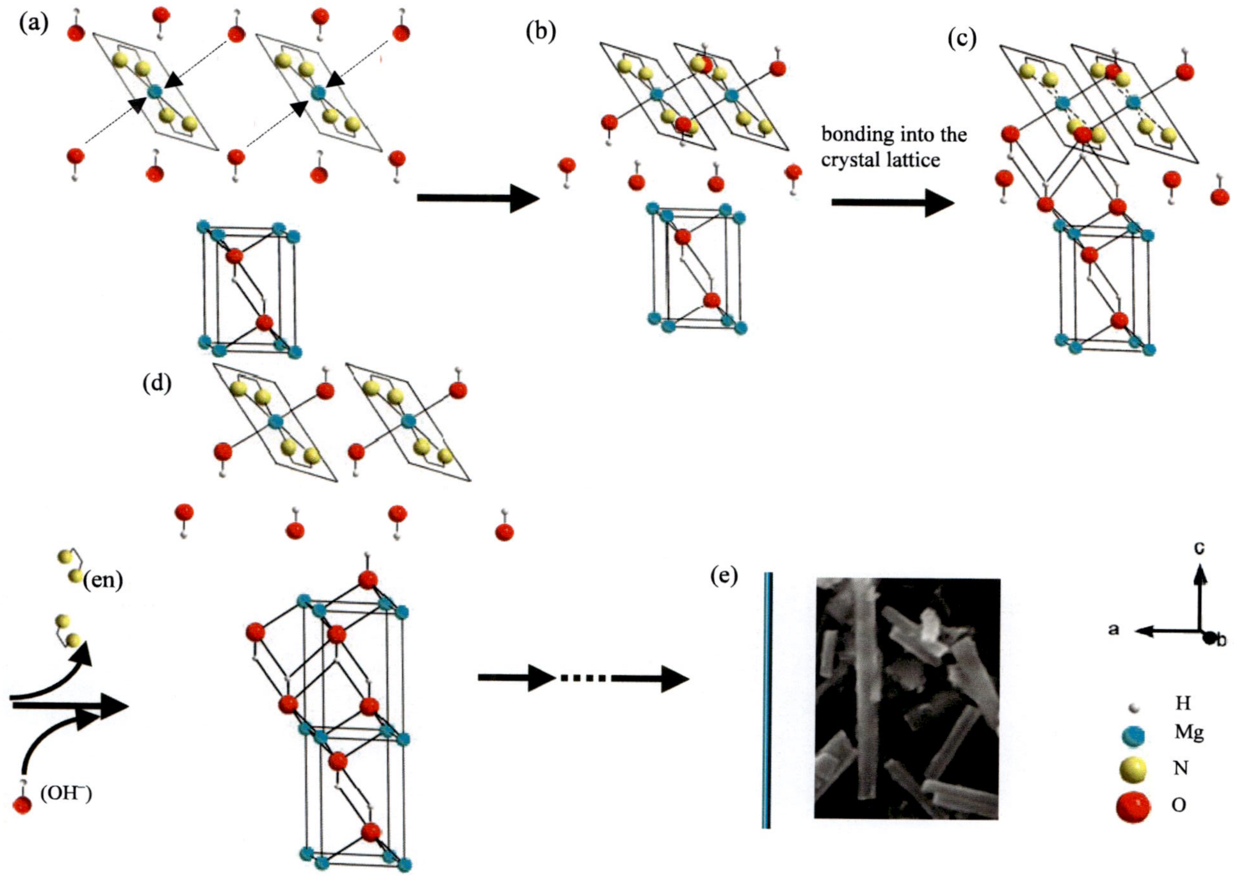

Figure 4. Schematic drawing of the growth process of rod-like $Mg(OH)_2$ using ethylenediamine as the solvent. (a) In the transition phase zone, it has the tendency to bond between the growth units due to the shorter distance of the growth units than that in liquid phase, and then they come into being the complex in the form of (b). Subsequently, the growth units bond into the crystal lattice along [001] direction (c), which leads that the bond strength of Mg–N becomes weaker and destroyed with the formation of hydrogen bond and makes $Mg(OH)_2$ growth along [001]. Therefore, rod–like $Mg(OH)_2$ forms at last. (e) SEM image of rod-like $Mg(OH)_2$.

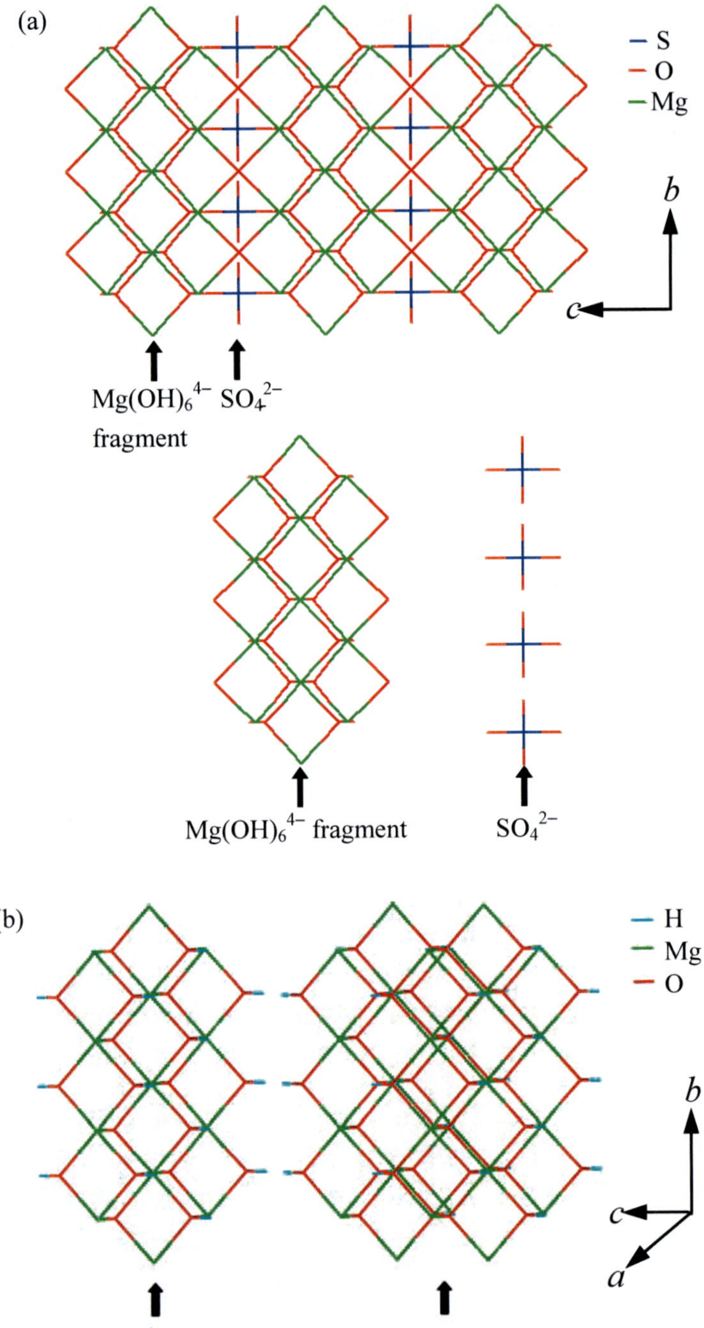

Figure 5. Crystallographic comparison of 512MHSH and $Mg(OH)_2$. (a) The top view is crystal structure of 512MHSH. All H atoms were omitted, and the adjacent $Mg(OH)_6^{4-}$ fragments are connected through SO_4^{2-} ions. The bottom view is the building block of 512MHSH, individually extracted from the top view to show involved $Mg(OH)_6^{4-}$ fragments and SO_4^{2-} ions. (b) The right is crystal structure of $Mg(OH)_2$, and the left is individually extracted from the right to show the involved $Mg(OH)_6^{4-}$ fragment. The adjacent $Mg(OH)_6^{4-}$ fragments are connected through hydrogen bonds in the crystallographic frame of $Mg(OH)_2$. It should be noted that the similarity of both 512MHSH and $Mg(OH)_2$ is the involved $Mg(OH)_6^{4-}$ fragments.

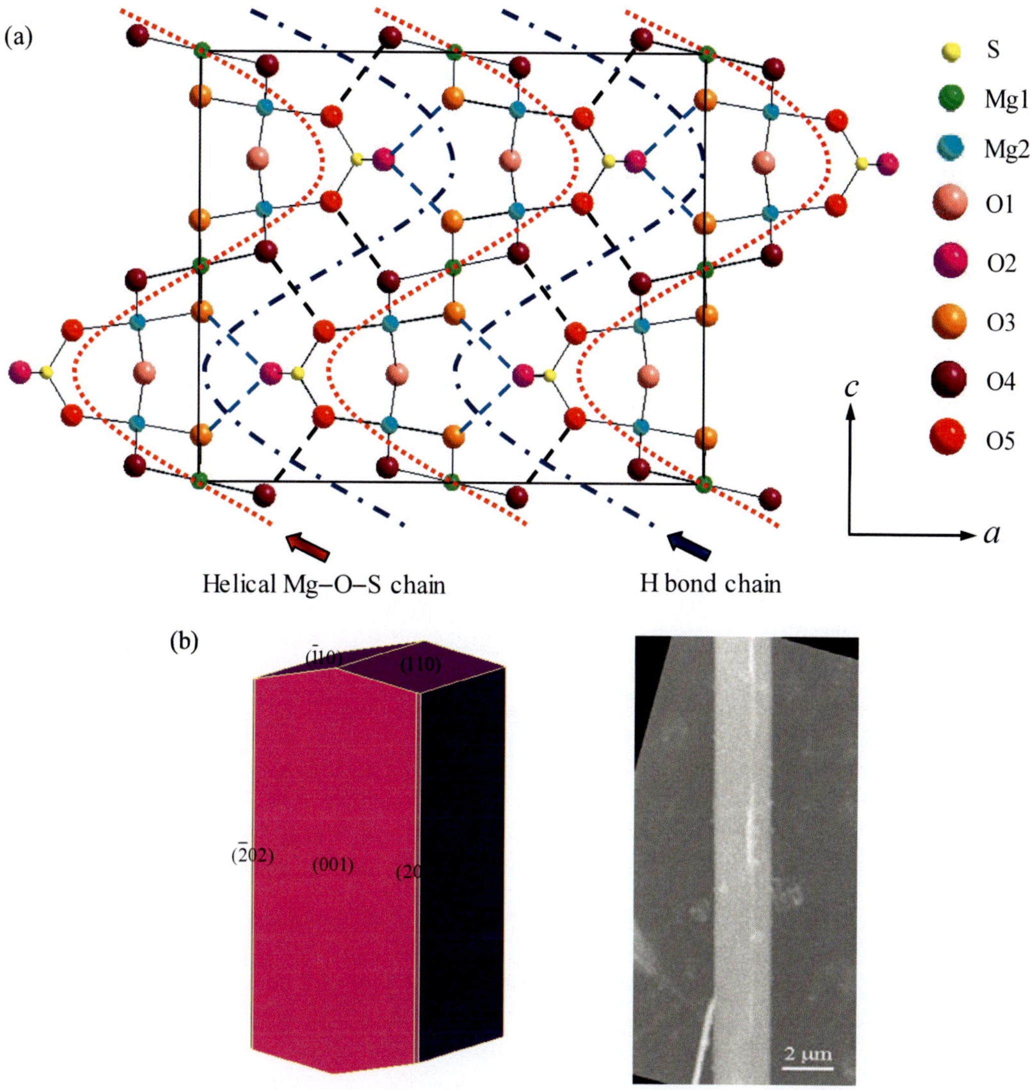

Figure 6. (a) Constituent atoms and chemical bonds of 512MHSH crystal. Those strong bonds such as Mg–O and S–O are shown as helical chain along the *c* axis in cyan dotted lines. All H atoms were omitted. The hydrogen bonds exist between neighboring helical Mg–O–S chains, and two kinds of hydrogen bonds are respectively drawn in black dashed lines and blue dotted lines. (b) The ideal morphology and experimental growth morphology of 512MHSH are shown for comparison.

It is well known that the morphology of the precursors also plays an important role in the final morphology of the corresponding products. The morphology of the precursors will be well maintained in the final samples by controlling the reaction conditions properly. Therefore, the research of the morphological evolution of these precursors becomes more important and valuable. Recently, using magnesium hydroxide sulfate hydrate (MHSH) as the precursor can synthesize $Mg(OH)_2$ whiskers. In our previous work, $Mg(OH)_2$ whiskers can be successfully obtained by using 512MHSH as the precursor, and 1D 512MHSH materials directed by SO_4^{2-} ions have been also systematically studied from both dynamic and thermodynamic viewpoint [5]. As shown in Figure 6(b), by utilizing the chemical bonding theory of single crystal growth, we can obtain the ideal morphology of 512MHSH, which has the tendency to grow along one dimension. From Figure 5a, it can be found that 512MHSH is composed of fragments of $Mg(OH)_6^{4-}$ and SO_4^{2-} ions, while the omitted hydrogen atoms link the neighboring helical Mg–O–S chains via hydrogen bonds (Figure 6(a)). Along the b-axis, the $Mg(OH)_6^{4-}$ fragments are compactly connected by the strong Mg–O bonds, while the total bond number is bigger and the total bond strength is stronger than those along both c and a axes, therefore, the growth rate along the b-axis is faster than that along both c- and a-axes, which benefits the formation of 1D 512MHSH along this direction. The other strong bond chain propagates along the c-axis, the $Mg(OH)_6^{4-}$ fragments are bonded by Mg–O bonds and SO_4^{2-} ions. Therefore, 512MHSH crystal tends to grow along the b-axis to form 1D morphologies such as the prism, wire, and belt, or along both b- and c-axes to form 2D morphologies such as the platelet. On the basis of the analysis of the structure in detail and theoretical calculation by the chemical bonding theory of single crystal growth, it can give an effective guide to control the morphology of 512 MHSH, and further to the $Mg(OH)_2$ crystal whiskers.

4. Conclusion

Chemical bonding theory of single crystal growth is proposed from the crystallographic viewpoint, and all theoretical relations of that are generally formulated by eq 16, in which both crystal structure and reaction condition (i.e. both thermodynamic and kinetic factors) have been taken into account. For any crystal the constituent chemical bonds at each particular crystallographic plane can be quantitatively studied on the basis of its detailed crystallographic structure. Under the ideal growth conditions, the bonds which are formed between the growth units in the crystal growth process, are not affected by the crystallizing environments, and the bonding strength can be expressed by the crystal constituent bond strengths. The ideal morphology of crystals can thus be quantitatively calculated by using this chemical bonding theory. The inorganic crystals/powders within seven crystal systems were respectively selected, and their ideal morphologies were quantitatively predicted. The results show that the ideal morphology of the inorganic crystals is in a good agreement with the experimental one, which demonstrates the general applicability of the chemical bonding theory for single crystals. We also take $Mg(OH)_2$ as the example to analyze and discuss the effect of both crystal structure and reaction conditions on the powder morphology. Our results indicated that the current theory can lead to a comprehensive understanding of the single crystal growth and the morphology control of various powders/particles, which can guide the control of crystal morphology from a kinetic viewpoint.

Acknowledgment

The authors would like to thank Dongli Xu, Xiaoxing Yan, Xu Zhang and Lei Wang for their important support in preparing the draft of this chapter. The financial support from a Program for New Century Excellent Talents in University (NCET-05-0278), the National Natural Science Foundation of China (Grant No. 20471012), a Foundation for the Author of National Excellent Doctoral Dissertation of P. R. China (Grant No. 200322), and the Research Fund for the Doctoral Program of Higher Education (Grant No. 20040141004) is greatly acknowledged.

References

[1] Bao, Z.; Weatherspoon, M. R.; Shian, S.; Cai, Y.; Graham, P.D.; Allan, S. M.; Ahmad, G.; Dickerson, M. B.; Church, B. C.; Kang, Z.; Abernathy, H. W.; Summers, C. J.; Liu, M.; Sandhage, K. H. *Nature* 2007, 446, 172–175.

[2] Tian, Z. R.; Voigt, J. A.; Liu, J.; Mckenzie, B.; Mcdermott, M. J.; Rodriguez, M. A.; Konishi, H.; Xu, H. *Nature Mater.* 2003, 2, 821–826.

[3] Manna, L.; Milliron, D.; Meisel, A.; Scher, E. C. *Nature Mater.* 2003, 2, 382–385.

[4] Li, H.; Bian, Z.; Zhu, J.; Zhang, D.; Li, G.; Huo, Y.; Li, H.; Lu, Y. *J. Am. Chem. Soc.* 2007, 129, 8406–8407.

[5] Yan, X.; Xu, D.; Xue D. *Acta Mater.* 2007, 55, 5747–5757.

[6] Piana, S.; Reyhani, M.; Gale, J. D. *Nature* 2005, 438, 70–73.

[7] Beyer, T.; Day, G. M.; Price, S. L. *J. Am. Chem. Soc.* 2001, 123, 5086–5094.

[8] Deij, M. A.; Cuppen, H. M.; Meekes, H.; Vlieg, E. *Cryst. Growth Des.* 2007, 7, 1936–1942.

[9] Deij, M. A.; Meekes, H.; Vlieg, E. *Cryst. Growth Des.* 2007, 7, 1949–1957.

[10] Auer, S.; Frenkel, D. *Nature* 2001, 409, 1020–1023.

[11] Bennema, P.; Meekes, H.; Boerrigter, S. X. M.; Cuppen, H. M.; Deij, M. A.; Van Eupen, J.; Verwer, P.; Vlieg, E. *Cryst. Growth Des.* 2004, 4, 905–913.

[12] Xu, D.; Xue, D.; Ratajczak, H. *J. Mol. Struct.* 2005, 740, 37–45.

[13] Xu, D.; Xue, D. *J. Synth. Cryst.* 2006, 35, 598–603. (in Chinese)

[14] Xu, D.; Xue, D. *J. Cryst. Growth* 2006, 286, 108–113.

[15] Xu, D.; Xue, D. *Physica. B* 2005, 370, 84–89.

[16] Hartman, P.; Perdok, W. G. *Acta Crystallogr.* 1955, 8, 49–52.

[17] H. J. Scheel, in *Crystal Growth Technology*, edited by H. J. Scheel and T. Fukuda; ISBN: 0-471-49059-8; John Wiley & Sons, Ltd: Chichester, West Sussex, 2003, Chap. 1, 3–13.

[18] Zhang, X.; Xue, D.; Xu, D.; Feng, X.; Wang, J. *J. Cryst. Growth* 2005, 280, 234–238.

[19] Zhang, X.; Xue, D.; Xu, D.; Wang, J.; Feng, X. *J. Cryst. Growth* 2006, 292, 505–509.

[20] Yan, C.; Xue, D. *Adv. Mater.* 2007, DOI: 10.1002/adma.200701752.

[21] Yan, C.; Xue, D. *J. Phys. Chem. B* 2006, 110, 7102–7106.

[22] Yan, C.; Xue, D. *J. Phys. Chem. B* 2006, 110, 1581–1586.

[23] Xu, J.; Xue, D. *J. Phys. Chem. B* 2006, 110, 7750–7756.

[24] Yan, C.; Xue, D. *J. Phys. Chem. B* 2006, 110, 25850–25855.

[25] Xu, J.; Xue, D. *J. Phys. Chem. B* 2006, 110, 17400–17405.

[26] Liu, M.; Xue, D.; Zhang, S.; Zhu, H.; Wang, J.; Kitamura, K. *Mater. Lett.* 2005, 59, 1095–1097.

[27] Yan, C.; Xue, D.; Zou, L.; Yan, X.; Wang, W. *J. Cryst. Growth* 2005, 282, 448–454.

[28] Liu, M.; Xue, D.; Luo, C. *J. Alloy. Compd.* 2006, 426, 118–122.

[29] Yan, C.; Xue, D. *J. Alloy. Compd.* 2007, 431, 241–245.

[30] Luo, C.; Xue, D. *Langmuir* 2006, 22, 9914–9918.
[31] Xu, J.; Xue, D. *Acta Mater.* 2007, 55, 2397–2406.
[32] Yan, C.; Xue, D. *J. Phys. Chem. B* 2005, 109, 12358–12361.
[33] Xu, J.; Xue, D. *J. Phys. Chem. B* 2005, 109, 17157–17161.
[34] Yan, C.; Xue, D.; Zou, L. *Mater. Res. Bull.* 2006, 41, 2341–2348.
[35] Yan, C.; Xue, D. *J. Phys. Chem. B* 2006, 110, 11076–11080.
[36] Xu, J.; Xue, D.; Yan, C. *Mater. Lett.* 2005, 59, 2920–2922.
[37] Liu, M.; Xue, D. *Mater. Lett.* 2005, 59, 2908–2910.
[38] Liu, M.; Xue, D. *Solid State Ionics* 2006, 177, 275–280.
[39] Yan, C.; Xue, D. *Electrochem. Commun.* 2007, 9, 1247–1251.
[40] Liu, M.; Xue, D. *J. Alloy. Compd.* 2007, 427, 256–259.
[41] Xu, J.; Xue, D. *J. Phys. Chem. B* 2006, 110, 11232–11236.
[42] Hammond, R. B.; Pencheva, K.; Ramachandran, V.; Roberts, J. *Cryst. Growth Des.* 2007, 7, 1571–1574.

In: Astrophysics and Condensed Matter
Editor: Thomas G. Hardwell, pp. 235-256
ISBN: 978-1-60456-332-0

Chapter 9

CANDIDATES FOR DARK MATTER IN THE $SU(3)_C \otimes SU(3)_L \otimes U(1)_N$ MODELS

H. N. Long[a], N. Q. Lan[b], D. V. Soa[b] and L. N. Thuc[b]
[a]Institute of Physics, VAST, P. O. Box 429, Bo Ho, Hanoi, Vietnam
[b]Department of Physics, Hanoi University of Education, Hanoi, Vietnam

Abstract

It has recently been pointed out that the 511 keV emission line detected by INTEGRAL/SPI from the bulge of our galaxy could be explained by annihilations of light dark matter particles into $e^+ \ e^-$. We present the possibility that dark matter could be made of scalar candidates, namely , of the Higgs bosons in the models based on $SU(3)_C \otimes SU(3)_L \otimes U(1)_N$ (3-3-1) gauge group.

These particles are singlet of the $SU(2)_L \otimes U(1)_Y$ group, so they do not interact with the ordinary particles, except the Higgs boson of the standard model. The Spergel-Steinhardt condition for self-interacting dark matter gives a bound on the mass of the candidates to be a few MeVs. Beside the scalar candidates, which exist in both non-SUSY and SUSY 3-3-1 models with right-handed neutrinos, the spin $\frac{1}{2}$ candidate exists in a variant 3-3-1 version with exotic neutral lepton.

In contrast to the singlet models, where an extra symmetry must be imposed to account the stability of the dark matter, in this case the decay of the candidates is automatically forbidden in all orders of perturbative expansion. This is because of that, these scalars are singlets, i.e., in bottom of the Higgs triplet. Therefore, the standard model fermions and the standard gauge bosons cannot couple with them. The 3-3-1 models are favor of the light dark matter.

1 Introduction

Observation of the cosmic microwave background, the primordial abundances of light elements and large scale structure have revealed that a great deal of the mass of our universe consists of dark matter. It is an amazing fact that even as our understanding of cosmology progresses by leaps and bounds, we remain almost completely ignorant about the nature of most of the matter in the universe [1]. It has recently been pointed out that the 511 keV emission line detected by INTEGRAL/SPI [2] from the bulge of our galaxy could be explained by annihilations of light dark matter particles into $e^+ e^-$. The nature of dark matter is still a challenging question in cosmology.

Cosmological models with a mixture of roughly 35% collisionless cold dark matter such as axions, WIMPs, or any other candidate interacting through the weak and gravitational forces only, and 65% vacuum energy or quintessence match observation of the cosmic microwave background and large scale structure on extra-galactic scales with remarkable accuracy [3,4]. It is known that only a fraction of the dark matter can be made of ordinary baryons and its enormous amount has unknown, nonbaryonic origin [5].

Until a few years ago, the more satisfactory cosmological scenarios were those ones composed of ordinary matter, cold dark matter and a contribution associated with the cosmological constant. To be consistent with inflationary cosmology, the spectrum of density fluctuations would be nearly scale-invariant and adiabatic. However, in recent years it has been pointed out that the conventional models of collisionless cold dark matter lead to problems with regard to galactic structures. They were only able to fit the observations on large scales ($\gg 1$ Mpc). Also, N-body simulations in these models result in a central singularity of the galactic halos [6] with a large number of sub-halos [7], which are in conflict with astronomical observations. A number of other inconsistencies are discussed in Refs. [8,9]. Thus, the cold dark matter model is not able to explain observations on scales smaller than a few Mpc.

It has recently been shown that an elegant way to avoid these problems is to assume the so called *self-interacting dark matter* [10]. One should notice that, in spite of all, self-interacting models lead to spherical halo centers in clusters, which is not in agreement with ellipsoidal centers indicated by strong gravitational lens observations [11] and by Chandra observations [12].

However, self-interacting dark matter models are self-motivated as alternative models. It is a well-accepted fact that the plausible candidates for dark matter are elementary particles. The key property of these particles is that, they must have a large scattering cross-section and negligible annihilation or dissipation. The Spergel-Steinhard model has motivated many follow-up studies [5, 13, 14]. Several authors have proposed models in which a specific scalar singlet that satisfies the self-interacting dark matter properties is introduced in the standard model in an *ad hoc* way [5, 14].

The standard model offers no options for dark matter. The first gauge model for self-interacting dark matter is the extended standard model based on the $\mathrm{SU}(3)_C \otimes \mathrm{SU}(3)_L \otimes \mathrm{U}(1)_N$ gauge group [15, 16]. The 3-3-1 model found by Fregolente and Tonasse [17] to be self-interacting dark matter differs from another one found by Long and Lan [18].

Models with $SU(3)_C \otimes SU(3)_L \otimes U(1)_N$ gauge symmetry (called 3-3-1 models for short) are interesting possibilities for the physics at the TeV scale [15, 16]. At low energies they coincide with the standard model and some of them give at least partial explanation to some fundamental questions that are accommodated but not explained by the standard model. For instance, in order to cancel the triangle anomalies, together with asymptotic freedom in QCD, the model predicts that the number of generations must be three and only three; *(ii)* the model of Ref. [15] predicts that $(g'/g)^2 = \sin^2\theta_W/(1-4\sin^2\theta_W)$, thus there is a Landau pole at the energy scale μ at which $\sin^2\theta_W(\mu) = 1/4$. According to recent calculations $\mu \sim 4$ TeV [19,20] ; *(iii)* the quantization of the electric charge [21] and the vectorial character of the electromagnetic interactions [22] do not depend on the nature of the neutrinos i.e., if they are Dirac or Majorana particles; *(iv)* as a consequence of item ii) above, the model possesses perturbative $\mathcal{N} = 1$ supersymmetry naturally at the μ scale [23, 24]; *(v)* the Peccei-Quinn [25] symmetry occurs naturally in these models [26]; *(vi)* since one generation of quarks is treated differently from the others this may be lead to a natural explanation for the large mass of the top quarks [27]. Moreover, if right-handed neutrinos are considered transforming non-trivially [16], 3-3-1 models [15] can be embedded in a model with 3-4-1 gauge symmetry in which leptons transform as $(\nu_l, l, \nu_l^c, l^c)_L \sim (\mathbf{1}, \mathbf{4}, 0)$ under each gauge factors [28, 29]. The $SU(3)_L$ symmetry is possibly the largest symmetry involving the known leptons (and $SU(4)_L$ if right-handed neutrinos do really exist). This make 3-3-1 or 3-4-1 models interesting by their own, and it has been the source of interest recently [30] because it requires that the number of fermion families be a multiple of the quark color in order to cancel anomalies, which suggest a path to the solution of the flavor problem. The Peccei-Quinn [25] symmetry occurs naturally in these models [26]. Another important feature of these models is that the SU$(2)_L$ group is totally embedded in SU$(3)_L$.

A subject that has not been given much attention by particle physicists in the past, could prove to be a remarkable powerful and precise probe of the properties of dark matter.

This paper is organized as follows: In Sec. 2 we briefly introduce necessary elements of the non-SUSY 3-3-1 models, and the candidates for dark matter. Sec. 3 is devoted to supersymmetric 3-3-1 model with right-handed neutrinos. We summarize our result and make conclusions in the last section.

2 The Non-SUSY 3-3-1 Models

To frame the context, it is appropriate to recall briefly some relevant features of the 3 - 3 - 1 models [15, 16]. We first introduce the 3-3-1 model with exotic lepton proposed by Pleitez and Tonasse [31]

2.1 The 3-3-1 Model with Exotic Lepton

The model treats the leptons as SU$(3)_L$ triplets [31]

$$f_L^a = \begin{pmatrix} \nu_L^a \\ e_L^a \\ P^a \end{pmatrix} \sim (1, 3, 0),\ e_R^a \sim (1, 1, -1),\ P_R^a \sim (1, 1, +1), \tag{1}$$

where $a = 1, 2, 3$ is the generation index. Each charged left-handed fermion field has its right-handed countepart transforming as a singlet of the SU$(3)_L$ group. Two of the three quark generations transform as antitriplets and the third generation is treated differently - it belongs to a triplet:

$$Q_{iL} = \begin{pmatrix} d_{iL} \\ -u_{iL} \\ D_{iL} \end{pmatrix} \sim (3, \bar{3}, -\frac{1}{3}), \tag{2}$$

$$u_{iR} \sim (3, 1, 2/3), d_{iR} \sim (3, 1, -1/3), D_{iR} \sim (3, 1, -4/3),\ i = 1, 2,$$

$$Q_{3L} = \begin{pmatrix} u_{3L} \\ d_{3L} \\ T_L \end{pmatrix} \sim (3, 3, 2/3), \tag{3}$$

$$u_{3R} \sim (3, 1, 2/3), d_{3R} \sim (3, 1, -1/3), T_R \sim (3, 1, 5/3).$$

Of the nine gauge bosons $W^a (a = 1, 2, ..., 8)$ and B of $SU(3)_L$ and $U(1)_N$, four are light: the photon (A), Z and $W^{\pm}$. The remaining five correspond to new heavy gauge bosons Z_2, $Y^{\pm}$ and the doubly charged bileptons $X^{\pm\pm}$.

Symmetry breaking and fermion mass generation can be achieved by three scalar $SU(3)_L$ triplets

$$\begin{array}{l} \mathrm{SU}(3)_C \otimes \mathrm{SU}(3)_L \otimes \mathrm{U}(1)_N \\ \quad \downarrow \langle\chi\rangle \\ \mathrm{SU}(3)_C \otimes \mathrm{SU}(2)_L \otimes \mathrm{U}(1)_Y \\ \quad \downarrow \langle\eta\rangle, \langle\rho\rangle \\ \mathrm{SU}(3)_C \otimes \mathrm{U}(1)_Q, \end{array}$$

where the minimally required scalar multiplets are summarized as

$$\begin{aligned} \chi &= \begin{pmatrix} \chi^- \\ \chi^{--} \\ \chi^0 \end{pmatrix} \sim (1, 3, -1), \\ \eta &= \begin{pmatrix} \eta^0 \\ \eta_1^- \\ \eta_2^+ \end{pmatrix} \sim (1, 3, 0), \\ \rho &= \begin{pmatrix} \rho^+ \\ \rho^0 \\ \rho^{++} \end{pmatrix} \sim (1, 3, +1), \end{aligned} \tag{4}$$

The vacuum expectation value (VEV) $\langle\chi^T\rangle = (0, 0, w/\sqrt{2})$ yields masses for the exotic quarks and lepton, the heavy neutral gauge boson (Z_2) and two charged gauge bosons (X^{++}, Y^+). The masses of the standard gauge bosons and the ordinary fermions are related to the VEVs of the other scalar fields, $\langle\eta^0\rangle = v/\sqrt{2}, \langle\rho^o\rangle = u/\sqrt{2}$.

In order to be consistent with the low energy phenomenology we have to assume that $w \gg u, v$.

By matching the gauge coupling constants we get a relation between g and g_N – the couplings associated with $\mathrm{SU}(3)_L$ and $\mathrm{U}(1)_N$, respectively:

$$\frac{g_N^2}{g^2} = \frac{6\, s_W^2(M_{Z_2})}{1 - 4s_W^2(M_{Z_2})}, \tag{5}$$

where $e = g\, s_W$ is the same as in the standard model. The most *economical*, gauge invariant and renormalizable Higgs potential is

$$V(\eta, \rho, \chi) = \mu_1^2\eta^\dagger\eta + \mu_2^2\rho^\dagger\rho + \mu_3^2\chi^\dagger\chi + a_1\left(\eta^\dagger\eta\right)^2 + a_2\left(\rho^\dagger\rho\right)^2 + a_3\left(\chi^\dagger\chi\right)^2$$

$$+\left(\eta^\dagger\eta\right)\left(a_4\rho^\dagger\rho + a_5\chi^\dagger\chi\right) + a_6\left(\rho^\dagger\rho\right)\left(\chi^\dagger\chi\right) + a_7\left(\rho^\dagger\eta\right)\left(\eta^\dagger\rho\right)$$
$$+a_8\left(\chi^\dagger\eta\right)\left(\eta^\dagger\chi\right) + a_9\left(\rho^\dagger\chi\right)\left(\chi^\dagger\rho\right) + \frac{1}{2}\left(f\epsilon^{ijk}\eta_i\rho_j\chi_k + \text{H. c.}\right). \quad (6)$$

Here the μ's and f are coupling constants with dimension of mass with $a_3 < 0$ and $f < 0$ from the positivity of the scalar masses [32].

Symmetry breaking is initiated when the scalar neutral fields are shifted as $\varphi = v_\varphi + \xi_\varphi + i\zeta_\varphi$, with $\varphi = \eta^0, \rho^0, \chi^0$. The details of the physical spectrum of the neutral scalar sector are crucial for our results. It is given in Refs. [32] and we summarize them here. Firstly we notice that real part of the shifted fields leads to the three massive physical scalar fields H_1^0, H_2^0, H_3^0 defined by

$$\begin{pmatrix} \xi_\eta \\ \xi_\rho \end{pmatrix} \approx \frac{1}{v_W}\begin{pmatrix} v & u \\ u & -v \end{pmatrix}\begin{pmatrix} H_1^0 \\ H_2^0 \end{pmatrix}, \qquad \xi_\chi \approx H_3^0, \quad (7)$$

where we are using $w \gg v, u$. The scalar H_1^0 is the one that we can identify with the standard model Higgs, since its squared mass,

$$m_1^2 \approx 4\frac{a_2u^4 - a_1v^4}{v^2 - u^2}, \quad (8)$$

carries no any feature from the 3-3-1 breakdown to the standard model. On the other hand H_3^0, with squared mass

$$m_3^2 \approx -4a_3w^2, \quad (9)$$

is a typical 3-3-1 scalar. So, there is no any massless Goldstone boson rising from the real part of the neutral sector. On the other hand, from the imaginary part we have two Goldstone and one massive physical state h^0 with eigenstate

$$\zeta_\chi \approx h^0 \quad (10)$$

and squared mass

$$m_h^2 = -f\frac{v_W^2 w^2 + v^2u^2}{vuw}. \quad (11)$$

It is important to notice that ζ_η and ζ_ρ are pure massless Goldstone states. The approximation in Eqs. (7), (8), (9) and (10), is valid for $w \gg v, u$. This condition leads to relations among the parameters of the scalar potential (6). One of them, which enters in the $H_1^0h^0h^0$ interaction, is

$$a_5v^2 + 2a_6u^2 \approx -\frac{vu}{2} \quad (12)$$

We must consider also the matter coupling through the scalar fields. In the model of Ref. [31], the full Yukawa Lagrangians that must be considered are

$$\mathcal{L}_\ell = -\sum_{ab}\left(\frac{1}{2}\epsilon^{ijk}G_{ab}^{(\nu)}\overline{\psi_{aiL}}^C\psi_{bjL}\eta_k + G_{ab}^{(\ell)}\overline{\psi}_{aL}\ell_{bR}^-\rho - G_{ab}^{(P)}\overline{\psi}_{aL}P_{bR}^+\chi\right) + \text{H. c.} \quad (13)$$
$$\mathcal{L}_Q = \overline{Q}_{1L}\sum_b\left(G_{1b}^{(U)}U_{bR}\eta + G_{1b}^{(D)}D_{bR}\rho + G^{(J)}J_{1R}\chi\right) +$$

$$\sum_{\alpha}\overline{Q}_{\alpha L}\left(F^{(U)}_{\alpha b}U_{bR}\rho^{*}+F^{(D)}_{\alpha b}D_{bR}\eta^{*}+\sum_{\beta}F^{(J)}_{\alpha\beta}J_{\beta R}\chi^{*}\right)+\text{H. c.,} \tag{14}$$

where $G^{(\nu)}_{ab}$, $G^{(\ell)}_{ab}$, $G^{(P)}_{ba}$, $G^{(U)}_{1b}$, $F^{(U)}_{\alpha b}$, $G^{(D)}_{1b}$, $F^{(D)}_{\alpha b}$, $G^{(J)}$ and $F^{(J)}_{\alpha\beta}$ are the Yukawa coupling constants. η^*, ρ^* and χ^* denote the η, ρ and χ antiparticle fields, respectively.

The main properties that a good dark matter candidate must satisfy are stability and neutrality. Therefore, we go to the scalar sector of the model, more specifically to the neutral scalars, and we examine whether any of them can be stable and in addition whether they can satisfy the self-interacting dark matter criterions [10]. In addition, one should notice that such dark matter particle must not overpopulate the Universe. On the other hand, since our dark matter particle is not imposed arbitrarily to solve this specific problem, we must check that the necessary values of the parameters do not spoil the other bounds of the model.

We can check through a direct calculation by employing the Lagrangians (6), (13) and (14) and by using the eigenstates (7) and (10) that the Higgs scalar h^0 and H^0_3 can, in principle, satisfy the criterions above. Remarkably they do not interact directly with any standard model field except for the standard Higgs H^0_1. However, h^0 must be favored, since we have checked that it is easier to obtain a large scattering cross section for it, relative to H^0_3, by a convenient choice of the parameters.

In contrast to the singlet models of the Refs. [5, 14], where an extra symmetry must be imposed to account the stability of the dark matter, here the decay of the h^0 scalar is automatically forbidden in all orders of perturbative expansion. This is because of the following features: i) this scalar comes from the triplet χ, the one that induces the spontaneous symmetry breaking of the 3-3-1 model to the standard model. Therefore, the standard model fermions and the standard gauge bosons cannot couple with h^0, ii) the h^0 scalar comes from the imaginary part of the Higgs triplet χ. As we mentioned above, the imaginary parts of η and ρ are pure massless Goldstone bosons. Therefore, there is not physical scalar fields which can mix with h^0. So, the only interactions of h^0 come from the scalar potential and they are $H^0_3h^0h^0$ and $H^0_1h^0h^0$. This latter has strength $2i\left(a_5v^2+a_6u^2\right)/v_W\equiv 2i\Theta$. We can check also that h^0 does not interact with other exotic particle.

Hence, if $v\sim u\sim(100-200)$ GeV and $-1\leq a_5\sim a_6\leq 1$, the h^0 can interact only weakly with ordinary matter through the Higgs boson of the standard model H^0_1. The relevant quartic interaction for scattering is $h^0h^0h^0h^0$, whose strength is $-ia_3$. Other quartic interactions evolving h^0 and other neutral scalars are proportional to $1/w$ and so we neglect them. The cross section of the process $h^0h^0\to h^0h^0$ *via* the quartic interaction is $\sigma=a_3^2/64\pi m_h^2$. The contribution of the trilinear interactions *via* H^0_1 and H^0_3 exchange are negligible. There is no other contribution to the process involving the exchange of vector or scalar bosons. A self-interacting dark matter candidate must have mean free path $\Lambda=1/n\sigma$ in the range 1 kpc $<\Lambda<$ 1 Mpc, where $n=\rho/m_h$ is the number density of the h^0 scalar and ρ is its density at the solar radius [7]. Therefore, with $a_3!=-1$, -0.208×10^{-7} GeV $\leq f\leq -0.112\times10^{-6}$ GeV, $w=1000$ GeV, $u=195$ GeV and $\rho=0.4$ GeV/cm^3, we obtain the required Spergel-Steinhardt bound, *i. e.*, 2×10^3 GeV$^{-3}\leq\sigma/m_h\leq 3\times10^4$ GeV^{-3} [10].

With this set of parameter values, we see from Eq. (11) that 5.5 MeV $\leq m_h\leq$ 29 MeV. This means that our dark matter particle is non-relativistic in the decoupling era (decoupling temperature $\sim$ 1 eV) and, for a standard model Higgs boson mass $\sim$ 100 GeV [33], it is produced by a thermal equilibrium density of the standard Higgs scalar to h^0h^0 pairs [14]. The density of the h^0 scalar from the H_1^0 decay can be obtained following the standard procedure, *i. e.*, we must solve the Boltzmann equation

$$\frac{dn_h}{dt} + 3Hn_h = \langle \Gamma_H \rangle n_H^{(\mathrm{eq})}, \tag{15}$$

where n_h is the number density of the h^0 scalar at the time t, H is the Hubble expansion rate,

$$\Gamma_H = \frac{\Theta^2}{4\pi E} \tag{16}$$

is the decay rate for the H_1^0 with energy E and

$$n_H^{(\mathrm{eq})} = \frac{1}{2\pi^2} \int_{m_1}^{\infty} \frac{E\sqrt{E^2 - m_1^2}}{\mathrm{e}^{E/T} - 1} \tag{17}$$

is the thermal equilibrium density of the standard H_1^0 at temperature T [34]. We are using the condition that the temperature is less than the electroweak phase transition $T_{\mathrm{EW}} \geq 1.5m_1$ [14]. The thermal average of the decay rate is given by

$$\langle \Gamma_H \rangle = \frac{\alpha\,(\Theta T)^2}{8\pi^3 n_H^{(\mathrm{eq})}} \mathrm{e}^{m_1/T}, \tag{18}$$

where α is an integration parameter that can be taken to be 1.87 [14]. We define $\beta \equiv n_h/T^3$ and in the radiation-dominated era we write the evolution equation (15) as

$$\frac{d\beta}{dT} = -\frac{\langle \Gamma_H \rangle \beta^{(\mathrm{eq})}}{KT^3} = -\frac{\alpha}{8\pi^3 K \mathrm{e}^{m_1/T}} \left(\frac{\Theta}{T^2}\right)^2, \tag{19}$$

where $K^2 = 4\pi^3 g\,(T)\,/45m_{\mathrm{Pl}}^2$, $\beta^{(\mathrm{eq})} = n_h^{(\mathrm{eq})}/T^3$ is the β parameter in the thermal equilibrium, $m_{\mathrm{Pl}} = 1.2 \times 10^{19}$ GeV is the Planck mass and $g\,(T) = g_B + 7g_F/8 = 136.25$ for the model of the Ref. [31]. g_B and g_F are the relativistic bosonic and fermionic degrees of freedom, respectively. Here we are taking $T = m_1$ since this regime gives the larger contribution to β [14]. Hence,

$$\beta = \frac{\alpha\Theta^2}{4\pi^3 K m_1^3}. \tag{20}$$

Now, the cosmic density of the h^0 scalar is

$$\Omega_h = 2g\,(T_\Gamma)\,T_\gamma^3 \frac{m_h \beta}{\rho_c g\,(T)}, \tag{21}$$

where $T_\gamma = 2.4 \times 10^{-4}$ eV is the present photon temperature, $g\,(T_\gamma) = 2$ is the photon degree of freedom and $\rho_c = 7.5 \times 10^{-47}h^2$, with $h = 0.71$, being the critical density of the Universe. Let us take $m_h = 7.75$ MeV, $v = 174$ GeV, $a_5 = 0.65$, $-a_6 = 0.38$ (actually in our calculations, we have used a better precision for a_5 and a_6) and $m_1 = 150$ GeV. Thus, from Eqs. (20) and (21) we obtain $\Omega_h = 0.3$. Therefore, without imposing any new fields or symmetries, the 3-3-1 model possesses a scalar field that can satisfy all the properties required for the self-interacting dark matter and that does not overpopulate the Universe.

The candidate for self-interacting dark matter that we propose here differs from the singlet models of Refs. [5,14] in an important point. As we have discussed above it comes from a gauge model proposed with another motivation that has an independent phenomenology. Therefore, the values of the parameters that we impose here must not spoil the preexisting bounds. We can obtain $m_1 \approx 150$ GeV from Eq. (8) with $a_1 = 1.2$ and $a_2 = 0.36$. From Eq. (9) we have $m_3 = 1$ TeV. On the other hand, one should notice that m_h has a small value since $-f \sim (10^{-7} - 10^{-6})$ GeV and $u \sim 195$ GeV. However, h^0 does not couple to the particles of the standard model for the Higgs boson. Thus, it evades the present accelerator limits. The constants a_5 and a_6 do not enter in the masses of the particles of the model and so, it is free in this work [32].

2.2 The 3-3-1 Model with Right-Handed Neutrinos

In this model the leptons are in triplets, and the third member is a right-handed neutrino:

$$f_L^a = (\nu_L^a, e_L^a, (\nu_L^c)^a)^T \sim (1, 3, -1/3), e_R^a \sim (1, 1, -1). \tag{22}$$

Each charged left-handed fermion field has its right-handed counterpart transforming as a singlet of the $\mathrm{SU}(3)_L$ group. The first two generations of quarks are in antitriplets while the third one is in a triplet:

$$Q_{iL} = (d_{iL}, -u_{iL}, D_{iL})^T \sim (3, \bar{3}, 0), \tag{23}$$

$$u_{iR} \sim (3, 1, 2/3), d_{iR} \sim (3, 1, -1/3), D_{iR} \sim (3, 1, -1/3),\ i = 1, 2,$$

$$Q_{3L} = (u_{3L}, d_{3L}, T_L)^T \sim (3, 3, 1/3), \tag{24}$$

$$u_{3R} \sim (3, 1, 2/3), d_{3R} \sim (3, 1, -1/3), T_R \sim (3, 1, 2/3).$$

The charged gauge bosons are defined as

$$\begin{aligned} \sqrt{2}\, W_\mu^+ &= W_\mu^1 - iW_\mu^2, \sqrt{2}\, Y_\mu^- = W_\mu^6 - iW_\mu^7, \\ \sqrt{2}\, X_\mu^o &= W_\mu^4 - iW_\mu^5. \end{aligned} \tag{25}$$

The *physical* neutral gauge bosons are again related to Z, Z' through the mixing angle ϕ.

The symmetry breaking can be achieved with just three $\mathrm{SU}(3)_L$ triplets

$$\begin{aligned} \chi &= \left(\chi^0, \chi^-, \chi'^0\right)^T \sim (1, 3, -1/3), \\ \rho &= \left(\rho^+, \rho^0, \rho'^+\right)^T \sim (1, 3, 2/3), \\ \eta &= \left(\eta^0, \eta^-, \eta'^0\right)^T \sim (1, 3, -1/3), . \end{aligned} \tag{26}$$

The necessary VEVs are

$$\langle\chi\rangle = (0, 0, \omega/\sqrt{2})^T,\ \langle\rho\rangle = (0, u/\sqrt{2}, 0)^T,\ \langle\eta\rangle = (v/\sqrt{2}, 0, 0)^T. \tag{27}$$

After symmetry breaking the gauge bosons gain masses

$$m_W^2 = \frac{1}{4}g^2(u^2+v^2),\ M_Y^2 = \frac{1}{4}g^2(v^2+\omega^2),\ M_X^2 = \frac{1}{4}g^2(u^2+\omega^2). \quad (28)$$

Eqn.(58) gives us a relation

$$v_W^2 = u^2 + v^2 = 246^2\ \text{GeV}^2. \quad (29)$$

In order to be consistent with the low energy phenomenology we have to assume that $\langle\chi\rangle \gg \langle\rho\rangle, \langle\eta\rangle$ such that $m_W \ll M_X, M_Y$.

The symmetry-breaking hierarchy gives us splitting on the bilepton masses [35]

$$|M_X^2 - M_Y^2| \leq m_W^2. \quad (30)$$

Our aim in this paper is to show that the 3-3-1 model with right-handed neutrinos furnishes a good candidate for (self-interacting) dark matter. The main properties that a good dark matter candidate must satisfy are stability and neutrality. Therefore, we go to the scalar sector of the model, more specifically to the neutral scalars, and we examine whether any of them can be stable and in addition whether they can satisfy the self-interacting dark matter criterions [10]. In addition, one should notice that such dark matter particle must not overpopulate the Universe. On the other hand, since our dark matter particle is not imposed arbitrarily to solve this specific problem, we must check that the necessary values of the parameters do not spoil the other bounds of the model.

Under assumption of the discrete symmetry $\chi \to -\chi$, the most general potential can then be written in the following form [36]

$$\begin{aligned} V(\eta,\rho,\chi) &= \mu_1^2\eta^+\eta + \mu_2^2\rho^+\rho + \mu_3^2\chi^+\chi + \lambda_1(\eta^+\eta)^2 + \lambda_2(\rho^+\rho)^2 + \lambda_3(\chi^+\chi)^2 \\ &+ (\eta^+\eta)[\lambda_4(\rho^+\rho) + \lambda_5(\chi^+\chi)] + \lambda_6(\rho^+\rho)(\chi^+\chi) + \lambda_7(\rho^+\eta)(\eta^+\rho) \\ &+ \lambda_8(\chi^+\eta)(\eta^+\chi) + \lambda_9(\rho^+\chi)(\chi^+\rho) + \lambda_{10}(\chi^+\eta + \eta^+\chi)^2. \end{aligned} \quad (31)$$

We rewrite the expansion of the scalar fields which acquire a VEV:

$$\eta^o = \frac{1}{\sqrt{2}}(v + \xi_\eta + i\zeta_\eta);\ \rho^o = \frac{1}{\sqrt{2}}(u + \xi_\rho + i\zeta_\rho);\ \chi^o = \frac{1}{\sqrt{2}}(w + \xi_\chi + i\zeta_\chi). \quad (32)$$

For the prime neutral fields which do not have VEV, we get analogously:

$$\eta'^o = \frac{1}{\sqrt{2}}\left(\xi'_\eta + i\zeta'_\eta\right);\ \chi'^o = \frac{1}{\sqrt{2}}\left(\xi'_\chi + i\zeta'_\chi\right). \quad (33)$$

Requiring that in the shifted potential V, the linear terms in fields must be absent, we get in the tree level approximation, the following constraint equations:

$$\begin{aligned} \mu_1^2 + \lambda_1 v^2 + \frac{1}{2}\lambda_4 u^2 + \frac{1}{2}\lambda_5 w^2 &= 0, \\ \mu_2^2 + \lambda_2 u^2 + \frac{1}{2}\lambda_4 v^2 + \frac{1}{2}\lambda_6 w^2 &= 0, \\ \mu_3^2 + \lambda_3 w^2 + \frac{1}{2}\lambda_5 v^2 + \frac{1}{2}\lambda_6 u^2 &= 0. \end{aligned} \quad (34)$$

Since dark matter has to be neutral, then we consider only neutral Higgs sector. In the $\xi_\eta, \xi_\rho, \xi_\chi, \xi'_\eta, \xi'_\chi$ basis the square mass matrix, after imposing of the constraints (34), has a quasi-diagonal form as follows:

$$M_H^2 = \begin{pmatrix} M_{3H}^2 & 0 \\ 0 & M_{2H}^2 \end{pmatrix}, \tag{35}$$

where

$$M_{3H}^2 = \frac{1}{2}\begin{pmatrix} 2\lambda_1 v^2 & \lambda_4 vu & \lambda_5 vw \\ \lambda_4 vu & 2\lambda_2 u^2 & \lambda_6 uw \\ \lambda_5 vw & \lambda_6 uw & 2\lambda_3 w^2 \end{pmatrix}, \tag{36}$$

and

$$M_{2H}^2 = \left(\frac{\lambda_8}{4} + \lambda_{10}\right)\begin{pmatrix} w^2 & vw \\ vw & v^2 \end{pmatrix}. \tag{37}$$

The above mass matrix shows that the prime fields mix themselves but do not mix with others. In the limit

$$\lambda_1 v,\ \lambda_2 u,\ \lambda_4 u \ll \lambda_5 w,\ \lambda_6 w, \tag{38}$$

we obtained physical eigenstates $H_1(x)$ and $\sigma(x)$

$$\begin{pmatrix} H_1(x) \\ \sigma(x) \end{pmatrix} = \frac{1}{(\lambda_5^2 v^2 + \lambda_6^2 u^2)^{1/2}}\begin{pmatrix} \lambda_6 u & -\lambda_5 v \\ \lambda_5 v & \lambda_6 u \end{pmatrix}\begin{pmatrix} \xi_\eta \\ \xi_\rho \end{pmatrix}, \tag{39}$$

with masses [36]

$$m_{H_1}^2 \approx \frac{v^2}{4\lambda_6}(2\lambda_1\lambda_6 - \lambda_4\lambda_5) \approx \frac{u^2}{4\lambda_5}(2\lambda_2\lambda_5 - \lambda_4\lambda_6), \tag{40}$$

$$m_\sigma^2 \approx \frac{1}{2}\lambda_1 v^2 + \frac{\lambda_4\lambda_6 u^2}{4\lambda_5} \approx \frac{1}{2}\lambda_2 u^2 + \frac{\lambda_4\lambda_5 v^2}{4\lambda_6}. \tag{41}$$

Eqs. (40) and (41) also give us relations among coupling constants and VEVs. Another massive physical state H_3 with mass:

$$m_{H_3}^2 \approx -\lambda_3 w^2. \tag{42}$$

The scalar $\sigma(x)$ is the one that we can identify with the standard model Higgs boson [36].

In the approximation $w \gg v$, mass matrix M_{2H}^2 gives us one Goldstone ξ'_χ and one physical massive field ξ'_η with mass

$$m_{\xi'_\eta}^2 = -\left(\frac{\lambda_8}{4} + \lambda_{10}\right) w^2. \tag{43}$$

In the pseudoscalar sector, we have three Goldstone bosons which can be identified as follows: $G_2 \equiv \zeta_\eta$, $G_3 \equiv \zeta_\rho$, $G_4 \equiv \zeta_\chi$ and in the ζ'^o_η, ζ'^o_χ basis

$$M_{2A}^2 = \left(\frac{\lambda_8}{4} + \lambda_{10}\right)\begin{pmatrix} w^2 & vw \\ vw & v^2 \end{pmatrix}. \tag{44}$$

We easily get one Goldstone G'_5 and one massive pseudoscalar boson ζ'_η with mass

$$m^2_{\zeta'_\eta} = -\left(\frac{\lambda_8}{4} + \lambda_{10}\right) w^2. \quad (45)$$

It is to be emphasized that, both ξ'_η and ζ'_η are in an singlet of the $SU(2)$. Therefore they do not interact with the standard model gauge bosons $W^\pm$, Z^0 and γ. Unlike the 3-3-1 model considered in [17], here we have two fields which can be considered as dark matter.

To get the interaction of dark matter to the standard model Higgs boson, we consider the following relevant parts

$$\begin{aligned} L_{int}(\sigma, \zeta_\eta) &= \frac{1}{4}\lambda_1 \left[v^2 + 2v\xi_\eta + \xi^2_\eta + \zeta^2_\eta + \xi'^2_\eta + \zeta'^2_\eta + 2\eta^+\eta^-\right]^2 \\ &+ \frac{1}{4}\lambda_4 \left[v^2 + 2v\xi_\eta + \xi^2_\eta + \zeta^2_\eta + \xi'^2_\eta + \zeta'^2_\eta + 2\eta^+\eta^-\right] \\ &\times \left[u^2 + 2u\xi_\rho + \xi^2_\rho + \zeta^2_\rho + 2\rho^+\rho^- + 2\rho'^-\rho'^+\right] \end{aligned} \quad (46)$$

Substituting (39) we get couplings of self-interacting dark matter with the standard model Higgs boson σ

$$\begin{aligned} L(\sigma, \zeta_\eta) &= \left[\frac{\sigma(x)}{\sqrt{\lambda_5^2 v^2 + \lambda_6^2 u^2}}\left(\lambda_1\lambda_5 v^2 + \frac{\lambda_4\lambda_6}{2}u^2\right) + \frac{H_1(x)\sigma(x)}{(\lambda_5^2 v^2 + \lambda_6^2 u^2)}\left(\lambda_1 - \frac{\lambda_4}{2}\right)\lambda_5\lambda_6 uv \right. \\ &+ \left. \frac{\sigma^2(x)}{2(\lambda_5^2 v^2 + \lambda_6^2 u^2)}\left(\lambda_5^2 v^2 + \frac{\lambda_6^2}{2}u^2\right)\right]\left(\xi'^2_\eta + \zeta'^2_\eta\right). \end{aligned} \quad (47)$$

From Yukawa couplings, we see that our candidates do not interact with ordinary leptons and quarks [37].

$$\begin{aligned} \mathcal{L}^\eta_{Yuk} &= \lambda_{3a}\bar{Q}_{3L}u_{aR}\eta + \lambda_{4ia}\bar{Q}_{iL}d_{aR}\eta^* + \text{h.c.} \\ &= \lambda_{3a}(\bar{u}_{3L}\eta^o + \bar{d}_{3L}\eta^- + \bar{T}_L\eta'^o)u_{aR} + \lambda_{4ia}(\bar{d}_{iL}\eta^{o*} - \bar{u}_{iL}\eta^+ + \bar{D}_{iL}\eta'^{o*})d_{aR} + \text{h.c.} \end{aligned}$$

We see that the candidates for dark matter in this model have not couplings with all the standard model particles except for the Higgs boson.

For stability of dark matter, we have to put mass of the standard model Higgs boson is twice bigger mass of the candidate

$$\frac{1}{2}\lambda_1 v^2 + \frac{\lambda_4\lambda_6 u^2}{4\lambda_5} \approx \frac{1}{2}\lambda_2 u^2 + \frac{\lambda_4\lambda_5 v^2}{4\lambda_6} \geq -\left(\frac{\lambda_8}{4} + \lambda_{10}\right) w^2. \quad (48)$$

To avoid the interaction of dark matter with Goldstone boson, we have

$$\lambda_1 = \frac{\lambda_4}{2} \quad (49)$$

The *wrong* muon decay ($\mu^- \to e^-\nu_e\bar{\nu}_\mu$) gives a lower limit for singly charged bilepton $M_Y \sim 230$ GeV. Combining Eqns. (58, 59) with (60) we obtain the following relation: $u \sim v \approx 100 - 200$ GeV and $w \approx (500 - 1000)$ GeV.

The cross section for $hh \to hh$ (where h stands for ξ'_η and ζ'_η) with quartic interaction is $\sigma = \lambda_1^2/4\pi m_h^2$. The requirement on the quality $\sigma_{el}/(m_h[GeV])$ denoting the ration of the DM elastic cross section to its mass (measured in GeV) is that [10, 14, 38]

$$2.05 \times 10^3 \text{ GeV}^{-3} \leq \frac{\sigma}{m_h} \leq 2.57 \times 10^4 \text{ GeV}^{-3} \tag{50}$$

Taking $\lambda_1 = 1$ we get 4.7 MeV $\leq m_h \leq$ 23 MeV. The self-interacting dark matter candidates interact with the standard model Higgs boson by strength 0.65 if $\lambda_5 = \lambda_6 = 1$ and $u = v = 175$ GeV are taken.

Now consider the cosmic density of the h scalar given by [17]:

$$\Omega_h = 2g(T_\gamma)T_\gamma^3 \frac{m_h \beta}{\rho_c g(T)}, \tag{51}$$

where $T_\gamma = 2.4 \times 10^{-4}$ eV is the present photon temperature, $g(T_\gamma) = 2$ is the photon degree of freedom and $\rho_c = 7.5 \times 10^{-47} h^2$ with $h = 0.71$, being the critical density of the Universe. Taking $m_h = 4.7$ MeV, we obtain $\Omega_h = 0.18$. This means that the self-interacting dark matter candidates do not overpopulate the Universe.

2.3 The 3-3-1 Model with Exotic Neutral Lepton

In this model the leptons are in triplets, and the third member is a *neutral exotic* lepton and left-handed fermion field has its right-handed counterpart transforming as a singlet of the SU(3)$_L$ group [39]:

$$f_L^a = (\nu_L^a, e_L^a, N_L^a)^T \sim (1, 3, -1/3), e_R^a \sim (1, 1, -1), N_R^a \sim (1, 1, 0). \tag{52}$$

The first two generations of quarks are in antitriplets while the third one is in a triplet:

$$Q_{iL} = (d_{iL}, -u_{iL}, D_{iL})^T \sim (3, \bar{3}, 0), \tag{53}$$

$$u_{iR} \sim (3, 1, 2/3), d_{iR} \sim (3, 1, -1/3), D_{iR} \sim (3, 1, -1/3),\ i = 1, 2,$$

$$Q_{3L} = (u_{3L}, d_{3L}, T_L)^T \sim (3, 3, 1/3), \tag{54}$$

$$u_{3R} \sim (3, 1, 2/3), d_{3R} \sim (3, 1, -1/3), T_R \sim (3, 1, 2/3).$$

The charged gauge bosons are defined as

$$\begin{aligned}\sqrt{2}\, W_\mu^+ &= W_\mu^1 - iW_\mu^2, \sqrt{2}\, Y_\mu^- = W_\mu^6 - iW_\mu^7, \\ \sqrt{2}\, X_\mu^o &= W_\mu^4 - iW_\mu^5.\end{aligned} \tag{55}$$

The *physical* neutral gauge bosons are again related to Z, Z' through the mixing angle ϕ.

The symmetry breaking can be achieved with three SU$(3)_L$ triplets and an sextet

$$\begin{aligned} \chi &= \left(\chi^0, \chi^-, \chi'^0\right)^T \sim (1,3,-1/3), \\ \rho &= \left(\rho^+, \rho^0, \rho'^+\right)^T \sim (1,3,2/3), \\ \eta &= \begin{pmatrix} \sigma^o_1 & \frac{h^-_1}{\sqrt{2}} & \frac{\sigma^0_2}{\sqrt{2}} \\ \frac{h^-_1}{\sqrt{2}} & H^{--}_1 & \frac{h^-_2}{\sqrt{2}} \\ \frac{\sigma^0_2}{\sqrt{2}} & \frac{h^-_1}{\sqrt{2}} & \sigma^0_3 \end{pmatrix} \sim (1,6,-2/3). \end{aligned} \quad (56)$$

The necessary VEVs are

$$\langle\chi\rangle = (0,0,\omega/\sqrt{2})^T,\ \langle\rho\rangle = (0,u/\sqrt{2},0)^T,\ \langle\eta\rangle = (v/\sqrt{2},0,0)^T. \quad (57)$$

The VEV of sextet has the form

$$\langle S\rangle = \begin{pmatrix} 0 & 0 & v_{\sigma_2} \\ 0 & 0 & 0 \\ v_{\sigma_2} & 0 & 0 \end{pmatrix}$$

After symmetry breaking the gauge bosons gain masses

$$m^2_W = \frac{1}{4}g^2(u^2+v^2+v_4),\ M^2_Y = \frac{1}{4}g^2(v^2+\omega^2+v_4^2),\ M^2_X = \frac{1}{4}g^2(u^2+\omega^2+v_4^2). \quad (58)$$

Eqn.(58) gives us a relation

$$v^2_W = u^2+v^2+v_4^2 = 246^2\ \ \mathrm{GeV}^2. \quad (59)$$

In order to be consistent with the low energy phenomenology we have to assume that $\langle\chi\rangle \gg \langle\rho\rangle,\ \langle\eta\rangle$ such that $m_W \ll M_X, M_Y$.

The symmetry-breaking hierarchy gives us splitting on the bilepton masses [35]

$$|M^2_X - M^2_Y| \leq m^2_W. \quad (60)$$

The Yukawa couplings are given

$$\mathcal{L}_Y = f_1\bar{L}_a\phi\nu_R + \frac{1}{2}f_1\bar{L}_aSL_b \quad (61)$$

The mass eigenstates $(\nu_e, \tilde{N})$ are related to the weak eigenstates $(\nu_e, \tilde{N})$

$$\tilde{\nu} = \nu\cos\theta - N\sin\theta,\ \tilde{N} = \nu\sin\theta + N\cos\theta \quad (62)$$

where θ is a mixing angle.

A mass distribution extended well beyond the visible galaxy [40]

$$M(R) = \int^R \rho(r)dV = v^2RG^{-1} \quad (63)$$

Figure 1: Two diagrams for the decay $N \to \nu_e \gamma$

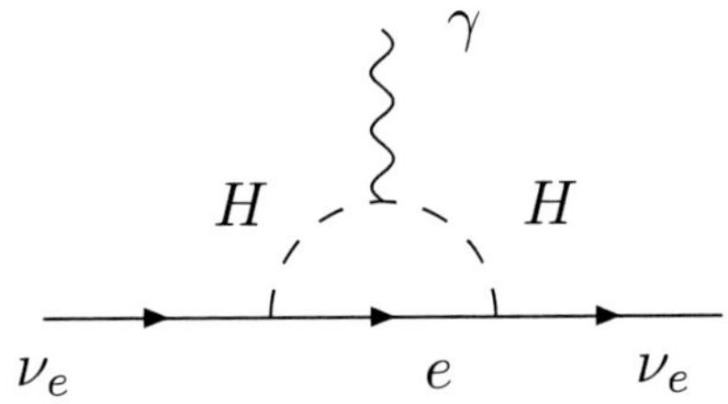

Figure 2: Diagram for Higgs contribution to the decay $N \to \nu_e \gamma$

The decay of N particle into neutrino and photon contains three graphs depicted in Fig.1 which give [39]

$$\Gamma(N \to \nu_e \gamma) = \frac{9}{4} \frac{m_N^5 G_F^2 \alpha}{512\, \pi^2} \sin^2(2\theta) \tag{64}$$

which corresponds to lifetime

$$\tau \approx 4.67 \times 10^{14} \sin^{-2}(2\theta)(1\mathrm{keV}/m_N c^2)^5 \mathrm{yr} \tag{65}$$

It is obvious that the N- particle's life can be longer than the age of the universe, if $m_N \sim 1$ kEV. It was shown that the Higgs contribution is much smaller than those of the two first, and the N- particle meets the constraints on dark matter from cosmology and galaxy formation.

3 Supersymmetric 3-3-1 Model with Right-Handed Neutrinos

Supersymmetric version on the 3-3-1 model with right-handed neutrinos has been proposed in [23, 41]. Here we will follow the usual notation writing for a given fermion f, the respective sfermions by $\tilde{f}$ *i.e.*, $\tilde{l}$ and $\tilde{q}$ denote sleptons and squarks respectively. Then, we have the following additional representations

$$\tilde{Q}_{\alpha L} = \begin{pmatrix} \tilde{d}_\alpha \\ \tilde{u}_\alpha \\ \tilde{d}'_\alpha \end{pmatrix}_L \sim (\mathbf{3}, \mathbf{3}^*, 0), \; \tilde{Q}_{3L} = \begin{pmatrix} \tilde{u}_3 \\ \tilde{d}_3 \\ \tilde{u}' \end{pmatrix}_L \sim (\mathbf{3}, \mathbf{3}, 1/3),$$

$$\tilde{L}_{aL} = \begin{pmatrix} \tilde{\nu}_a \\ \tilde{l}_a \\ \tilde{\nu}^c_a \end{pmatrix}_L \sim (\mathbf{1}, \mathbf{3}, -1/3), \tag{66}$$

$$\begin{aligned} &\tilde{l}^c_{aL} \sim (\mathbf{1}, \mathbf{1}, 1), \\ &\tilde{u}^c_{iL}, \tilde{u}'^c_L \sim (\mathbf{3}^*, \mathbf{1}, -2/3), \; \tilde{d}^c_{iL}, \tilde{d}'^c_{\alpha L} \sim (\mathbf{3}^*, \mathbf{1}, 1/3), \end{aligned} \tag{67}$$

with $a = e, \mu, \tau$; $i = 1, 2, 3$; and $\alpha = 1, 2$. However, when considering quark (or squark) singlets of a given charge we will use the notation u^c_{iL}, d^c_{iL} ($\tilde{u}_{iL}, \tilde{d}^c_{iL}$ with $i(j) = 1, 2, 3$.

The supersymmetric partner of the scalar Higgs fields, the higgsinos, are

$$\begin{aligned} \tilde{\eta} &= \begin{pmatrix} \tilde{\eta}^0_1 \\ \tilde{\eta}^- \\ \tilde{\eta}^0_2 \end{pmatrix}, \tilde{\chi} = \begin{pmatrix} \tilde{\chi}^0_1 \\ \tilde{\chi}^- \\ \tilde{\chi}^0_2 \end{pmatrix} \sim (\mathbf{1}, \mathbf{3}, -1/3), \\ \tilde{\rho} &= \begin{pmatrix} \tilde{\rho}^+_1 \\ \tilde{\rho}^0 \\ \tilde{\rho}^+_2 \end{pmatrix} \sim (\mathbf{1}, \mathbf{3}, 2/3), \end{aligned} \tag{68}$$

and the respective extra higgsinos, needed to cancel the chiral anomaly of the higgsinos in Eq. (68), are

$$\begin{aligned} \tilde{\eta}' &= \begin{pmatrix} \tilde{\eta}'^0_1 \\ \tilde{\eta}'^+ \\ \tilde{\eta}'^0_2 \end{pmatrix}, \tilde{\chi}' = \begin{pmatrix} \tilde{\chi}'^0_1 \\ \tilde{\chi}'^+ \\ \tilde{\chi}'^0_2 \end{pmatrix} \sim (\mathbf{1}, \mathbf{3}^*, 1/3), \\ \tilde{\rho}' &= \begin{pmatrix} \tilde{\rho}'^-_1 \\ \tilde{\rho}'^0 \\ \tilde{\rho}'^-_2 \end{pmatrix} \sim (\mathbf{1}, \mathbf{3}^*, -2/3), \end{aligned} \tag{69}$$

and the corresponding scalar partners denoted by η',χ', ρ', with the same charge assignment as in Eq. (69), and with the following VEVs: $v' = \langle \eta'^0_1 \rangle/\sqrt{2}$, $w' = \langle \chi'^0_2 \rangle/\sqrt{2}$ and $u' = \langle \rho'^0 \rangle/\sqrt{2}$. This complete the representation content of this supersymmetric model.

Concerning the gauge bosons and their superpartners, if we denote the gluons by g^b the respective superparticles, the gluinos, are denoted by λ^b_C, with $b = 1, \ldots, 8$; and in the electroweak sector we have V^b, the gauge boson of $SU(3)_L$, and their gauginos partners λ^b_A; finally we have the gauge boson of $U(1)_N$, denoted by V', and its supersymmetric partner λ_B. This is the total number of fields in the minimal supersymmetric extension of the 3-3-1 model of Refs. [16,31].

3.1 Superfields

The superfields formalism is useful in writing the Lagrangian which is manifestly invariant under the supersymmetric transformations [42] with fermions and scalars put in chiral superfields while the gauge bosons in vector superfields. As usual the superfield of a field ϕ

will be denoted by $\hat{\phi}$ [43]. The chiral superfield of a multiplet ϕ is denoted by

$$\begin{aligned}\hat{\phi} \equiv \hat{\phi}(x, \theta, \bar{\theta}) &= \tilde{\phi}(x) + i\, \theta\sigma^m\bar{\theta}\, \partial_m\tilde{\phi}(x) + \frac{1}{4}\, \theta\theta\, \bar{\theta}\bar{\theta}\, \Box\tilde{\phi}(x) \\ &+ \sqrt{2}\, \theta\phi(x) + \frac{i}{\sqrt{2}}\, \theta\theta\, \bar{\theta}\bar{\sigma}^m\partial_m\phi(x) \\ &+ \theta\theta\, F_\phi(x), \end{aligned} \tag{70}$$

while the vector superfield is given by

$$\begin{aligned}\hat{V}(x, \theta, \bar{\theta}) &= -\theta\sigma^m\bar{\theta}V_m(x) + i\theta\theta\bar{\theta}\bar{\tilde{V}}(x) - i\bar{\theta}\bar{\theta}\theta\tilde{V}(x) \\ &+ \frac{1}{2}\theta\theta\bar{\theta}\bar{\theta}D(x). \end{aligned} \tag{71}$$

The fields F and D are auxiliary fields which are needed to close the supersymmetric algebra and eventually will be eliminated using their motion equations.

For fermion superfields we use the notation

$$\hat{L}_{aL},\ \hat{l}^c_{aL},\ \hat{Q}_{\alpha L},\ \hat{Q}_{3L},\ \hat{u}^c_{iL},\ \hat{d}^c_{iL},\ \hat{u}'^c_L,\ \hat{d}'^c_{\alpha L}. \tag{72}$$

For scalar superfields we write: $\hat{\eta}$, $\hat{\chi}$, $\hat{\rho}$ and similar expressions for $\hat{\eta}'$, $\hat{\chi}'$, $\hat{\rho}'$ and we must change (field) by (field)$'$.

The vector superfield for the gauge bosons of each factor $SU(3)_C$, $SU(3)_L$ and $U(1)_N$ are denoted by $\hat{V}_C, \hat{\bar{V}}_C$; $\hat{V}, \hat{\bar{V}}$; and $\hat{V}'$, respectively, where we have defined $\hat{V}_C = T^b\hat{V}^b_C$, $\hat{V} = T^b\hat{V}^b$; $\hat{\bar{V}}_C = \bar{T}^b\hat{V}^b{}_C$, $\hat{\bar{V}} = \bar{T}^b\hat{V}^b$; $T^b = \lambda^b/2$, $\bar{T}^b = -\lambda^{*b}/2$ are the generators of triplet and antitriplets representations, respectively, and λ^b are the Gell-Mann matrices.

The Lagrangian of the model has the following form

$$\mathcal{L}_{331S} = \mathcal{L}_{SUSY} + \mathcal{L}_{\text{soft}}, \tag{73}$$

where $\mathcal{L}_{SUSY}$ is the supersymmetric part and $\mathcal{L}_{\text{soft}}$ the soft terms breaking explicitly the supersymmetry.

3.2 The Supersymmetric Lagrangian

The supersymmetric part of the Lagrangian is decomposed in the lepton, quark, gauge, and the scalar sectors as follow:

$$\mathcal{L}_{SUSY} = \mathcal{L}_{\text{Lepton}} + \mathcal{L}_{\text{Quark}} + \mathcal{L}_{\text{Gauge}} + \mathcal{L}_{\text{Scalar}}, \tag{74}$$

where

$$\mathcal{L}_{\text{Leptons}} = \int d^4\theta \left[\hat{\bar{L}}_{aL} e^{2g\hat{V} - \frac{g'}{3}\hat{V}'} \hat{L}_{aL} + \hat{\bar{l}}^c_{aL} e^{g'\hat{V}'} \hat{l}^c_{aL} \right], \tag{75}$$

in the lepton sector, we have omitted the sum over the three lepton family for simplicity, and

$$\begin{aligned}\mathcal{L}_{\text{Quarks}} &= \int d^4\theta \Big[\hat{\bar{Q}}_{\alpha L} e^{[2(g_s\hat{V}_C + g\hat{\bar{V}})]} \hat{Q}_{\alpha L} + \hat{\bar{Q}}_{3L} e^{[2(g_s\hat{V}_C + g\hat{V}) + \frac{g'}{3}\hat{V}']} \hat{Q}_{3L} \\ &+ \hat{\bar{u}}^c_{iL} e^{[2g_s\hat{\bar{V}}_C - \frac{2g'}{3}\hat{V}']} \hat{u}^c_{iL} + \hat{\bar{d}}^c{}_{iL} e^{[2g_s\hat{\bar{V}}_C + \frac{g'}{3}\hat{V}']} \hat{d}^c_{iL} \\ &+ \hat{\bar{u}}'^c_L e^{[2g_s\hat{\bar{V}}_C - \frac{2g'}{3}\hat{V}']} \hat{u}'^c_L + \hat{\bar{d}}'^c{}_{\alpha L} e^{[2g_s\hat{\bar{V}}_C + \frac{g'}{3}\hat{V}']} \hat{d}'^c_{\alpha L} \Big], \end{aligned} \tag{76}$$

in the quark sector, and we have denoted g_s, g, g' the gauge coupling constants for the $SU(3)_C, SU(3)_L, U(1)_N$ factors, respectively. In the gauge sector we have

$$\begin{aligned}\mathcal{L}_{\text{Gauge}} &= \frac{1}{4}\int d^2\theta \; [\mathcal{W}_C\mathcal{W}_C + \mathcal{W}\mathcal{W} + \mathcal{W}'\mathcal{W}'] \\ &+ \frac{1}{4}\int d^2\bar{\theta} \; [\bar{\mathcal{W}}_C\bar{\mathcal{W}}_C + \bar{\mathcal{W}}\bar{\mathcal{W}} + \bar{\mathcal{W}}'\bar{\mathcal{W}}'] ,\end{aligned} \tag{77}$$

where $\mathcal{W}_C$, $\mathcal{W}$ e $\mathcal{W}'$ are fields that can be written as follows

$$\begin{aligned}\mathcal{W}_{\zeta C} &= -\frac{1}{8g_s}\bar{D}\bar{D}e^{-2g_s\hat{V}_C}D_\zeta e^{2g_s\hat{V}_C}, \\ \mathcal{W}_\zeta &= -\frac{1}{8g}\bar{D}\bar{D}e^{-2g\hat{V}}D_\zeta e^{2g\hat{V}}, \\ \mathcal{W}'_\zeta &= -\frac{1}{4}\bar{D}\bar{D}D_\zeta\hat{V}', \;\; \zeta = 1, 2.\end{aligned} \tag{78}$$

Finally, in the scalar sector we have

$$\begin{aligned}\mathcal{L}_{\text{Escalar}} &= \int d^4\theta \; \Big[\hat{\bar{\eta}}\, e^{[2g\hat{V} - \frac{g'}{3}\hat{V}']}\hat{\eta} + \hat{\bar{\chi}}\, e^{[2g\hat{V} - \frac{g'}{3}\hat{V}']}\hat{\chi} + \hat{\bar{\rho}}\, e^{[2g\hat{V} + \frac{2g'}{3}\hat{V}']}\hat{\rho} \\ &+ \hat{\bar{\eta}}'\, e^{[2g\hat{\bar{V}} + \frac{g'}{3}\hat{V}']}\hat{\eta}' + \hat{\bar{\chi}}'\, e^{[2g\hat{\bar{V}} + \frac{g'}{3}\hat{V}']}\hat{\chi}' + \hat{\bar{\rho}}'\, e^{[2g\hat{\bar{V}} - \frac{2g'}{3}\hat{V}']}\hat{\rho}'\Big] \\ &+ \int d^2\theta W + \int d^2\bar{\theta}\bar{W},\end{aligned} \tag{79}$$

where W is the superpotential.

3.3 The scalar Potential

In the present model the scalar potential is written as

$$V_{331} = V_F + V_D + V_{\text{soft}}, \tag{80}$$

where

$$\begin{aligned}V_F &= -\mathcal{L}_F = \sum_m F_m^\dagger F_m \\ &= \sum_{ijk}\Big[\left|\frac{\mu_\eta}{2}\eta'_i + \frac{f_2}{3}\epsilon_{ijk}\rho_j\chi_k\right|^2 + \left|\frac{\mu_\chi}{2}\chi'_i + \frac{f_2}{3}\epsilon_{ijk}\eta_j\rho_k\right|^2 + \left|\frac{\mu_\rho}{2}\rho'_i + \frac{f_2}{3}\epsilon_{ijk}\chi_j\eta_k\right|^2 \\ &+ \left|\frac{\mu_\eta}{2}\eta_i + \frac{f'_2}{3}\epsilon_{ijk}\rho'_j\chi'_k\right|^2 + \left|\frac{\mu_\chi}{2}\chi_i + \frac{f'_2}{3}\epsilon_{ijk}\eta'_j\rho'_k\right|^2 + \left|\frac{\mu_\rho}{2}\rho_i + \frac{f'_2}{3}\epsilon_{ijk}\chi'_j\eta'_k\right|^2\Big]\end{aligned} \tag{81}$$

and

$$\begin{aligned}V_D &= -\mathcal{L}_D = \frac{1}{2}(D^aD^a + DD) = \frac{g'^2}{18}(-\eta^\dagger\eta + \eta'^\dagger\eta' - \chi^\dagger\chi + \chi'^\dagger\chi' + 2\rho^\dagger\rho - 2\rho'^\dagger\rho')^2 \\ &+ \frac{g^2}{8}(\eta_i^\dagger\lambda_{ij}^b\eta_j - \eta_i'^\dagger\lambda_{ij}^{*b}\eta'_j + \chi_i^\dagger\lambda_{ij}^b\chi_j - \chi_i'^\dagger\lambda_{ij}^{*b}\chi'_j + \rho_i^\dagger\lambda_{ij}^b\rho_j + \rho_i'^\dagger\lambda_{ij}^{*b}\rho'_j)^2,\end{aligned} \tag{82}$$

finally,

$$\begin{aligned} V_{\text{soft}} &= -\mathcal{L}_{\text{soft}} = m_\eta^2 \eta^\dagger \eta + m_\rho^2 \rho^\dagger \rho + m_\chi^2 \chi^\dagger \chi + m_{\eta'}^2 \eta'^\dagger \eta' \\ &+ m_{\rho'}^2 \rho'^\dagger \rho' + m_{\chi'}^2 \chi'^\dagger \chi' - \epsilon_{ijk}(k_1 \rho_i \chi_j \eta_k + k_1' \rho_i' \chi_j' \eta_k' \\ &+ H.c.), \end{aligned} \tag{83}$$

where we have used the scalar multiplets in Eqs. (57) and (69).

With Eqs. (81)-(83) we can work out the mass spectra of the scalar and pseudoscalar fields by making the usual shift $X^0 \to \frac{1}{\sqrt{2}}(v_X + H_X + iF_X)$. The analysis is similar to that of Ref. [23] and we will not write the constraints equation, etc.

By using as input the following values for the parameters: $\sin^2\theta_W = 0.2314$, $g = 0.6532$, $g' = 1.1466$; $f_2 = 2$, $f_2' = 10^{-3}$; $k_1 = k_1' = 10$ GeV; $\mu_\eta = \mu_\rho = \mu_\chi = -10^3$ GeV; $m_\eta = 15$ GeV, $m_\rho = 10$ GeV. $m_\rho = 244.99$ GeV; $m_{\chi 2} = m_{\chi_2'} = 10^3$ GeV and $m_{\rho'} = 13$ GeV, we obtain the masses

$$\begin{aligned} &m_1 \approx 1702, m_2 \approx 1449, m_3 \approx 387, \\ &m_4 \approx 380, m_5 \approx 361, m_6 \approx 130, \end{aligned} \tag{84}$$

for the scalar sector (all masses are in GeV). Note that the lightest neutral scalar is heavier than the lower limit of the Higgs scalar of the standard model, i.e., $m_H \gtrsim 114$ GeV. For the pseudoscalar sector we obtain

$$\begin{aligned} &M_1 \approx 1702, M_2 \approx 1449, M_3 \approx 363, \\ &M_4 \approx 5, M_5 = 0, M_6 = 0, \end{aligned} \tag{85}$$

only the two massless pseudoscalars are exact values, i.e., there are two Goldstone bosons as it should be. Notice that there is a light pseudoscalar.

A carefully study shows that there is a Higgs boson satisfies the conditions for self-interacting dark matter [44].

4 Conclusions

In conclusion, it is a remarkable fact that the 3-3-1 models have an option for self-interacting dark matter without the need of imposing any new symmetry to stabilize it. We have shown that the Spergel-Steinhardt bound for self-interacting dark matter [31] can be realized in the 3-3-1 model with a reasonable choice of the values of the parameters. The 3-3-1 model with right-handed neutrinos provides two Higgs bosons: one is scalar or CP-even and another is pseudoscalar or CP- odd particle having properties of candidates for dark matter. In difference with the previous candidate which introduced by hand, our self-interacting dark matter arises without imposing new properties to satisfy all the criteria. From the conditions for self-interacting dark matter one get the bounds for scalar Higgs bosons: $m_h = 7.75$ MeV in the 3-3-1 model with exotic leptons [17] and $4.7\ MeV \leq m_h \leq 23$ MeV in the 3-3-1 version with right-handed neutrinos [18]. Scalar dark matter candidates have been recently investigated in [63].

It means that the candidate for self-interacting dark matter has not to be introduced *ad hoc* as in other models [5]. This feature remains valid in the supersymmetric 3-3-1 model with right-handed neutrinos. The spin $\frac{1}{2}$ particle exists in the version with exotic neutral lepton also satisfies the dark matter criteria.

This would turn out to be particularly interesting if observations reveal to be in favor of light dark matter.

This work was supported in part by National Council for Natural Sciences of Vietnam contract No: KT - 04.1.2.

References

[1] M. S. Turner, *Phys. Rep.* **333**-334 (2000) 619.

[2] J. Knodlseder et al, [astro-ph/0309442] to appear in *Astronomy & Astrophysics*.

[3] N. A. Bahcall et al, *Science* **284** (1999) 1481.

[4] L. Wang et al, *Astrophys. J.* **530** (2000) 17.

[5] C. P. Burgess et al, *Nucl. Phys. B* **619** (2001) 709; M. C. Bento et al, *Phys. Rev. D* **62** (2000) 041302; M. C. Bento et al, *Phys. Lett.* **B 518** (2000) 276; D. E. Holz and A. Zee, *Phys. Lett.* **B 517** (2001) 239.

[6] S. Ghigna et al, *Astrophys. J.* **544** (2000) 616; J. F. Navarro et al, ibid 462 (1996) 563; B. Moore et al, *Mon. Not. R. Astron. Soc.* **310** (1999) 1147.

[7] B. Moore et al, *Astrophys. J.* **524** (1999) L19; A. Klypin et al, *Astrophys. J.* **522** (1999) 82.

[8] R. Davé et al, *Astrophys. J.*, **547** (2001) 574.

[9] J. S. Bullock et al, Astrophys. J. 539 (2000) 517; R. A. Swaters et al, *Astrophys. J.* **531** (2000) L107; F. C. van den Bosch et al, *Astrophys. J.* **119** (2000) 1579; J. F. Navarro and M. Steinmetz, *Astrophys. J.* **528** (2000) 607; C. Firmani et al, *Mon. Not. R. Astron. Soc.* **315** (2000) L29; V. P. de Battista and J. A. Sellwood, *Astrophys. J.* **493** (1998) L5.

[10] D. N. Spergel and P. J. Steinhardt, *Phys. Rev. Lett.* **84** (2000) 3760.

[11] N. Yoshida et al., *Astrophys. J.* **544** (2000) L87.

[12] D. A. Buote et al, *Astrophys. J.* **577** (20020 183.

[13] J. P. Ostriker, *Phys. Rev. Lett.* **84** (2000) 5258; A. Burkert, astro-ph/0002409.

[14] J. McDonald, *Phys. Rev. Lett.* **88** (2002) 091304.

[15] F. Pisano and V. Pleitez, *Phys. Rev. D* **46**, (1992) 410; P. H. Frampton, *Phys. Rev. Lett.* **69**, (1992) 2889; R. Foot et al, *Phys. Rev. D* **47**, (1993) 4158.

[16] R. Foot, H. N. Long, and Tuan A. Tran, *Phys. Rev. D* **50**, (1994) R34; H. N. Long, *Phys. Rev. D* **54**, (1996) 4691.

[17] D. Fregolente and M. D. Tonasse, *Phys. Lett.* **B555** (2003) 7.

[18] H. N. Long and N. Q. Lan, *Europhys. Lett.* **64**, 571 (2003).

[19] P. H. Frampton and T. W. Kephart, *Phys. Lett.* **B585**, 24 (2004); P. H. Frampton, Plenary Talk at SUGRA 20, Northeastern University, March 2003, hep-th/0306029.

[20] A. G. Dias, R. Martinez and V. Pleitez, hep-ph/0407074.

[21] C. A. de S. Pires and O. Ravinez, *Phys. Rev. D* **58**, 035008 (1998).

[22] C. A. de S. Pires, *Phys. Rev. D* **60**, 075013 (1999).

[23] H. N. Long and P. B. Pal, *Mod. Phys. Lett.* **A13**, 2355 (1998); T. V. Duong and E. Ma, *Phys. Lett.* **B316**, 307 (1993); J. C. Montero, V. Pleitez and M. C. Rodriguez, *Phys. Rev. D* **65**, 035006 (2002); R. A. Diaz, R. Ramirez, J. Mira and J. A. Rodriguez, *Phys. Lett* **B552**, 287 (2003).

[24] M. Capdequi-Peyranere and M. C. Rodriguez, *Phys. Rev. D* **65**, 035006 (2002).

[25] R. D. Peccei and H. Quinn, *Phys. Rev. Lett.* **38**, 1440 (1977); Phys. Rev. D **16**, 1791 (1977).

[26] P. B. Pal, *Phys, Rev. D* **52**, 1659 (1995); A. G. Dias, V. Pleitez, and M. D. Tonasse, *Phys. Rev. D* **67**, 095008 (2003); A. G. Dias and V. Pleitez, *Phys. Rev. D***69**, 015007 (2004); A. G. Dias, C. A. de S. Pires, and P. S. Rodrigues da Silva ,*Phys. Rev. D* **68**, 115009 (2003).

[27] H. N. Long and V. T. Van, *J. Phys.* **G25**, 2319 (1999).

[28] R. Foot et al in [16].

[29] F. Pisano and V. Pleitez, *Phys. Rev. D* **51**, 3865 (1995).

[30] G. Tavares- Velasco and J. J. Toscano, *Phys. Rev.* **D 65**, 013005 (2004);

[31] V. Pleitez and M. D. Tonasse, *Phys. Rev.* **D48**, (1993) 2353.

[32] M. D. Tonasse, *Phys. Lett. B* **381** (1996) 191 [See also N. T. Anh, N. A. Ky and H. N. Long, *Int. J. Mod. Phys.* **A 16** (2001) 541 and references cited therein].

[33] Particle Data Group, S. Eidelman et al, *Phys. Lett.* **B592**, 1 (2004).

[34] See for example, E. W. Kolb and M. S. Turner, The early universe (Addison-Wesley Publishing Co., Reading, 1990).

[35] H. N. Long and T. Inami, *Phys. Rev. D* **61**, 075002 (2000).

[36] H. N. Long, *Mod. Phys. Lett. A* **13** (1998) 1865.

[37] H. N. Long: Right-handed neutrino currents in the $\mathrm{SU}(3)_L \otimes \mathrm{U}(1)_N$ electroweak theory. [hep-ph/9603258].

[38] B. D. Wandelt et al, in Marina de Rey 2000, *Sources and detection of dark matter and dark energy in the Universe*, pp. 263 - 274, [astro-ph/0006344].

[39] F. Z. Chen, *Phys. Lett. B* **442** (1998) 223.

[40] A. De Rujula and S. L. Glashow, *Phys. Rev. Lett.* **45** (1980) 942.

[41] J. C. Montero, V. Pleitez and M. C. Rodriguez, Supersymmetric 3-3-1 model with right-handed neutrinos, preprint IFT-P.027/2004 [hep-ph/0406299].

[42] J. Wess e J. Bagger, Supersymmetry and Supergravity 2nd edition, Princeton University Press, Princeton NJ, (1992).

[43] H. E. Haber and G. L. Kane, *Phys. Rep.* **117**, 75 (1985).

[44] H. N. Long, N. Q. Lan and D. T. Huong, work in progress.

[45] C. Boehm and P. Fayet, *Nucl. Phys.* **B683** (2004) 219.

[46] J. C. Montero, V. Pleitez, and M. C. Rodriguez, Contribution to XXII Encontro Nacional de Partículas e Campos, São Lorenço (MG), Brazil, Oct. 22-26/2002; hep-ph/0204130, and references therein.

[47] See http://www.slac.stanford.edu/exp/e158+.

[48] P. L. Anthony et al. (E158 Collaboration), *Phys. Rev. Lett.* **92**, 181602 (2004).

[49] J. C. Montero, V. Pleitez, and M. C. Rodriguez, *Phys. Rev. D* **58**, 097505 (1998).

[50] E. M. Gregores, A. Gusso, and S. F. Novaes, *Phys. Rev. D* **64**, 015004 (2002); D. V. Soa, T. Inami, and H. N. Long, *Eur. Phys. J.* **C34**, 285 (2004).

[51] G. Tavares-Velasco and J. J. Toscano, Europhys. Lett. **53**, 465 (2001); *Phys. Lett.* **B472**, 105 (2000).

[52] B. Dion, T. Grégoire, D. London, L. Marleau, and H. Nadeu, *Phys. Rev. D* **59**, 075006 (1999).

[53] P. Das, P. Jain and D. W. McKay, *Phys. Rev. D* **59**, 055011 (1999).

[54] T. Kitabayashi and M. Yasue, *Phys. Rev. D* **63**, 095006 (2001); T. Kitabayashi, *Phys. Rev. D* **64**, 057301 (2001).

[55] T. Kitabayashi and M. Yasue, *Phys. Lett.* **B508**, 85 (2001).

[56] J. C. Montero, C. A. de S. Pires, and V. Pleitez, *Phys. Rev. D* **65**, 095001 (2002).

[57] M. Kuze and Y. Sirois, *Prog. Part. Nucl. Phys.* **50**, 1 (2003).

[58] P. H. Frampton and M. Harada, *Phys. Rev. D* **58**, 095013 (1998); H. N. Long and T. Inami, *Phys. Rev. D* **61**, 075002 (2000).

[59] D. Ng, *Phys. Rev. D* **49**, 4805 (1994); P. Jain and S. D. Joglekar, *Phys. Lett. B,* **407**, 151 (1997).

[60] Y. Hayato et al. (SuperKamiokande Collaboration), *Phys.Rev. Lett.* **83**, 1529 (1999).

[61] C. A. de Pires and P. S. Rodrigues da Silva, [hep-ph/0307253]; A. G. Dias and V. Pleitez, [hep-ph/0308037].

[62] J. T. Liu and D. Ng, Z. *Phys. C* **62** (1994) 693; N. A. Ky, H. N. Long and D. V. Soa, *Phys. Lett. B* **486** (2000) 140.

[63] C. Boehm and P. Fayet, *Nucl. Phys.* **B683** (2004) 219.

[64] C. B. Netterfield et al, *Astrophys. J.* **571** (2002) 604.

[65] S. Weinberg, Phys. Rev. D **26**, 287 (1982); P. Nath, A. H. Shamseddine, and R. Arnowitt, *Phys. Rev. D* **32**, 2348 (1985); I. Hinchliffe and T. Kaeding, *Phys. Rev. D* **47**, 279 (1993); T. Goto and T. Nihei, *Phys. Rev. D* **59**, 115009 (1999).

[66] H. Murayama and A. Pierce, *Phys. Rev. D* **65**, 055009 (2002); B. Bajc, P. F. Perez, and G. Senjanovic, *Phys. Rev. D* **66**, 075005 (2002), and references therein.

[67] R. Martinez, W. A. Ponce, L. A. Sanchez, *Phys.Rev.D* **65**, 055013 (2002); L. A. Sanchez, W. A. Ponce, R. Martinez, Phys.Rev.D **64**, 075013 (2001).

[68] M. C. Rodriguez, work in progress.

INDEX

A

B

C

D

E

F

G

N

O

P

Q

R

S

T

U

V

W

X

Y

Z